PRINCIPLES OF MANAGEMENT

By

SHIKHA TANEJA
B.E. in Information Technology
M.B.A. in IT-HR
Lecturer, Applied Science & Humanities
Echelon Institute of Technology,
Faridabad (Haryana)

RENU RANA
M.B.A. (HR & Finance)/M.A. (Eco.) Pursuing
Asst. Professor, Humanities Department
Echelon Institute of Technology,
Faridabad (Haryana)

S.K. KATARIA & SONS®
Publisher of Engineering Books
4885/109, Prakash Mahal, Dr. Subhash Bhargav Lane,
Opposite Delhi Medical Association, Daryaganj, New Delhi-110002
Phone: +91-11-43551243; Mobile: +91-9871775858, 7678425209
e-mail: katariabook@gmail.com, katariabooks@yahoo.com
Website: www.skkatariaandsons.com
(e-books, subscriptions and bulk buying also available on our website)

Published By :

S.K. KATARIA & SONS®
4885/109, Prakash Mahal, Dr. Subhash Bhargav Lane,
Opposite Delhi Medical Association, Daryaganj, New Delhi-110002
Phone: +91-11-43551243, Mobile: +91-9871775858, 7678425209
e-mail: katariabook@gmail.com, katariabooks@yahoo.com

First Edition: 2012; Second Edition: 2014; Reprint: 2026

Laser Typesetting at: Goswami Associates, Delhi.

Preface

It is matter of immense pleasure to release this book **Principles of Management.** The contents of the book have been organized to meet the requirement of the students preparing for Management, Business Administration, Engineering and other management courses. This book will help the students to learn the aspects of competitive business world.

This book is a modest attempt to present the basic concepts, principles, and functions of management for the benefit of students to learn the importance of management in all the organization and also help them to occupy managerial positions in multi-national corporations. This book has been written in a very easy language and it makes extensive use of diagrams for the benefit of the students. It is hoped that the book will be found to be of great use by the concerned students and also the practicing managers.

The text of the book is designed under six units as follows:

Unit 1: Management

Unit 2: Management Information System (MIS)

Unit 3: Organization Structure and Organization Analysis

Unit 4: Motivation and Productivity

Unit 5: HRD and Organization

Unit 6: Organizational Behavior

The first unit is all about the principles, nature, characteristics, evolution, theories and functions of management. Additionally contain the role of administration in the organization. The second unit contains the portion of management information system and its importance in the organization as decision making system. The third unit deals with the organization structure and organization analysis and the fourth unit include the motivation and productivity of the employees and the last unit contain HRD and Organization.

With simple and easy language the book has case studies included in every chapter to illustrate the concepts. The book provides chapter end exercises with concept review and critical thinking questions.

We are deeply indebted to the authors and researchers whose publications have contributed towards in our understanding of the subject matter. We are confident that this edition of the book will only help to maintain the popularity but prove it worthwhile for the students. We shall greatly appreciate the suggestion for further improvements of the book from teachers and students of the subject.

In the end we owe our deep sense of gratitude to Dr. Chavi Sharma (Director FMS, MRIU), Dr. Renu Sethi (Assistant Professsor, YMCA), Ms. Smruti Mohanty, Ms. Neha Goyal, Mr. Tushar Banerjee and Mr. Sengupta for providing us the opportunities for interacting with them regarding the topics covered in this book.

Also we are thankful to our publishers S.K Kataria and Sons, New Delhi for their wonderful work in bringing this book in an attractive manner.

AUTHORS

Contents

2. MANAGEMENT INFORMATION SYSTEM 79–116

3. ORGANIZATION STRUCTURE AND ORGANIZATION ANALYSIS 117–167

4. MOTIVATION AND PRODUCTIVITY 168–216

CHAPTER 1

Management

1.0 INTRODUCTION

One can easily perform various activities and get the results but the question arises—**Is the results effective?** This is the main concern of every organisation (*e.g.*, a family a college, a manufacturing firm, a business concern, an insurance company, a governmental agency, social organisations, hospital, a university, or a religious trust). And to make the results or task effective human beings join together in the form of groups, teams and organisations. These groups or teams join together **"to manage"** the activities to achieve the goals.

A central organ or agency is required to coordinate the activities and efforts of various individuals working together in the organisation so that they can work collectively as a team. Management is an essential part of any goal oriented activity. And managing has become one of the most important areas of human activity because of increasing role of large and complex organisations in the society. Each and every organisation requires planning of activities, organisation of resources, leading and motivation of people and various other function for the development of overall organisation. This chapter is the modest attempt to present the meaning. Objectives, nature, roles, features and levels of management. Additionally theories of management, starting with traditional approach to modern approach and how organisation work as a system, is the main concern of this chapter.

1.1 MEANING OF MANAGEMENT

As one author has rightly said "Management is the art of getting things done through others". And this definition has become popular because it is very brief and concise. It stresses that management is basically an art and that a manager directs the efforts of other people rather than performing the task himself. This definition also highlights the attainment of objective. **But this definition suffers from several deficiencies.**

- First, it is simplistic and not sufficiently broad-based as it excludes scientific elements in management.

- Secondly, it does not point out the managerial functions involved in getting work done from others.
- Thirdly, it is silent on the behavioral and participative aspects of management.
- Fourthly, it gives an impression that a manager obtains results by treating his subordinates as were tools and a manager is not supposed to get people to perform by hook or crook, Lastly, this definition overstresses enterprise objectives and does not point out that a formal organisation structure is required for management.

"Management is the art of getting things done through people."

–Mary Parkery Follett

"Management consists of getting things done through others—A manager is one who accomplishes organisational objectives by directing the efforts of others."

–C.S. Gorge

Modern Definition: Modern management involves creating a conducive environment in the organisation whereby individuals are motivated to work efficiently for the achievement of organizational goals.

According to Harold Knootz and Cyril O'Donnell: Management is the creation and maintenance of an internal environment in an enterprise where individuals working in groups, can perform efficiently and effectively towards the attainment of group goals.

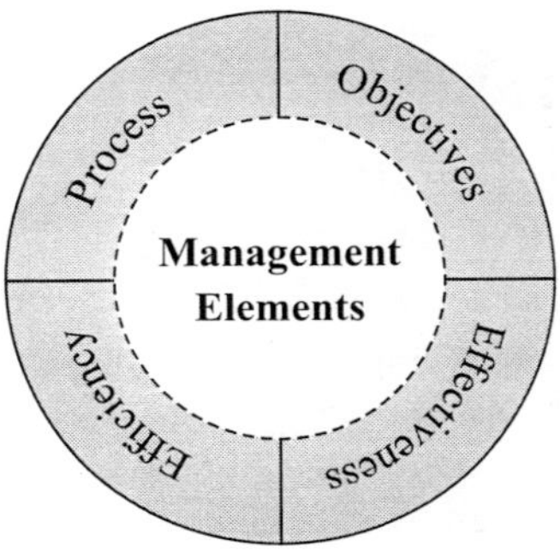

Fig. 1.1 Elements of Management

The basic elements of this definition are discussed below:

1. **Process:** The term process in this definition means the primary functions or activities that management performs to get things done. These functions are planning, organizing, staffing, directing and controlling.
2. **Organizational Objectives:** The concept of management insists that will be the activities of managers must be directed towards the achievement of organisational goals and these goals should be considered as common goals of all employees.
3. **Effectiveness:** Being effective or doing works effectively means finishing the given task. Effectiveness in management is concerned with doing the rights things completing activities result.
4. **Efficiency:** It means optimum utilization of resources in performing the given task. The achievement of the end result with the least amount of resources. It signifies the relationship between inputs and outputs. Efficiency would be greater if less outputs are used to produce the required amount of goods or if more goods are produced with the given output.

Some other definitions of management are given below:

According to Kim Ball: "Management is defined as the art of applying the economic principles that underlie the control of men and materials in the enterprise under consideration."

This definition looks after the various resources of organisation *i.e.,* men and materials and their co-ordination and control for the achievement of the organisation goals.

According to Haimann: "Management is the function of getting things done through people and directing efforts of individuals towards a common objective."

This definition contains three main elements: (*i*) Managers are responsible for creating an environment conducive to good performance (*ii*) Management operates through individuals and groups of persons, and (*iii*) the aim of management is to achieve efficiency and effectiveness in the accomplishment of common objectives.

According to Henry L. Sisk: "Management is the co-ordination of all resources through the process of planning, organising, directing and controlling in order to attain stated goals.''.

This definition appears to be closer to reality for three reasons: First, it highlights the essential nature of manager's job, *i.e.*, coordinating human and physical resources. Secondly, it points out the functions of a manager. Thirdly, it reveals the fact that management strives to achieve the specific objectives of the organisation.

1.2 CONCEPT OF MANAGEMENT

Management comprises different meanings depending upon the context in which it is used. Some of the important concepts of management are given below:

1.2.1 Management as a Process

Management refers to a series of inter-related functions or elements to achieve stated objectives through the effective utilisation of human and other resources.

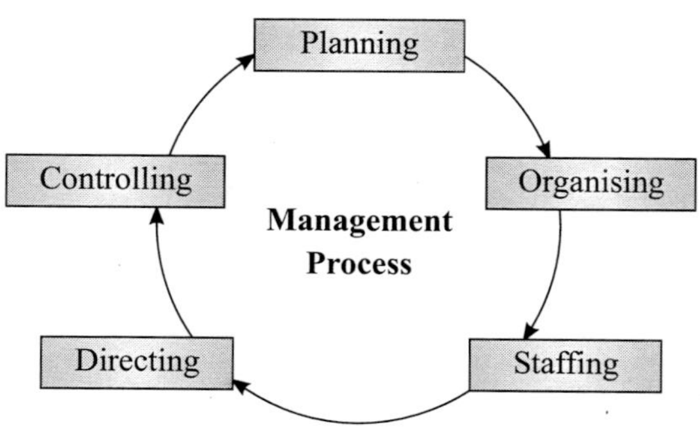

Fig. 1.2 Management Process

It involves five managerial functions:

1. Planning: To take decisions and to prepare plans, policies, procedures, rules, etc.

2. Organising: To divide work among individuals and to create authority and responsibility.

3. Staffing: To employ people to manage various positions in the organisation and to provide them necessary training.

4. Directing: To give instructions to the subordinates, to motivate them and to provide them leadership.

5. Controlling: To compare actual performance with predetermined standards and to take corrective action.

Management is called a process because it comprises a series of function that leads to the achievement of certain objective. As shown in Fig. 1.2 management process involves planning, organising, staffing, directing and controlling. These are the basic functions which every manager performs for the achievement of certain goals. The manager draws plan to translate these goals into reality. He organises, directs and controls the activities of the people continues as long as the business unit is in operation. In fact, management is a continuous process.

Management also called as social process because it deals with various social groups such as shareholders supplies, workers, customers, in particular and the society in general. And also major task of management is motivating and leading the human resources for the effective use of physical and financial resources.

1.2.2 Management as an Economic Resource

Resources like land, labours, capital and entrepreneurship, management is a vital factor of production. The inputs of labour, capital and materials do not by themselves ensure growth. They require the catalyst of management to produce results. It is management that coordinates various factors of production. Therefore, management occupies a central place among productive factors. For the growth of economy the development of these resources is must which is only possible with management. Management plays a vital role in the development of organisation, society and Nation (economy). The collective function of management supports the life cycle of organisation followed by economy. Due to these factors we can say that management is an economic resource.

1.2.3 Management as a Team

Management consists of all those individuals who guide and direct the efforts of other individuals to achieve specified objectives. Managers occupy positions at different levels of authority but perform the same basic functions. Top-level managers have great authority than middle-level managers, who, in turn, have great authority than operating managers. In this way, a system of authority (called chain of command) is created in every organisation. The management group consists of different types of managers, *e.g.* professional managers (formally trained in management and divorced

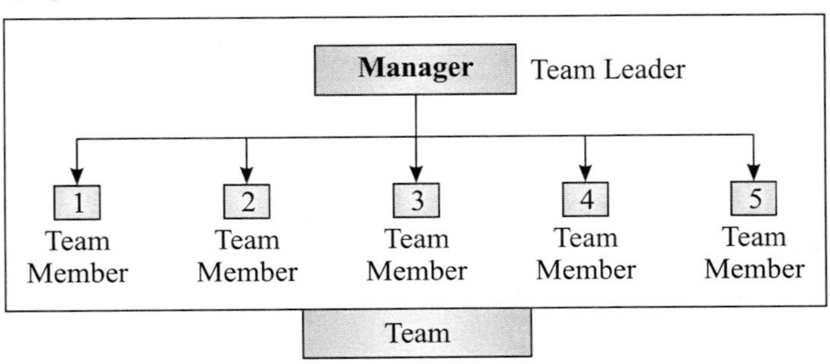

Fig. 1.3 Management as a Team

from ownership), family managers (owner managers having little formal training) and civil servant managers (IAS persons occupying managerial positions in public sector), etc.

1.2.4. Management as an Art

Art involves the practical application of personal skills and knowledge to achieve the concern results. It is a practical way of doing specific things. If a science is learnt, an art is practiced. Stated differently, science is to seek knowledge and art is to apply knowledge.

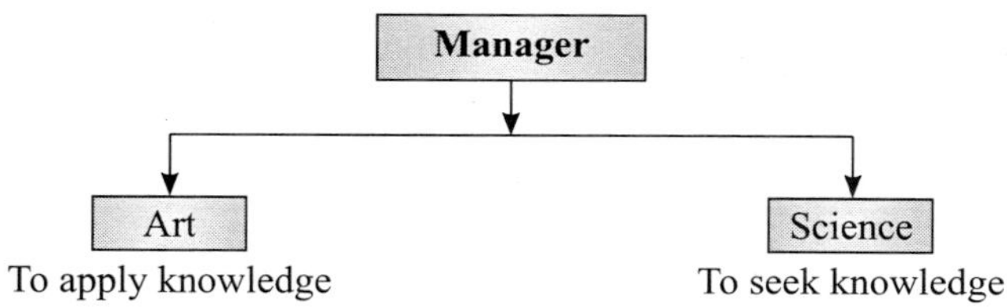

Fig. 1.4

An art has the following features:

(*i*) ***Body of knowledge*:** Art is based on theoretical knowledge of concepts, Principles and application of knowledge about a particular field such as music, painting, etc.

(*ii*) ***Personalised Application of Knowledge*:** Art implies personalized application of knowledge and skills about a particular field, say, music or painting. Every artist or practitioner develops his personal skills and style of producing concrete results.

(*iii*) ***Practice*:** Art is learned and refined through continuous practice.

(*iv*) ***Creativity*:** Art is creative in nature. An artist uses his skills and style to create better results.

Management is considered an art because of the following reasons:

(*i*) The implementation of management process does involve the use of knowledge of management concepts, principles and techniques.

(*ii*) Management seeks to achieve concrete practical results, *e.g.* profits, growth, social service, etc. in a given situation.

(*iii*) Every manager has to apply his personal skills to deal with various problems of the unit he is managing. At times, he may have to use personal judgement to take managerial decision.

(*iv*) Expertise in management requires a sufficiently long period of experience in managing. The managerial art can be refined through continuous practice.

(*v*) The art of management can be learnt and mastered through continuous practice.

As an art, management calls for a corpus of abilities and judgement and a continuous practice of management theories and principles. The main purpose of every manager is to mould the attitude and behaviour of people at work towards the achievement of certain goals is an art of the highest order. Art implies the various method of doing work and so as management because it also involves various ways of managing the work with different principles and theories.

1.2.5 Management as a Science

Science is a body of systematised knowledge accumulated and accepted with reference to the understanding of general truths concerning a particular phenomenon, subject or object of study. It contains concepts, theories, hypotheses and principles to explain cause and effect relationship between two or more factors.

In order to be organized as a science, a discipline should have following characteristics:

- Systematized Body of Knowledge
- Scientific Observation
- Experimentation
- Verifiable principles and Universal Application.

It is a well-known fact that the management has systematised body of knowledge pertaining to its field. The researchers in management use scientific techniques to collect and analyses data about human cause and effect relationship. These principles have been verified by many researchers. More or less these principles have universal application in different type of organisations in different countries. That is only, management is called science.

Nature of Management Science: Management is a social science like Economics because it involves the study of human behaviour. Its principles are not hard and fast laws like those of physics or chemistry. Rather, they are flexible guidelines to be modified in different situations. This is because living and complex human beings in dynamic situations cannot be studied under controlled laboratory conditions. Validity of the principles cannot be tested through experiments. Moreover, management is a comparatively young and growing science which developed in the 20th century. Therefore, management is called an in exact science or soft science. Management is also an inter-disciplinary science as it draws on freely the knowledge from several other sciences like Economics, Mathematics, Sociology, Psychology and Anthropology.

However, **management is not a perfect Science** like other physical sciences such as physics, chemistry, biology, etc. The main reasons for the inexactness of science of management are as follows:

(*i*) Many of the principles of management are not supported by research.

(*ii*) In management, it is difficult to establish cause and effect relationship as in chemistry or biology.

(*iii*) Application of management principles depends upon situations and factors.

(*iv*) Management deals with people at work and it is very difficult to predict their behaviour accurately. Since it is a social process, it is also called a Social Science.

1.2.6 Management as a Profession

'Profession' refers to an occupation backed by specialised body of knowledge and training and to which entry is regulated by a representative body. The essential requirements of a profession are as follows:

Profession

- Specialised field of knowldege
- Restricted entry based on education and training
- Representative on professional association
- Ethical code of conduct for self-regulation
- Social recognition
- Professional fees.

Table 1.1 Management as a Profession

Professional Criteria	*Management*
(*i*) Specialised body of knowledge	Management has a systematised body of knowledge and principle.
(*ii*) Entry based on education and Training	Entry to this field does not require any prescribed qualifications or training.
(*iii*) Professional Association	There is no professional association of which membership is essential for practising manager. In India, We have AIMA, but is not essential to hold its membership to be a manager.
(*iv*) Code of Ethics	Managers are not bound by any common code of conduct. Some association have framed code of conduct for members, but they exist on paper only.
(*v*) Service Responsibilities	At present, service is not the only goal of management. It is fast moving in direction. There is a growing concern for social responsibility of management.

Management can be considered a profession because of the following reason:

(*i*) The field of management is supported by a well-defined body of knowledge that can taught and learnt.

(*ii*) Management of modern organizations requires competent application of management principles, techniques and skills. Thus, their is a need for formal eduction and training in management. Many institutions of management have come up now is India and abroad which offer courses like MBA, PGDBM to create competent managers.

(*iii*) Certain association of managers have been formed in different countries of the world. These association have prescribed standards of education and training for their members.

(*iv*) Several association of managers [such as All India Management Association (AIMA) have prescribed code of conduct for their members.

(*v*) Managers are aware of their social responsibilities towards various groups of the society including customers, labour suppliers, government, etc. They are guided by service motive. The managers enjoy higher status in the society.

Management can't be fully regarded as a profession it does not fulfill all the requirements of a profession as is the case with medical or accountancy profession. Arguments against professional status of management are as follows:

(*i*) Entry to management profession is not restricted. Their is no prescribed standard qualification (*e.g.*, MBA) and training programme to become a manager.

(*ii*) Management does not have all-India representative association like Medical Council of India or Institute of Chartered Accountants of India to prescribe professional standards and enforce them.

(*iii*) There is no ethical code of conduct for the managers as is the case with doctors and C.As.

Thus, Professionalisation of management is not complete yet. However, it is moving in that direction.

1.3 NATURE AND CHARACTERISTICS OF MANAGEMENT

To understand the nature of management, it is essential to know the features or characteristics of management. The basic features or characteristics of management are as follows:

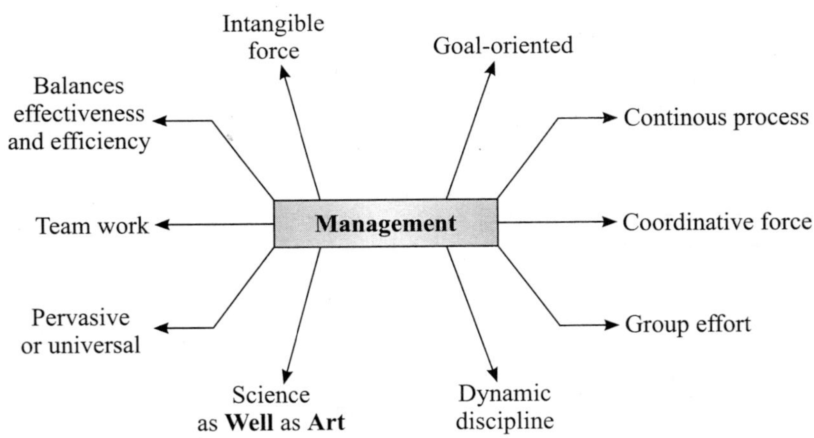

Fig. 1.5 Characteristics of management

(*i*) ***Management is Goat Oriented*****:** The main purpose of management is to achieve the goals of the organisation. For example: Management of a business is to achieve the goals of the organisation by satisfaction of customers, earning of profits and increasing the goodwill and image of business. The basic aim of management is to achieve maximum efficiency of the organisation.

(*ii*) ***Management is a Continuous Process*:** The five functions of management– Planning, organizing, staffing, directing and controlling, works as a process. These functions are repeated time and again and does not stop anywhere. It is an ongoing process of planning the activities and execution of plans by the other functions.

(*iii*) ***Management is Coordinative Force*:** The coordination or integration of human and other resources for effective performance. It brings physical and financial resources together and leads the human resources for the efficient use of non-human resources. All these resources are properly organised and dividend into various work-units for the purpose of achieving greater co-ordination. Management set as an effective tool in getting minimum productivity.

(*iv*) ***Management is an Intangible Force*:** The intangible force refers to unseen force and management has been called as intangible force because its presence is evident by the results of its efforts. Thus, feeling of management is results are apparently known. It is basically the efforts of human resources and its proper utilization towards the achievement of goals of an organisation.

(*v*) ***Management is the Part of Group Effort*:** Each and every task accomplished by management is going to achieved by the group efforts. As management is an integral part of any activity. It involves the use of group efforts in the pursuit of well defined goals or objectives. It cannot exist independent of the group or organisation it manages.

(*vi*) ***Management Balances Effectiveness and Efficiency*:** Effective management requires that all the activities related to management are performed effectively and efficiently. An effective organisation is the one who is able to accomplish its objectives by making optimum use of resources.

For example, it a target is given to the managers of a company to produce 2000 pairs of shoes per month. Both attain their targets and are so effective. But the cost of production per shoe of the first production manager is Rs. 1000 per pair and that of the second is Rs. 1200 per pair. In this case, the first production manager will be termed as both effective and efficient.

(*vii*) ***Management is a Dynamic Discipline*:** Now-a-days management is a field of study which is taught in universities and management institutes. So, it is also called as multi-disciplinary in nature. It contains principles drawn many social sciences like psychology, anthropology, sociology, etc. Management has become a very popular subject of study as is evident from the great rush for admissions into institutions and universities imparting education and training in management. Management offers a very fruitful and rewarding career.

(*viii*) ***Management is a Science as well as Art*:** "Science is a body of systematised knowledge accumulated and accepted with reference to the understanding of general truths concerning a particular subject or object of study." And management has an organized body of knowledge consisting of distinct concepts, principles and techniques which have wide application. So, it is treated as science.

The application of these concepts, principles and techniques requires specialized knowledge and skills on the part of the manager. Since the skills acquired by a manger are his personal possession, management is viewed as an art. The skills can be learnt through training and experience.

(*ix*) ***Management is Pervasive or Universal*:** For an effective performance of an organized activity management is essential. Thus, it is universal in nature. The principles and techniques of management have universal application. They can be applied to all types organization—business, social, educational and religious. But the principles and techniques should not be applied blindly as they are not rigid laws. And due to this, management principles and techniques should be modified to match the given situation and the type of organization. Thus it is pervasive.

(*x*) ***Management Accomplishes Results Through the Cooperation of Others*:** Putting together a team is rather like designing a mini-organization to achieve the goal. Also managers cannot do everything themselves. They must have the necessary ability and skills to get work accomplished through the efforts of others. They must motivate the subordinates for the accomplishment of the tasks assigned to them. It is through motivation that managers can influence the behaviour of their subordinates.

1.4 SIGNIFICANCE OR IMPORTANCE OF MANAGEMENT

Management refers to as the dynamic life-giving element of every business enterprise. Management is the thinking organ that provides vision to the business. It is also the integrating force for the accomplishment of business objectives. The importance of management to a modern business is discussed below:

Fig. 1.6 Importance of Management

1. **Accomplishment of Goals:** It is the management which determines the goals of the organizations and of various departments and functional groups. The goals are communicated to the employees to seek their cooperation. All Organizational activities are directed towards the organizations objectives. Clear-cut definition of goals is essential for the success of any organizations.
2. **Effective Utilization of Resources:** Management ensures optimum utilization of resources. Through planning and organizing, management eliminates all types of wastages and achieves efficiency in all business operations. Management motivates workers to put in her best performance.
3. **Order to Endeavour or Efforts:** Management deals with integration of human and non-human resources in order to achieve organisational objectives. It directs and coordinates the activities of individuals and groups in the use of materials, methods and machines. It thus, brings order to endeavors of different groups.
4. **Sound Organisation:** Management establishes sound organisation for the accomplishment of the desired objectives. It clarifies authority-responsibility relationships among various positions in the enterprise. It fills various positions with persons having the right qualification and training. Management also provides the workers with proper environment and encourages the spirit of cooperation.
5. **Provides vision and Foresight:** Management keeps itself in touch with the external environment and supplies vision and foresight to the enterprise. It helps in predicting what is going to happen in future which will influence the working of the enterprise. It also take steps to ensure that the enterprise is able to meet the demands of changing environment.
6. **National Growth and Prosperity:** At national level, the efficient management of resources is equally important at the national level. According to Peter Drucker, "Management is the crucial factor in economic and social development." The development of a country virtually depends on the quality of management of its resources. Efficient management of resources is a key to growth of the economy and prosperity and better standard of living of the general public.

1.5 MANAGEMENT VS ADMINISTRATION

Some authors suggest that there is no fundamental difference between management and administration, whatever difference between the two exists, exists only in terms of their usage in different activities.

Various views expressed in this regard have led to the emergence of three approaches:

(*i*) Administration is above management.

(*ii*) Administration is a part of management.

(*iii*) Management and Administration are same.

- Administration is concerned with laying down objectives of the enterprise, formulating its policies, determining the broad objectives structure and overall control of the undertaking.

- "Administration defines the goals management strives towards it."
- Administration is a determinative function, where as management is an executive function concerned with implementation of policies laid down by administration.

(*i*) **American Viewpoint:** Administration is a higher level function where as management is a lower level function.

(*ii*) **British Viewpoint:** Management is a generic term that includes administration.

(*iii*) **Modern Viewpoint:** Both the terms are used interchangeably.

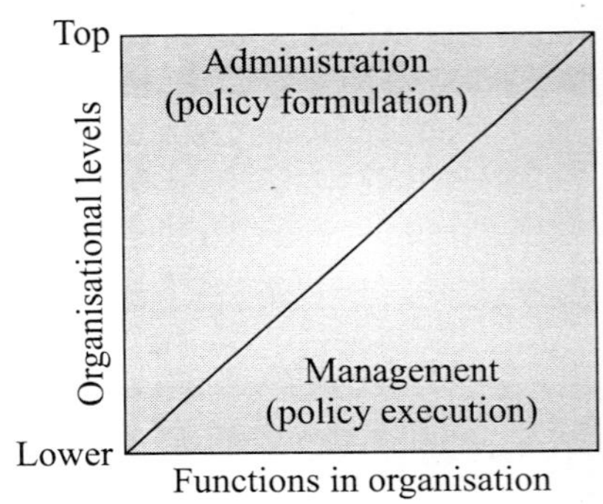

Fig. 1.7 Administration and management functions

1.5.1 Administration is Above Management

According to this viewpoint, administration is a top-level function while management is a lower-level function. Administration is a determinative (thinking) function concerned with laying down basic objectives and broad policies of an organisation. On the other hand, management is an executive (doing) function involving the direction of human effort towards the realization of such objectives. This view is held by eminent American experts on management.

American experts such as Florence, Lansburg, Theo Harimann, G.E. Milward, D.E. McFarland, W.R. Spriegel, Ordway Tead and Schulze also hold similar views. According to them, administration involves policy formulation whereas management is concerned with the execution of policies and supervision of day-to-day operations. Thus, this view considers administration as superior to management.

1.5.2 Administration is a Part of Management

According to E.F. L. Brech, "Management is the generic term for the total process of executive control involving responsibility for effective planning and guidance of the operations of an enterprise. Administration that part of management which is concerned with the installation and carrying out of the procedures by which the programme is laid down and communicated and the progress of activities is regulated and checked against plans."

Kimball and Kimball, Richman and Copen also hold similar views. In their opinion, management comprises both policy-making and execution. Administration is that part of management which involves doing routine things in a known setting. Thus, the European viewpoint is exactly opposite to the American viewpoint.

1.5.3 Administration and Management are Same

Herri Fayol, William Newman, Chester Barnard, George R. Terry, Louis, Allen and many other writers make no distinction between administration and management.

According to William H. Newman, management or administration is "the guidance, leadership and control of the efforts of a group of individuals towards some common goals." Fayol said, all undertakings require the same functions and all must observe the same principles. There is one common science which can be applied equally well to public and private affairs. In practice, the two terms are used interchangeably because both involve the same principles and functions. Somehow the word 'management' has become popular in business enterprises where economic performance is of primary importance. On the other hand, the term administration is preferred in government departments, hospitals, religious trusts, educational institutions and other non-business originations.

Management Vs. Administration

Table 1.2 Distinguish between Management and Administration

Basic of Difference	*Administration*	*Management*
• Definition	Administration is concerned with the formulation of objective, plans and policies of the organisation.	Management means getting the works done through and with others by leading and motivating them.
• Level in organisation	Top level in the organisation.	Middle or lower level in the organisation.
• Major Focus	Policy formulation and objective determination.	Policy executive for objective achievement.
• Nature of function	Determinative	Executive
• Scope of function	Broad and conceptual	Narrow and operational
• Decision-making	Administration determines what is to be done and when it is to be done?	Management decides who shall implement the administrative decisions.
• Usage	The term administration is associated with non-business organisations such as government departments, public enterprises, military organisations, sociocultural organisation.	The term management is widely used in business organisation in the private sector.
• Employer-employee relations	Entrepreneurs and owners.	Employees.
• Designations in organisation	Minister, secretary, commissioner, director, vice-chancellor, registrar, etc.	General manager, managing director, plant superintendent, branch manager, controller, etc.

1.6 FUNCTIONS OF MANAGEMENT

The elements of management process are known as functions of management. There is, however, no single list of functions acceptable to all. Various authors have classified these functions differently. Henri Fayol has classified them into planning, organizing, staffing, directing and controlling, *(P, O, S, D, C)*.

Table 1.3 Managerial Functions and Sub-functions

Functions	*Sub-functions*
• **Planning**	Forecasting, decision-making, strategy formulation, policy-making, programming, scheduling, budgeting, problem-solving, innovation, investigation and research.
• **Organizing**	Functionalisation, divisionalisation, departmentalisation, delegation, decentralisation, activity analysis, task allocation.
• **Staffing**	Manpower planning, recruitment, selection, training, placement, compensation, promotion, appraisal, etc.
• **Directing**	Supervision, motivation, communication, leadership, activating, etc.
• **Controlling**	Fixation of standards, recording, measurement, reporting, corrective action.

Thus, for the sake of analysis of the management process we can classify the management functions into the following categories:

1. Planning
2. Organising
3. Staffing
4. Directing
5. Controlling.

1.6.1 Planning

Planning lays down the objectives to be achieved and the step to be followed to achieve them. It is a mental process requiring the use of intellectual facilities, foresight and sound judgement. It involves selecting and relating the future in the visualisation and formation of proposed activities believed necessary to achieve the desired results. It involves deciding in advance what to do, when to do, where to do and who is to do and how the results are to be evaluated. Thus, planning is the systematic thinking about the ways and the means for the accomplishment of predetermined objectives.

Planning involves the following steps:

(*i*) Determination of objectives

(*ii*) Forecasting

(*iii*) Formulation of policies and programmes

(*iv*) Preparation of schedules

(*v*) Budgeting

Planning includes determination of the objectives, policies, rules, programmes, strategies, budgets and procedures. The purpose of planning is to minimise the risk and to make at the same time full advantage of the available opportunities. It bridges the gap from where we are and where we want to go. Good planning is flexible in nature as it can adjust to changes in business conditions. It covers both short-term and long-term periods.

1.6.2 Organising

Organising is an important function of management by which management brings together the human and material resources. This function must be performed when an activity involves two or more persons. Organising involves determining the activities to be done, grouping the activities, assigning the grouped activities to individuals and creating a structure of authority and responsibility among the people to achieve the objectives of the enterprise.

Urwick define organising as determining what activities are necessary to achieve any purpose and arranging these in groups which may be necessary to assign to the individual. The process of organisation involves the determination of authority and responsibility relations in the organisation. An important function of every manager is to determine the nature of the activities required to attain the group goals, the grouping of these activities and the assignment of the activities to the individuals with necessary delegation of authority.

The process of organisation involves the following steps:

(*a*) Determination of objectives,

(*b*) Identification and grouping of activities,

(*c*) Assignment of duties to individuals, and

(*d*) Development of relationships.

1.6.3 Staffing or Human Resource Management

After organising the various activities to be performed, management is in position to know the manpower requirements of the enterprise at different levels in the organisation structure. After determining the number and types of personal to be selected to fill different jobs, management proceeds with resulting, selecting and training the people to fulfil the requirements of the enterprise. In a running enterprise, staffing is a continuous process because new jobs are created in the enterprise and existing employees leave the enterprise.

Staffing comprises of those activities which are essential to keep manned the positions created by the organisation structure. It includes the task of determining the requirements with regard to number and types of people for the jobs to be done, laying down qualifications for various jobs and recruiting, selecting and training people to perform those jobs efficiently.

Staffing is concerned with both managers and non-managers. It is a function performed by managers at all levels. Earlier, staffing was considered a part of organising. But with the recognition of the importance of the human factor in industry and business, it began to be considered as a separate function. Staffing usually includes the following activities:

(*i*) Human resource planning

(*ii*) Deciding sources of recruitment

(*iii*) Receiving application

(*iv*) Testing and interviewing

(*v*) Final selection and appointment letter

(*vi*) Orientation and placement

(*vii*) Training and development

1.6.4 Directing

Direction is the process of guiding, supervising, leading and motivating the subordinates to work in a way that is beneficial to the enterprise. The manager not only shows the right path but also leads the subordinates to achieve the objective of the enterprise. He creates a sense of belongingness faith and loyalty among the subordinates.

Fig. 1.8

The direction functions involves the following four elements:

(*i*) **Leadership:** Leadership is the process by which a manager guides and influences the work of others in choosing and attaining specified goals. According to Chaster Bachard, leadership is the quality of the behaviour of the individuals whereby they guide people towards the accomplishment of some common goal.

(*ii*) **Communication:** A manager has to tell the workers that they are required to do, how to do and when to do it. It has to create an understanding in the minds of the subordinates of work to be done.

(*iii*) **Motivation:** It is the function of a manager to motivate the people working under him to perform the work assigned. A successful manager has to make proper use of motivation to enthuse the people to work harmoniously for the attainment of the desired objectives.

(*iv*) **Supervision:** Supervision is the process by which conformity between planned and actual results is maintained. Effective supervision ensures greater output of high quality. It teaches the subordinates the way their are to be performed.

1.6.5 Controlling

According to Fayol, "In an understanding control consists of verifying whether everything occurs in conformity with the plan adopted, the instructions issued and the principles established." Controlling is that management activity whereby the managers compare actual performance against the planned one, find out the deviations, take corrective action to remove the deviations, incorporate positive deviations, in the plans and help ensure the realisation of the specific goals.

In a running concern, planning and control go together because planning seeks to set consistent, integrated and articulated events conform to plans. The most notable feature of the process of control is that it is forward-looking. A manager cannot control the past but can avoid the problems in future by taking actions in the light of past experiences. The control process consists of the following steps:

(*i*) Setting up of standards.

(*ii*) Measuring performance.

(*iii*) Comparing performance with standards.

(*iv*) Taking corrective action.

Although planning must precede controlling, plans are not self-achieving. Plans guide managers in the use of resources to accomplish specific goals; then activities are checked to determine whether they conform to the plans.

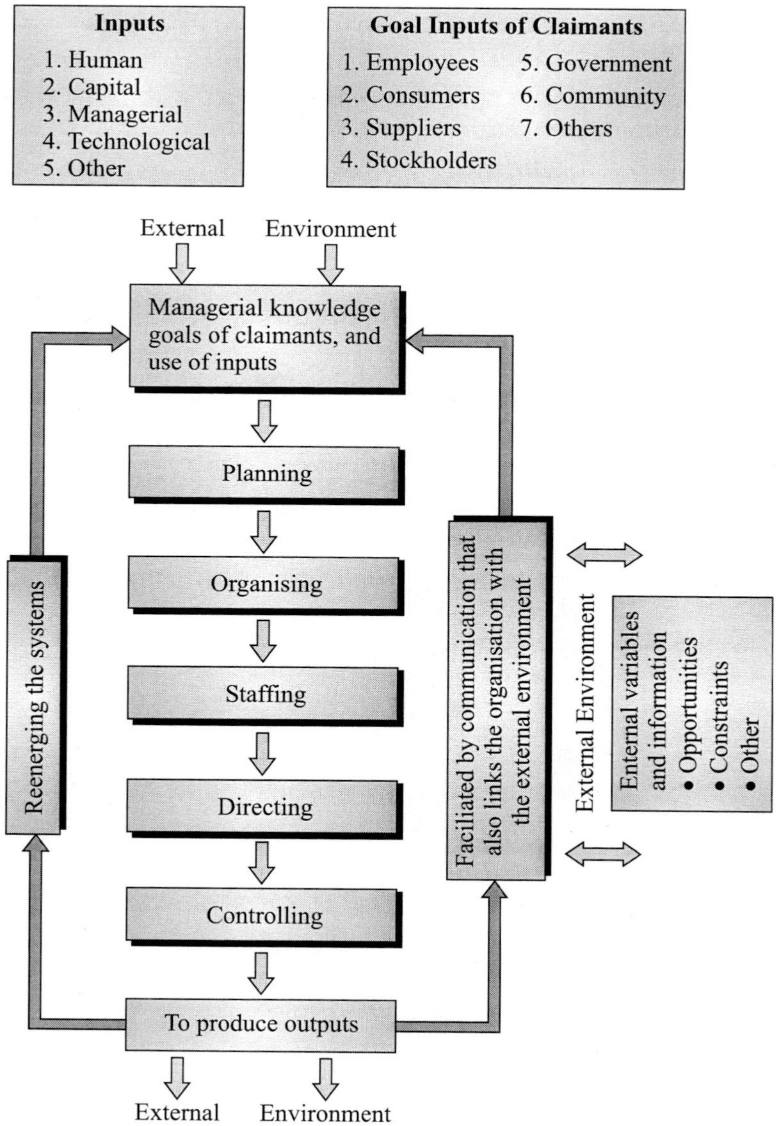

Fig. 1.9 Process of Management

1.7 INTER-RELATIONSHIP BETWEEN MANAGERIAL FUNCTION

Managing is a continuous process consisting of

(*i*) Planning (*ii*) Organising
(*iii*) Staffing (*iv*) Directing
(*v*) Controlling

When a manager reaches the last step, viz. control, his job does not end here. Actually, he will again start with planning on the basis of his past observations and experience and his projection about the future. One cannot say that a manager will organise only when the job of planning is over or will direct only when the stage of organising is completed. In practice, a manager has to perform these activities simultaneously to achieve the desired objectives.

Thus, for theoretical purposes, it may be convenient to separate management functions, but practically they defy such categorisation. They are highly inseparable. Each functions blends into other functions and affects the performance of others and is also affected by the others. In other words, the functions of management are interdependent and inter related. This interaction between the functions of management has been shown by dotted arrows in the Fig. 1.10.

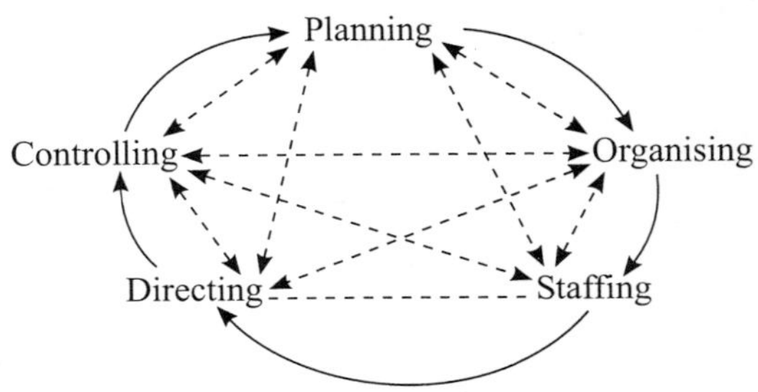

Fig. 1.10 Inter-relationship between Management Functions

1.8 PRINCIPLES OF MANAGEMENT

Management Principle

A principle is a basic statement that provides understanding and guide thinking and action. Principle furnish the structural framework of a science. Principle are referred to as being descriptive, perspective or normative.

1.8.1 The Need and Importance of Management Principles

The principles of management have a tremendous impact upon the practice of management in increasing the efficiency of an organisation. The need and importance of management principle can be visualised as follows:

(*i*) ***To increase Efficiency*:** The established principle of management provide managers guidelines as to how they should work in different situations. These principle increase managerial efficiency. Today, a management graduate, who has acquired the knowledge of management principles, definitely puts better efficiency and effectiveness in the organisation. Though, there is a serious limitation of management principles, that is these have to be modified according to situations as these deal with human beings to diverse nature, these enable a manager to understand the different situations is a better way and save him from costly trial-and-error method.

(*ii*) ***To Crystalline the Nature of Management*:** Lack of understanding of management principles makes it difficult to analyse the management job and to define the exact scope of managerial functions. Thus, individuals cannot be trained effectively for managerial positions. According to Henry Fayol, management principles, rules, methods and procedures are necessary to train and educate future managers.

(*iii*) ***To carry on Researches*:** If, in any subject, certain fundamental principles are developed, the scope and limitations defined, these become the basis for future researches. In the absence of these principles, researches become difficult and future horizons of knowledge cannot be expanded. It is scarcely too much to say that the most important index of the state of maturity of science is state of it's systematic theory. This includes the character of the general conceptual scheme in use in the filed, the kinds and degrees of logical integration of the different elements which make it up, and the ways in which it is actually used in empirical research. The recent emphasis on management researches has increased the quantum of knowledge in this field.

(*iv*) ***To attain Social Objectives*:** Management itself is part of the society and it takes the inputs from the society and gives the output to the society. Thus the standard of the society depends upon the quality of the management. If the management is efficient, the resources of the society are better utilized thereby giving more satisfaction of the society and improving the quality of life of people. In this context, management principles plays an important role.

1.8.2 Nature of Management Principles

(*i*) ***General Statement*:** management principle have been derived out of experience and analysis only. They have not been tested under laboratory or controlled conditions. That is why they have been expressed in the form of general statement. They cannot be stated as rigorously as the principles to physical science. The obvious season is that management principle deal with human beings who have different background, perceptions and needs.

(*ii*) ***Dynamic Guidelines*:** The principles have been derived from management are finable in nature. Refinements in principles take place continuously. This is made possible by the ever increasing research in the field of management.

(*iii*) ***Based on Situations*:** The application of management principle depends upon the specific situation being faced by the manager. Thus, they can be modified by the manager to meet the needs of the situation. The principles are neither absolute nor stable for all times to come. They are flexible in the sense that they can undergo a change according to the changed conditions. However, it does not mean that they may be overhalved according to the whims and experiences of even those who have not natured themselves in managing. If this is allowed, this principles will not withstand the test of time.

(*iv*) ***Careful Application*:** Management principle are relative and not absolute laws which can be applied blindly in all situations. They must be applied carefully depending an organisational requirements and situational demands. According to G.R Teryy, "Management principles are capsules of selected management wisdom to be used carefully and discretely."

(*v*) ***Universal Application*:** Most of the management principles can be applied to all kinds of organisations. In other words, they are applicable to business organisations, social, political, religious and non-profit organisation. Moreover, they have utility in all kinds of economics of the world. The universality of these principle has been established beyond doubt as they are being applied by various people managing different organisations. They are applicable to any kind of organisation wherever there is a need of coordinated efforts of human being.

1.8.3 Henri Fayol's Principles of Management

Fayol has given fourteen principles of management. He has made distinction between management principles and management elements. While management principle is a fundamental truth and establishes cause-effect relationship, management donates the function performed by a manager.

Various principles of Management are as follows:

- Division of Work
- Authority and Responsibility
- Discipline
- Unity of Command
- Unity of Direction
- Subordination of Individual Interest to General Interest
- Remuneration of Personal
- Centralisation
- Scalar Chain
- Order
- Equity
- Stability of Ensure of Personnel
- Initiative
- Esprit de corps (Union of Strength)

1. **Division of Work:** Fayol has advocated division of work to take the advantage of specialisation. According to him, "specialisation belongs to natural order. The workers always work on the same part the managers concerned always with the same matters, acquire an ability, surness and accuracy which increase their output. Each change of work brings in it training and adaptation which reduces output, yet division of work has its

limits which experience and a sense of proportion teach us may not be exceeded. This division of work can be applied at all levels of the organisation.

2. **Authority and Responsibility:** The authority and responsibility are related, authority denotes the right or power to give orders to the subordinates. Responsibility means the duty which the subordinates is expected to perform by virtue of his position in the organisation. Responsibility must be expressed either in terms of functions or in terms of objectives. When a subordinate is asked to control the working of a machine, the responsibility is in terms of function and when a subordinate is asked to produce a number of units of a product, the responsibility is created in terms of objectives.

3. **Discipline:** Discipline means getting obedience to rules and regulations of the organisation. According to Fayol, discipline is obedience application, energy and outwards marks of respect. Discipline is necessary for the smooth running of the organisation. Maintenance of discipline in the organisation depends upon the quality of leadership, clear and fair agreements and a judicious application of rules.

   According to Fayol, discipline can be maintained by:

   (*i*) having good superiors at all levels.

   (*ii*) entering into agreement either with the individual employees or with the union, as the case may be, that are as clear and fair as possible.

   (*iii*) Ensuring that penalties are judiciously imposed.

4. **Unity of Command:** This principle conveys that a subordinate should receive orders from one superior only. If he receives orders from more than one superior, he will not be able to carry out the orders in a proper manner. Fayol observed that if this principle is violated, authority will be undermined, discipline will be in a jeopardy, order will be disturbed and stability will be threatened. Dual command is a permanent source of conflict. Therefore in every organisation, each subordinate should have one superior whose command he has to obey.

   By observing the principle of unity of command, the following benefits may be achieved.

   (*i*) It helps to clarify authority-responsibility relationships in the organisation.

   (*ii*) There will be no possibility of the subordinate receiving conflicting orders.

   (*iii*) The organisation structure will be simple and management will be more effective because there will be no confusion as to who is responsible to whom.

5. **Unity of Direction:** According to this principle, each group of activities with the same objective must have one head and one plan. Unity of direction is different from unity of command in the sense that the formers is concerned with functioning of the organisation in respect of its grouping of activities or planning while the latter in terms of reporting relationship. Unity of direction provides better coordination among various activities to be undertaken by an organisation.

6. **Subordination of Individual Interest to General Interest:** Common interest is above the individual interest. Individual interest must be subordinate to general interest when there is conflict between the two. However, factors like ambition, laziness, weakness, etc. tend to reduce the importance of general interest. Therefore, superiors should set an examples in fairness. and goodness. The agreement between the employers and the employees should be fair and there should be constant vigilance and supervision.
7. **Remuneration of Personnel:** Remuneration of employees should be fair and provide minimum possible satisfaction to employees and employers. Fayol did not favour profit sharing plans for workers but advocated it for managers. He was also in favour of non-financial benefit though these were possible only in the case of larger-scale organisation. The method of employee remuneration should accord satisfaction to both the employers and the organisation.
8. **Centralisation:** Everything which goes to increase the importance of subordinate role is decentralisation, everything which goes to reduce it is centralisation. Without using the term 'centralisation of authority' Fayol refers the extent to which authority is or decentralised. Centralisation and decentralisation is the natural order, but in large firms, a series of intermediaries is required. Share of authority and initiative left to intermediaries depends on the personal character of the manager, his moral worth, the reliability of his subordinate and also on the relative values of managers and employees are constantly changing, it is desirable that the degree of centralisation or decentralisation may itself vary constantly.
9. **Scalar Chain:** There should be a scalar chain of authority and of communication ranging from the highest to the lowest. It suggests that each communication going up or coming down must flow through each position in the line of authority. It can be short-circuited only is special circumstances when its rigid following would be detrimental to the organisation. For this purpose, Fayol has suggested 'gang plank' which is used to present the scalar chain from bogging down action. His scalar chain and gang plank can be presented in Fig. 1.10.

   In Fig. 1.10, A is the man having immediate subordinates B and L. In turn B and L are having immediate subordinates C and M. This continues to be level of G and Q. Ordinarily, the communication must flow from A to B to C to D, and so on while coming from the top to down. Similarly, it must flow from F to A via E, D, C and B and coming down to P via L, M, N and O. Fayol suggests that this scaler chain system takes time and therefore, can be substituted by gang plank (dotted line) without weakening the chain of command. In order to maintain authority it is desirable that superiors of F and P authorise them to deal directly provided each informs his superiors of any action taken. Fayol suggested that this system allows F and P to deal in a few hours with some questions or other which via the scalar chain would pass through twenty transmissions, inconvenience people, involve masses of paper, lose weeks or months to get to a conclusion.

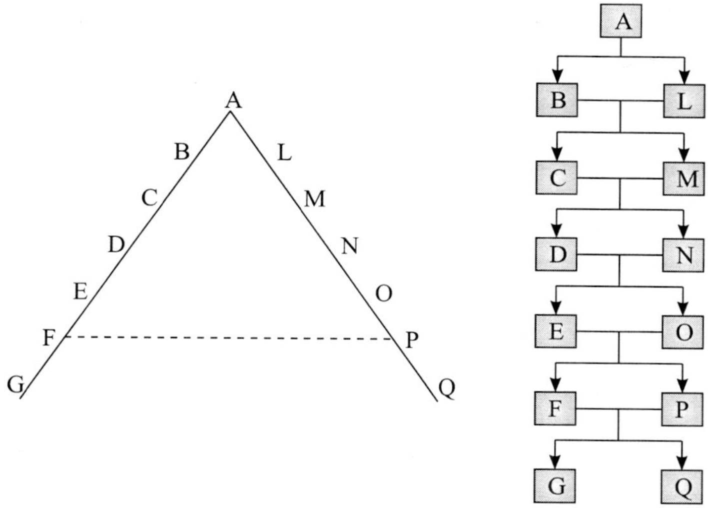

Fig. 1.11 Scalar chain and gang plank

10. **Order:** There must be "a place for everything and everything in its place." This is what is meant by order. Fayol deal with order in material things and also social order. In order to achieve order for material things, there must be a place specified for everything and everything must be in its specified place. For social order to the prevail in the organisation, there must be an appointed place for every employee and every employee must be in his/her appointed place. There will be perfect social order in the organisation if there is right person in the right place.
11. **Equity:** Equity refers to the fair judgement in dealing with human resources. Personnel must be treated with kindness and equity if devotion and loyalty are expected of them. Equity does not exclude forcefulness or sterness. The superiors in the organisation should be experienced and good-natured so as to deal with the subordinates in a proper manner.
12. **Stability of Tensure of Personnel:** This principle calls for lowest possible turnover of personnel for a well-being of the enterprise. Moreover, the employees should not be rotated at different jobs very frequently because considerable time is required to learn each job. According to Fayol, "time is required for an employee to get used to new work and succeed in doing it well, always assuring that he possesses the requisite abilities. If, when he has got used to it or before that, he is removed, he will not have had time to render worthwhile service."
13. **Initiative:** Fayol wanted that the subordinates should be given an opportunity to take some initiative in thinking out and executing the plans. Employees get satisfaction when they are allowed to take initiative. Initiative on their part can be a great source of organisational strength.
14. **Esprit de Corps (Union is Strength):** Esprit de corps means the spirit of loyalty and devotion which units the members of group. It also means the regard for the honour of the group to which they belongs. This principle

calls for harmonious human relations in the organisation so that the employees are loyal to the organisation. Harmony among organisational personnel is source of strength. Unit among personnel can be accomplished by proper communication. Verbal contacts with the personnel should be encouraged as for as possible.

Fayol Warned Against Two Enemies of Spirit de Corps:

(*i*) Divide and Rule

(*ii*) Abuse of Written communication.

It will be dangerous for the firm to divide its workers. They should rather be welded in cohesive and highly interacting work-groups. Over reliance on written communication also tends to disrupt the team spirit. Written communication where necessary, should always be supplemented by oral communication because face-to-face contacts tend to promote speed, clarity and harmony.

Table 1.4 Distinguish between Unity of Command and Unity of Direction

Unity of command	*Unity of direction*
1. It implies that a subordinate should receive orders and instruction from one boss only.	It means "one unit and on plan" for the group of activities having the same objective. This activities of a group should be directed towards its common goal.
2. This principle is related to the functioning.	This principle is related to the functioning of a department or the organisation as whole.
3. Unity of command is necessary of firm responsibility of each subordinate.	Unity of direction is necessary for sound organisation.
4. Unity of command avoids conflict in the undertaking.	Unity of direction avoids duplication of efforts and wastages of resources.

1.9 LEVELS OF MANAGEMENT

Every organisation consists of managerial hierarchy or chain of command which have several levels of authority. In a big company, the management levels may be classified as follows:

1. Top management,
2. Middle management, and
3. Supervisory or operating management.

In a very large enterprise, the middle management level may be sub-divided into upper middle management and lower middle management as shown in Fig. 1.12.

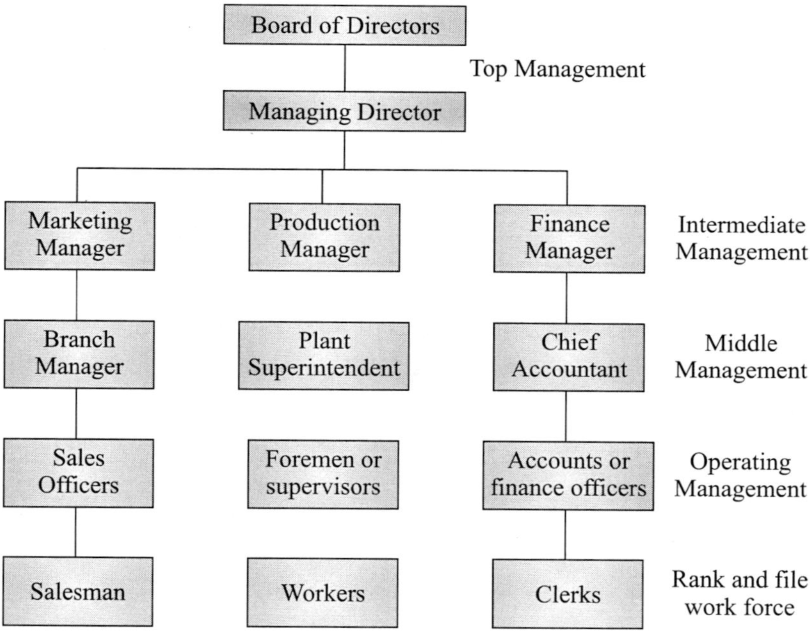

Fig. 1.12 Levels of Managment

1.9.1 Top Management

Top management consists of the Board of Directors and the chief executives. Board of Directors is accountable to the shareholders in the annual general meting of the company. It is basically an organ of overall review and control. Chief executive is concerned with the overall management of the company's operations. He maintains coordination among different departments/subsidiaries/units of the company.

Thus, the main functions of top-management are:

(*i*) To design broad organisation structure.

(*ii*) To analyse and interpret changes in the external environment of the company.

(*iii*) To establish long-term corporate plans (goals, policies and strategies) of the company.

(*iv*) To formulate and approve the master budget and departmental budget.

(*v*) To appoint departmental heads and key executives.

(*vi*) To coordinate and integrate the activities of different departments and divisions of the company.

(*vii*) To provide overall direction and leadership to the company.

(*viii*) To decide the distribution of profits, etc.

(*ix*) To exercise overall review and control of the financial and operating results of the company.

(*x*) To represent the company to the outside world.

1.9.2 Intermediate Management

The intermediate or upper middle management comprises departmental or divisional heads, *e.g.*, works manager, marketing manager, personnel manager, finance manager, etc. It is also known as departmental or functional management. Every divisional or departmental head is the overall incharge of one particular division or department. He is accountable for the performance of his division or department to the chief executive. He coordinates and controls the activities of all personnel working in different branches or sections of his particular department.

1.9.3 Middle Management

Middle management or lower middle management level consists of sectional heads, *e.g.* plant manager, area sales manager, branch manager, office manager, etc. These executives serve as a link between intermediate or top management and the operating management. The main function of middle level managers are:

(*i*) To interpret and explain the plans and policies formulated by top management.

(*ii*) To monitor and control the operating performance.

(*iii*) To cooperate among themselves so as to integrate the various activities of a department.

(*iv*) To train, motivate and develop supervisory personnel, and

(*v*) To lay down rules and regulations to be followed by supervisory personnel.

1.9.4 Supervisory or Operating Management

This is the lowest or first level of management in an organisation. It consists of supervisors, foreman, sales officers, accounts officers, purchase officers, etc. The distinguishing feature of supervisory management is that it deals with non-managers, *i.e.*, workers, where all other levels deals with executives.

The functions of supervisory management are as follows:

(*i*) To plan day-to-day production within the goals laid down by higher authorities.

(*ii*) To assign jobs to workers and to make arrangements for their training and development.

(*iii*) To issue orders and instructions.

(*iv*) To supervise and control workers operations and to maintain personal contacts with them.

(*v*) To arrange materials and tools and to maintain machinery.

(*vi*) To advise and assist workers by explaining work procedures, solving their problems, etc.

(*vii*) To maintain discipline and good human relations among workers.

(*viii*) To report feedback information and workers problems to the higher authorities.

1.10 SKILLS OF A MANAGER

The job of a manager has become very challenging several skills are, therefore required in order to be a successful manager. The skills of an effective manager may be classified into four categories:

- Technical skills
- Human skills
- Conceptual skills
- Diagnostic skills

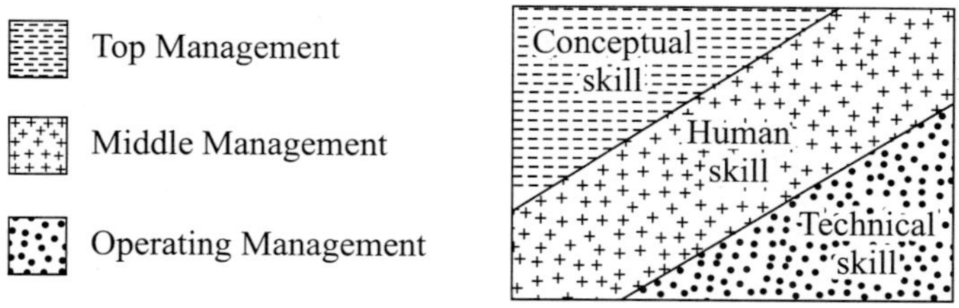

Fig. 1.13 Management skills at various levels

- **Technical skills:** Technical skills refer to the ability and knowledge in using the equipment, techniques and procedures involved in performing specific tasks.
- **Human skills:** Human skills consists of the ability to work effectively with other people both as individuals and as members of a group.
- **Conceptual skills:** Conceptual skills comprise the ability to see the whole organisation and the inter-relationships between its parts. Such skills help the manager to conceptualise the environment, to analyse the forces working in a situation and to take a broad and foresighted view of the organisation.
- **Diagnostic skills:** Diagnostic skills include the ability to determine, by analysis and examination, the nature and circumstances of a particular condition. It is the ability to cut through unimportant aspects and quickly get to the heart of the problem, *i.e.,* logical thinking, analytical reasoning and creativity.

Table 1.5 Managerial Skills

Skills	*Description*	*Example*
• Technical	The use of tools, techniques and procedures in a specialised manner.	• Pathologist analysing a blood sample. • An engineer designing a bridge.
• Human	Interpersonal relationships dealing with selecting, motivating and leading other employees.	• An accounting manager supervising a group of audit accountants. • A manufacturing manager resolving conflict of his team members.

(Contd.)

• Conceptual	The ability to see the total organisational picture by integrating and coordinating large number of activities.	• An analysis by the executive vice-president of the potential effects of a merger with another firm. • Examining the total impact of a proposed labour contract in the firm by the personnel vice-president.
• Diagnostic	The ability to quickly get at the true cause of a certain situation through a maze of data, observation and facts.	• Analysing the causes of employee turnover. • Anticipating changes in consumer buying habits by a marketing manager.

1.11 ROLES OF MANAGER

Some of the *characteristics* that are common most of the managers are as follows:

(*i*) Managers spend a major portion of their time in achieving coordination between human and non-human resources.

(*ii*) Managers do much work at an unrelenting pace.

(*iii*) Managerial tasks are characterised by beauty, variety, and fragmentation.

(*iv*) Managers prefer live action–brief, specific, well-defined.

(*v*) Managers prefer oral to written communication.

(*vi*) Managers maintain a vast number of contacts, spending most time with subordinates, linking them with supervisor and others in a complex network.

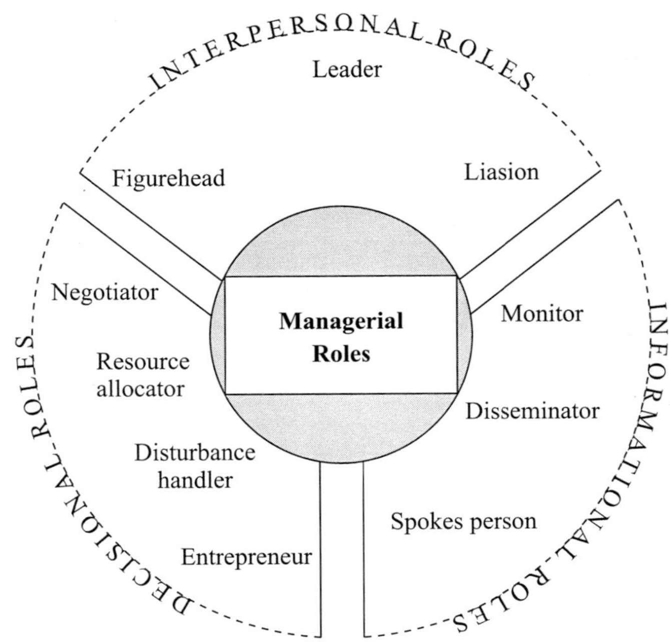

Fig. 1.14 Roles of Manager

1.11.1 Inter Personal Roles

The first set of behaviour concerns interpersonal roles, which include the following.

(*i*) ***Figurehead***: Executive managers perform a number of ceremonial duties such as representing their firm at public affairs and overseeing official functions. Lower level managers have ceremonial duties as well, perhaps on a lesser scale, including attending employees weddings, greeting visitors, and hosting customers.

(*ii*) ***Leader***: This encompasses a range of duties suggested earlier including motivating workers, guiding work-related behaviour, and encouraging activities that help achieve organisational objectives.

(*iii*) ***Liasion***: Managers find themselves acting as liasion (bond) between groups and individuals which are part of, or come in contact with, an organisation. The liasion role is important for establishing contacts with suppliers, coordinating activities among work groups, and encouraging harmony needed to assure effective performance.

1.11.2 Informational Roles

Informational roles refers to the role of manager where he is responsible for gathering and using information to help make effective decisions. Additionally he should be able to communicate to transmit the information and articulate decisions.

Mintzberg's three informational roles are follows:

(*i*) ***Monitor***: Managers monitor activity, solicit information, gather data, and observe behaviour. Well-informed managers are prepared for decision-making and can redirect behaviour to improve organisational performance.

(*ii*) ***Disseminator***: Here communications are reversed. Rather then receive information, managers transmit information. Obviously, this is a crucial aspect of management. Subordinates, superiors and managers of similar work groups rely on timely information disseminated with clarity.

(*iii*) ***Spokesperson***: Top executives find themselves more involved as spokesperson than lower-level managers. A firm's policy on competition, its philosophy of customer care, and its commitment to safety are topics common in executive speeches. However, managers at all levels are spokespersons who may be called upon to represent their groups.

For example*:* When department heads meet to discuss operating budgets, they must be prepared to present information and support budget requests of their respective departments.

1.11.3 Decisional Roles

Mintzberg identified four roles within the list of his behaviourial sets. These are as follows:

(*i*) ***Entrepreneur***: In recent years, entrepreneurs have been identified with a commitment to innovation. Managers in complex organisations let in entrepreneurial way, by constantly trying to improve their operations. They seek new ways of using resources, new technologies for enhanced performance, and new systems of organising human resources.

(*ii*) ***Disturbance Handler*:** Historically, this may be the best understood role of managers because they have always had the primary responsibility for resolving problems. It may also be the most stressful role as managers seem to find themselves constantly faced with disturbances that threaten the harmony and effectiveness of their organisation.

(*iii*) ***Resource Allocator*:** The third role links planning and organising functions. Managers must plan to meet their objectives and distribute resources accordingly. There will be sufficient time, money, materials, or manpower to accomplish all that is expected, so resource allocation often involves carefully assigning scarce resources.

Table 1.6 Roles of Manager

Interpersonal Roles		
Figure head	***Leader***	***Liasion***
Performs symbolic duties: Bestows honours, make speeches, serves on committees.	Motivates, inspires, sets an example, legitimizes the power of subordinates.	Maintains relationship with other organisations, governments, industry groups, etc.

Informational Roles		
Monitor	***Disseminator***	***Spokes person***
Observes, collects and reviews data on the meeting of standards. Notes compliance.	Transmits information about internal and external environments	Speaks for the organisation lobbies and defends. Engages in public relations.

Decisional Roles			
Entrepreneur	***Disturbance Handler***	***Resource Allocator***	***Negotiator***
Initiates changes Authorizes action. Set goals. Formulate plans.	Handles conflicts and complaints. Counters action of competitors.	Approves budgets, schedules and promotions. Sets priorities.	Works out agreement with customers, suppliers and agencies.

1.12 QUALITIES OF A SUCCESSFUL MANAGER

Manager has to direct, inspire and motivate his men at work. He has to clearly set forth the objectives for them and inject in them enthusiasm to achieve the objectives. He has to be a competent person in the area of his specialisation. He must be a man of open mind, moral and emotional integrity and capable to understand and solve problems.

Fig. 1.15 Qualities of a successful manager

- **Ability to Think:** It is difficult to think and think clearly and purposefully. Ability to thinks in clear terms and with definite purpose in view is a must for manager to succeed. He must be able to thinks rationally.
- **Broad vision:** A manager must be able to take into consideration the overall effect of his functioning on the company as a whole. Sectarian view, thinking in terms of 'my department', is not conducive to the smooth working of the company. Therefore, a manager must take a comprehensive view of each of his actions.
- **Clear Expression:** Another function of a manager is to direct his subordinates. It means he must communicate with them. Effective communication depends upon clarity of expression. Even the best conceived idea or instruction is worthless unless it can be communicated effectively. Therefore, the manager must be able to express his ideas clearly.
- **Technical Ability:** Though a manager need to be a technician, it is necessary that he possesses the necessary degree of technical competence relating to his field. This would help him to handle his subordinates effectively. Technical skills are more important at the supervisory level.
- **Human relations skills:** A manager must keep in mind that he has to lead his subordinates and not to drive them. He should possess adequate knowledge of the factors and forces which go to ensure good relations, motivate people for best of their performance and generate cooperative and competent human behaviour.
- **Organisational Ability:** A manager must be a good organiser. Ability to organise well is a quality of vital importance to make a successful manager.
- **Dynamic Personality:** A manager must possess the desire to move ahead, to introduce change for better, to do something new. He must always looks for doing something bigger and better.
- **Emotional stability:** A manager must be able to keep his personal likes and dislikes away from his organisational responsibility. It means that in solving

the business problems of his company, he must not allow his personal feelings to influence his divisions. This requires stable emotional feelings.

WHY MANAGERS FAIL

- They are driven by needs for higher salary.
- They are preoccupied with status symbols.
- They are concerned primarily with self.
- They cut themselves off from others.
- They withhold ideas and feelings.
- Lack of knowledge and skills of managing.
- Ignoring the values of the subordinates.

1.13 STYLES OF MANAGERS

1. **Entrepreneurial Style:** Entrepreneurial style is the style of manager where he acts as a decision-maker and owner for every task.
   - Manager behaves like an owner.
   - Takes bold decisions based on intuition.
   - Is pragmatic and result-oriented.
   - Seeks compliance of orders and standards.
   - Is unstructured and sentimental.
   - Prefers loyal employees.
2. **Conservative Style:** Conservative style of a manager refers to when he sets simple goals and works in a closed setting, involving each and every member of his term in decision-making.
   - Adopts classical approach.
   - Works in a closed setting.
   - Prefers continuity and status quo.
   - Uses tried and tested methods.
   - Set simple goals.
3. **Methodical Style:** Manager who brings the people together for the purpose of working in the organisation to achieve the goal and follows scientific approach. The establishment of formal organisation structure, formal lines of authority had to be undertaken for coordinated effort.
   - Follows scientific approach.
   - Seeks precise and optimum results.
   - Take team members together.
   - Is methodical and systematic.
4. **Quasi-Professional Style:** Manager with reactive and proactive skills, advanced and knows modern management techniques. Cope-up with the changing environment and manages work effectively.
   - Knows modern management techniques.

- Adopts open and dynamic approach.
- Is both reactive and proactive.
- Combines rationality with reality.
- Initiates and manages change effectively.

1.14 FUNCTIONAL AREAS OF MANAGEMENT

There is a very wide scope of management. The management function varies in accordance of its use. Every organisation have different functional area of management require for planning of activities, organisation of resources, establishment of communication system, leading and motivation of people, and control of operations for the realisation of its goals or objectives. The various functional areas of management may be classified into the following categories:

- Production management
- Marketing management
- Financial management
- Personnel management.

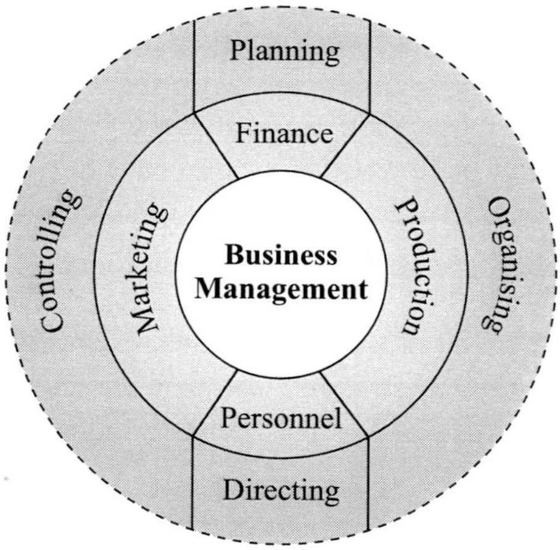

Fig. 1.16 Components of Management

1.14.1 Production Management

Production simply means the creation of utilities, *i.e.*, when raw materials are converted into finished products, it is known as creation of utilities.

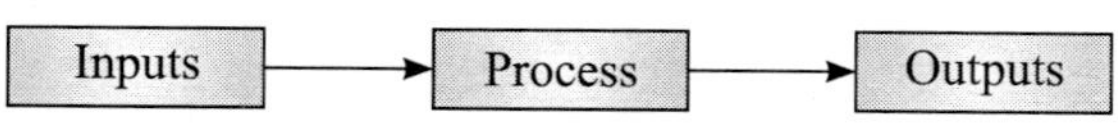

The main aim of any production system is to produce economically the goods and services required by the customers. In order to achieve this aim, it is essential to plan, organise, direct and control the production system. These activities comprise production management and are put under the charge of a 'production manager'. The production manager should have both the technical and managerial qualifications and skills.

> Production management deals with managerial functions related to the design of the production system and operation and control of the production system, i.e., production planning and control.

According to Elwood S. Buffa, "Production management deals with decision-making related to production process, so that resulting goods or service is produced according to the specification, in the amounts and by the schedule demanded, and at minimum cost. If accomplishing these objective, production management is associated with two broad areas of activities—the design and control of production system".

Concept of Production Management

The basic concept of production management is related with **5P's**.

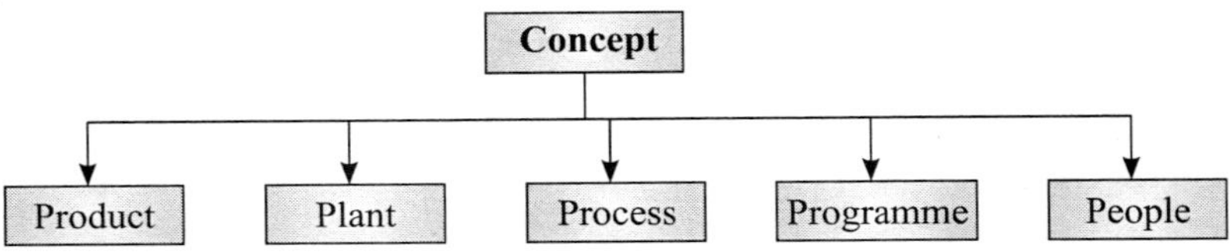

1. **Product:** Production is impossible with the very concept of product. Any produced is to meet the customer specification is terms of performance, quality, quantity reasonable price. In diciding a product, the production manager cannot afford to skip the effect of external factors like technological, economical, legal, social, etc. But internal factors are also important to the considered by the production manager.
2. **Plant:** For the making of a product, plant is required in the form of building and equipment. And like product, this plant should have concern for the level of future demand, what will be performance and reliability of equipment, what will be the safety norms.
3. **Processes:** The process is a sequence of operations and production process is related with those operations which transform the material into finished goods. A process is a series of activities leading to transform one thing to another.
4. **Programmes:** A programme is a schedule of operations arranged in a logical sequence. So in a production programme activities are arranged in series and time to move from one activity to another activity is also determined. Proper time tables are set out for delivery of finished goods. But this can only be possible if proper schedules of purchasing the material, manufacturing of the products are already made.
5. **People:** Production is the act which largely depend on people from beginning to end. It is the people who are responsible for the success of the organisation which includes the production department also. All human beings are different in nature, attitude and behaviour. And production manager has to formulate his policies keeping in mind the needs and requirements of the people.

Functions of Production Management

Production management involves a wide range of activities from the plant location to the packaging of products to be distributed by the marketing department of the enterprise. Production management has a very wide scope and it includes the following operations:

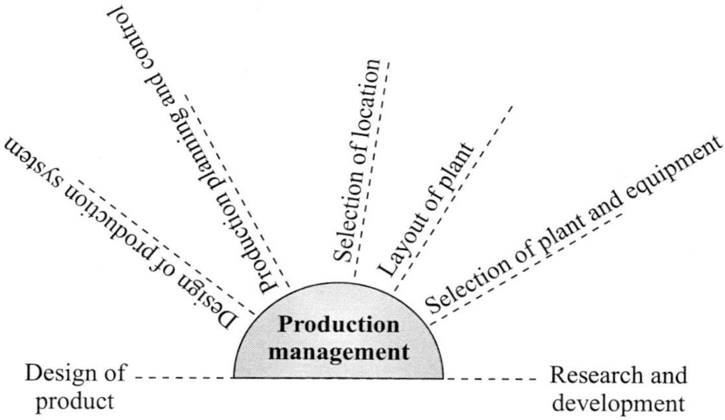

1. **Design of product:** It is the top management which determines the product to be produced by the firm. But the designing of the product is the responsibility of the production manager. The product designing deals with form and function. The form design deals with the product's shape and appearance whereas the functional design deals with its working.
2. **Design of Production system:** Production system is the framework within which the conversion if inputs into output occurs. There are three basic kind of production system, namely, process production, job production and intermittent production. The choice of a production system will depend upon the type of product to be produced and the scale of production carried on by the firm. The suitability of the production systems has been defined as the manufacture in production.
3. **Production planning and control:** It deals with the determination and regulation of production processes. After the production system has been designed and activated, the production manager is concerned with production planning. He establishes the exact sequence of operation of each individual item, part or assembly and lays down the schedule of its completion. Production planning is followed by production control. Production control is a process by which actual performance is compared with the predetermined standards. The production manager has to apply to control in these important areas:

   (*a*) Control of inventories,

   (*b*) Control of flow of raw materials into the plant, and

   (*c*) Control of work-in-process.
4. **Selection of Location:** Plant should be located at such place where production and distribution costs are the minimum. Costs which influence the location

decision include cost of land, rental value, transportation costs, labour cost, cost of water and power, etc. Other factors which influence the selection of location are: process inputs, process outputs, process requirements. Government policy, availability of site, personal, etc.

5. **Layout of plant:** The plant layout represents an arrangement of machines and facilities. The plant layout should be efficient to achieve economy and efficiency in operations of the production department. An efficient layout is one that allows materials to move through the necessary operations rapidly and in the most direct way possible.
6. **Selection of plant location:** The choice of plant and equipment depends upon:
   (*a*) What is available or what can be made available? That is technological feasibility constraints, and
   (*b*) What is economically reasonable: That is cost constraints.
   The quality of output, life of the machine and adaptability of the facilities are also important considerations.
7. **Research and Development:** Research means critical investigation in order to acquire new knowledge. Applied research explores information for the practical problems in mind and thus is directed to achieve immediate solution to practical problems. Development comes after applied research. It involves design and fabrication of new products and the testing their usefulness keeping in view the needs and demands of customers.

1.14.2 Marketing Management

Marketing management is a business discipline which is focused on the practical application of marketing techniques and the management of a firm's marketing resources and activities. It is an important functional area of business which generate revenues through the sale of goods and services to the customers.

The marketing management process consists of analyzing market opportunities, researching and selecting target markets, and implementing and controlling the marketing efforts. Marketing management is responsible for taking the decisions for the product, price, place and promotion keeping in view the requirements of the business.

Functions of Marketing Management

The major function of marketing department is to generate revenue for the business by selling want satisfying goods and services to the customers. In order to achieve this purpose, the marketing manager performs the following functions:

1. **Marketing research:** Marketing research is the function that links the customer, consumer and public to the marketer through information used to identify and define marketing opportunities and problems, generate, refine, and evaluate marketing actions, monitor marketing performance and improve understanding of marketing as a process.

   Marketing research specifies the information required to address these issues, designs the method for collecting information, manages and implements the data collection process.

2. **Product planning and development:** A product is something which is offered by a business firm to customers to satisfy their needs. It has great importance in all other areas of marketing management. For instance, marketing research is mainly directed towards knowing the needs of the customers and increasing the sale of the product, storage and transportation activities depend upon the nature of the product. Therefore, it is necessary to plan and develop products which meet the specification of the customers.
3. **Buying and Assembling:** Procurement of raw materials, semi-finished or finished products has gained great importance for the modern industrial and commercial enterprises. Raw materials are purchased for production by the industrial enterprises and finished goods are purchased for resale by the commercial enterprises.

   Purchasing is different from assembling. Purchasing involves determination of requirements, finding the sources of supply, planning the order and receiving the goods. But assembling means collection of goods already purchased from different sources at a common point.
4. **Selling:** Selling is providing the customer with the products that they want. **Examples**: Retail stores and sales people. Sale may take the form of (*i*) a negotiated sale, and (*ii*) an auction sale. In case of negotiated sale, the terms and conditions between the buyer and the seller are arrived at by bargaining or haggling. But in case of auction sale, there is no scope for negotiation between the seller and buyer. The buyers assembled at the place of auction and bid against one another for the goods are sale.
5. **Financing:** Financing involves getting the money that is necessary to operate a business. Business raise money by selling shares, selling their products and taking out loans. This is the preliminary function. Total budgeting is going to be done while planning for marketing the goods and services.
6. **Pricing:** Pricing a product is perhaps the most important part of the marketing functions. Pricing a product involves choosing a price where the profit generated from a product would be at its maximum. This is based on the competitors, prices and how much the consumer would want to pay for the product. Price of a product is influenced by the cost of product and services offered. Profit margin desired, prices faced by the rival firms and government policy, sound pricing policy is an important factor for selling the products to the customers.
7. **Promotion:** Promotion of a product is crucial to a product's success in the market. Efforts to try and convince consumers through communication to buy the product are generally called promotions. Promotion is one of the four elements of marketing mix (product, place, promotion place).

   It is the communication link between sellers and buyers for the purpose of influencing, informing or persuading a potential buyer's purchasing decision. There are three basic objective of promotion.

   (*i*) To present information to consumers as well as others.

   (*ii*) To increase demand.

   (*iii*) To differentiate product.
8. **Advertising:** Advertising has become an important functions of marketing in the competitive world. It helps to spread the message about the product and

thus promote its sale. It facilitates creation of a non-personal link between the advertiser and the receivers of the massage. The importance of advertising has increased in the modern era of large scale production and tough competition in the market.

Business firms use several media of advertisement to sell their product. These include newspapers, magazines, radio, television, cinema halls, hoardings, windows displays, etc.

So, these are the eight functions or ways, which is going to be considered by the marketing manager during the decision making process while marketing any product or services.

1.14.3 Financial Management

Financial management seeks to ensure the right amount and type of funds to business at the right time and at reasonable cost. Financial management means planning, organizing, directing and controlling the financial activities such as procurement of funds and utilisation of funds of the enterprise. It means applying general management principles of financial resources of the enterprise.

The financial manager of an organisation or business enterprise has to manage various aspect of the finance which determine the scope of his duties.

The various functions are discussed below:

(*i*) ***Formulation of objective*: ** The foremost function of the finance department is the formulation of objective. These objective refers to those activities which are correlated with organisation objectives and helps in the fulfillment of goals of the organisation for the accomplishment of finance department goals, it is necessary to have coordination with other various functional departments.

(*ii*) ***Estimation of capital requirements*: ** A finance manager has to make estimation with regards to capital requirements of the company. This will depend upon expected costs and profits and future programmes and policies of a concern. Estimations have to be made in an adequate manner which increases the earning capacity of enterprise.

(*iii*) ***Determination of capital structure*: ** Once the estimation have been made, the capital structure have to be decided. This involves short-term and long-term deft equity analysis. This will depend upon the proportion of equity capital a company is possessing and additional funds which have to be raised from outside parties.

(*iv*) ***Choice of sources of funds*: ** For additional funds to be procured, a company has many choices like:

(*a*) Issue of shares and debentures.

(*b*) Loans to be taken from banks and financial institutions.

(*c*) Public deposits to be drawn like in form of bonds. Choice of factor will depend upon the relative merits and demerits of each source and period of financing.

(*v*) ***Investment of funds*: ** The finance manager has to decide to allocate funds into profitable ventures so that there is safety on investment and regular returns is possible. The funds procured from various sources are to be prudently invested in various a sets so as to optimise the return on investment. While investment decision, management should be guided by three important principles– safety, profitability and liquidity.

(*vi*) ***Disposal of surplus*:** The net profits decision have to be made by the finance manager. This can be done in two ways:

(*a*) *Dividend declaration:* It includes identifying the rate of dividends and other benefits like bonus.

(*b*) *Retained profit:* The volume has to be decided which will depend upon expansional, innovational, diversification plans of the company.

(*vii*) ***Management of cash*:** Finance manager has to make decisions with regards to cash management. Cash is required for many purpose like payment of wages and salaries, payment of electricity and water bills, payment to creditors, meeting current liabilities, maintenance of enough stock, purchase of raw materials, etc. In order to know the need of cash during different periods, the management should prepare a cash flow statement in advances.

(*viii*) ***Financial controls*:** The finance manager has not only to plan, procure and utilize the funds but he also has to exercise control over finances. This can be done through many techniques like ratio analysis, financial forecasting, cost and profit control, etc.

1.14.4 Personnel Management

Personnel management involves planning, organising, directing and controlling the procurement, development, compensation, maintenance, etc. of the human resources in an enterprise. It consists of various functions for managing the human resource of an organisation.

Edwin B. Flippo has defined Personnel Management as that area of management which deals with planning the activities, maintain and utilising the workforce in the enterprise to help in the achievement of its objectives.

Personnel management consists of the following activities:

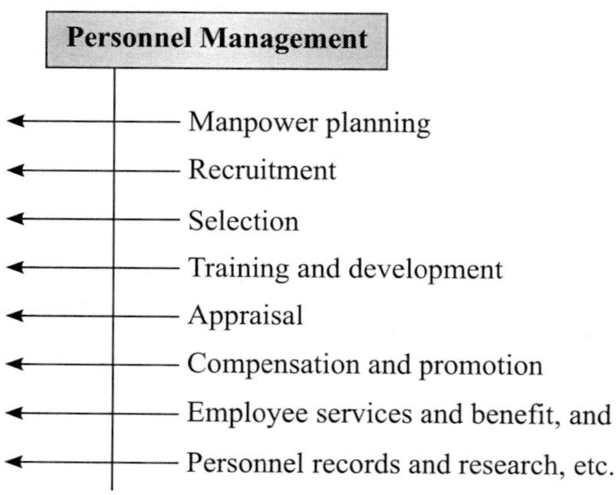

Manpower or Human Resource Planning

Human resource planning may be defined as a process by which the management ensures the right number of people and the right kinds of people at the right place at the right time doing the right things for which they are best suited for the achievement of organisational objectives.

The process of manpower planning involves:

- Forecasting manpower requirements
  - – Work load analysis
  - – Work-force analysis
- Manpower inventory preparation
- Manpower gap Identification
- Formulating manpower plans

Recruitment

Recruitment is the process of identifying the sources for prospective candidates to stimulate there to apply for jobs in the organisation.

It includes seeking and attracting a pool of people from which qualified candidates for job vacancies can be choosen. The process of identification different sources of personnel is known as recruitment. It is a positive process as it attracts suitable applicants who apply for available jobs.

The process of recruitment:

(*i*) identifies the different sources of labour supply.

(*ii*) assesses their validity.

(*iii*) chooses the most suitable source or sources, and

(*iv*) invities applicants from the prospective candidate for the vacant jobs.

Recruitment is concerned with listing the candidates for consideration of selection to various jobs. It enables the management to select suitable employees for different jobs. Recruitment is done before selection or employment of workers. Recruitment is a positive process of searching the prospective employees and attracting them to apply for vacancies.

Sources of Recruitment

Internal sources

1. Transfers
2. Promotions
3. Re-employment of ex-employees

External sources

1. Factory gate
2. Advertising
3. Employment exchange
4. Employment agencies
5. Educational institution
6. Recommendations
7. labour contractor

Selection

Selection is the process of choosing the best person for a particular job. It leads to employment of workers. Selection is a negative process it involves rejection of unsuitable candidates. More candidates are rejected than are selected. Selection involves several steps to weed out the unsuitable candidates for the job under consideration. Criteria are laid down at a each stage. Those who do not fulfil these criteria are rejected.

Training and Development

"Training is the act of increasing the knowledge and skills of an employee for doing a particular job."

Training refers to the acquisition of knowledge, skills and competencies as a result of the teaching of vocational or practical skills and knowledge that relate to specific useful competencies. Whereas training increases job skills, development shapes attitudes of employees.

Development involves growth of an employee in all respects. It denotes the process by which the employees acquire skills and competence to do their present jobs and increase their capabilities for handling higher jobs in the future.

Appraisal

Appraisal refers to the rating or evolution of the worth, merit or effectiveness. Performance appraisal (earlier known as merit rating) implies the formal and systematic evaluation of performance on the job. Almost every large organisation has a formal system of evaluating work performance of its employees because its success depends upon such performance. Performance refers to the degree of accomplishment of the tasks and it is measured in terms of results. According to Flipo, "Performance appraisal is the systematic, periodic and an impartial rating of an employee's excellence in matters pertaining to his present job and of his potentialities for a better job.

Compensation and Promotion

Compensation may be defined as money received in the performance of work, plus the many kind of benefits and services that organisation provide to their employees. 'Money' is included under direct compensation (popularly known as wages, *i.e.*, gross pay); while benefits come under indirect compensation, and may consists of life, accident and health insurance, the employer's contribution to retirement, pay for vacation or illness, and employee's required payments for employee welfare as social security.

Promotion means the transfer of an employee to a job that pays more money or one that enjoys some *preferred status*. A promotion involves reassignment of an employee to a position having higher pay, increased responsibilities, more privileges, increased benefits and greater potential. According to Pigors and Myers' "Promotion is the advancement of an employee to a better jobs in terms of greater responsibilities, more prestige of status, greater skill, and especially higher scale of pay or salary."

Personnel Records and Research

Personnel department maintains the personnel records of the employees working in the enterprise. It keeps full records about their training, achievements, transfer, promotions, etc. It also preserves many other records relating to the behaviour of personnel like absenteeism and labour turnover and the personnel programmes and policies of organisation. Also the research upon the various functions of personnel is necessary for managing the personnel of organisation with the changing environment.

1.15 PROFESSIONAL MANAGEMENT IN INDIA

We often hear the terms 'professional management' and 'family management', the distinguishing features of family or hereditary management are (*a*) both ownership and control of the business are in the hands of members of a family, (*b*) the main emphasis is an profit-maximisation, and (*c*) the family members may or may not have formal education and training in management. Under family management or patrimonial management, the authority of decision-making rests with the members of a family group. Apart from family orientation there is a predominant element of caste or class orientation wherein blood ties and caste ties provide the pass port to the top level of management. Outsiders are debarred from the top posts and serve mainly as yesmen of the family boss.

Family management was said to be effective in the early stages of industrialisation. But India has now reached a stage where professional management is needed. Our inability to make best possible use of modern technology is considered to be mainly due to task of professional management. In the public sector too the bereaucratisation of management has been the main cause of poor performance.

Even top positions in private sector companies are being offered to the professionals having no family ties. In owner-managed companies to management graduates are being preferred to fill managerial positions. This shows that owner-management and professional management can go together. Owner-managed firms are adopting scientific techniques of management to meet pressures of growing competition and modern technology. There is increasing awareness of social complication and modern technology. There is increasing awareness of social responsibility and business ethics. Thus, management in India is gradually getting professionalised.

In India the need for efficient professional management is particularly high due to the following reasons:

1. A large part of the economy is unorganised.
2. The infrastructure is comparatively poor.
3. There is a sharp divergence in technology and traditions requiring tough adjustment and balancing.
4. Productive efficiency is low.
5. The goal of the country is to balance economic growth and social justice.
6. There is widespread poverty and unemployment.

1.16 WHY STUDY MANAGEMENT THEORIES?

What is meant by theory? As dictioniary meaning it is an idea or an explanation which has not yet been proved to be correct.

Theories are perspectives with which people make sense of their world experiences. A theory is a coherent group of assumptions put forth to explain the relationship between two or more observable facts.

Firstly, theories provide a stable focus for understanding what we experience. A theory provides criteria for determining what is relevant.

Secondly, theories enable us to communicate efficiently and thus move into more and more complex relationship with other people. Imagine the frustration a person would encounter if, in dealing with other people, he always had to define even the most basic assumptions he make about the world in which he live!

Thirdly, theories challenge us indeed make it possible to keep learning about our world. As we all know that, theories have boundaries, there is only so much that can be covered by any one theory. Once we are aware of this, we are able to ask ourselves if there are alternative ways of looking at the world especially when our theories no longer seen to fit our experiences and to consider the consequences of adopting alternative beliefs.

1.17 THE EVOLUTION OF MANAGEMENT THOUGHT

Management practice is as old as human civilisation when people started living together in groups. For every human, group requires management and the history of human being is full of organisational activities. However, the study of how managers achieve result is predominantly a twentieth century phenomenon. In the earlier years, management could not get the attention of researchers because the field of business in which the management concepts were applied was held low, unworthy of study, indifferent approach of other social scientists like economists, sociologists, psychologists, etc. towards management and business organisation, treatment of management as an art not as a science, and the commonly held belief that managers are born and not made. These factors created the situation where the need for a systematic study of management was not felt.

Although academic writers and theorists contributed little to the study of management until the early 1950's, previous writings having come largely from practitioners, the past four to five decades have seen a veritable deluge of writing from the academic halls. The variety of approaches to management analysis, the amount of research, and the number of differing views have resulted in much confusion as to what management is, what management theory and science is, and how managerial events should be analyzed. Koontz some years ago called this situation "**the management theory jungle**." Since that time, the vegetation in this jungle has changed somewhat—new approaches have developed, and older approaches have taken on some new meanings with some new words attached, but the developments of management science and theory still have the characteristics of a jungle.

1.17.1 Approaches to Management Analysis

Starting with the oldest school of management theory, different approaches can be listed as follows which is shown in Fig. 1.17:

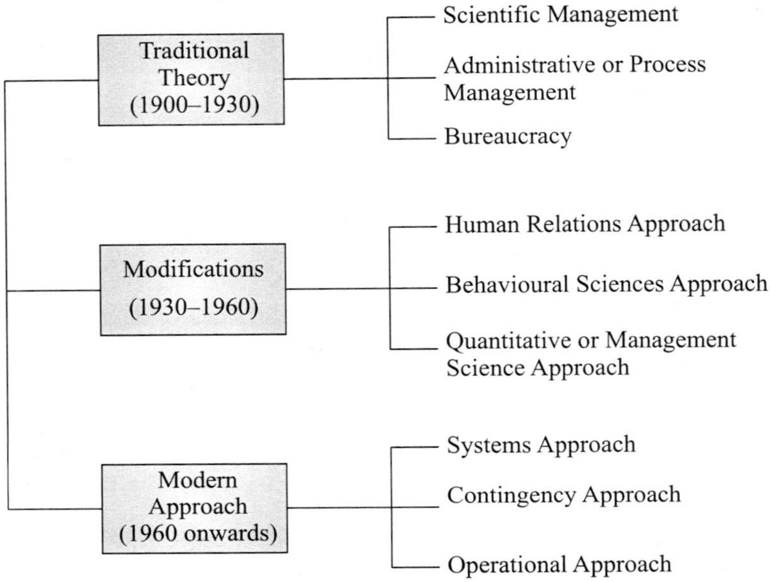

Fig. 1.17

Traditional School of Management Thought: Classical and neo-classical approaches constitute the "traditional school of management thought". These approaches are briefly discussed below:

1. **Classical Theory:** The classical theory represents the traditionally accepted views about the organisations. It signifies the beginning of the systematic study of organizations, that is why, it is said to be the oldest school of thought about organisations and their management. The classical theorists focuses on the organisation structure for the achievement of organisational goals and also developed certain principles of management. Writers who contributed to the classical thought in the early years of the 20th century include Taylor, Fayol, Luther, Weber, Gulick, Urwick, Mooney and many others. This theory is also known as "**Structural Theory of Organisation**" as all the writers are concentrated on finding sound principles of organisation.

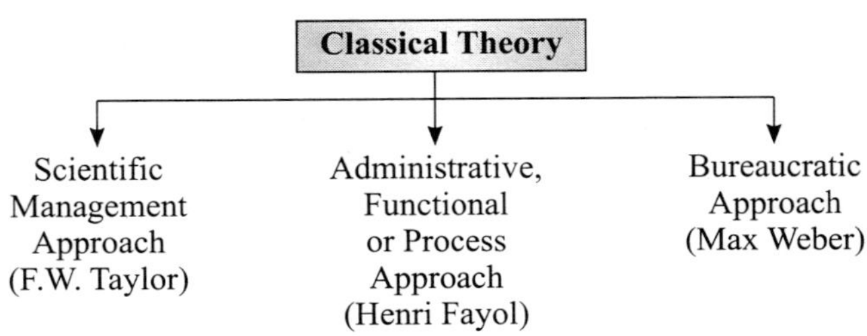

Fig. 1.18

Figure 1.18 incorporates three viewpoints of classical theory:

1. Taylor's Scientific Management
2. Fayol's Administrative Management
3. Weber's Ideal Bureaucracy.

Features of Classical Theory

The salient features of classical approach are as follows:

1. The classical theory placed emphasis on the planning of work, the technical requirements of the organisation, principles of management and the assumption of rational and logical behaviour. Thus, it deals with the anatomy of formal organisation *i.e.*, analysis of the organisation structure. This theory also laid emphasis on division of labour, and specialization, structure, scalar and functional process and span of control.
2. The classical theorists ignored the role of human element whereas emphasized organisation structure for coordination of various activities.
3. The classical theory treated the organisation as closed systems. As this theory ignored the impact of external environment on the working of the organisation.
4. The organisation's efficiency can be increased by making each individual efficient.
5. The integration of the organisation is achieved through the authority and control of the central mechanism. Thus, it is based on **centralization of authority**.
6. The people at work could be motivated by the economic rewards as they are supposed to be 'rational economic persons'.
7. There is no conflict between the individuals and the organisations. In case of any conflict, the interests of the organisation should prevail.

Contributions/Advantages of Classical Theory

Classical Theory leads to various contributions such as:

1. Classical theory provides a scientific basis for management practices. It also provides a starting point for researchers to verify the validity and to improve the applicability of management knowledge and such knowledge about management is effectively presented.
2. In order to motivate the workers, this approach is completely dependent on money or financial incentives.
3. This approach focussed on the functions of managers in different types of organisation.
4. Managers can modify the management principles (as they are flexible in nature) in order to suit the given situation.
5. It gave supreme importance to the organisation and its objectives. People are supposed to follow the organisation's policies, procedures and to obey the orders of their superior in order to achieve the organizational objectives.

Criticism/Disadvantages of Classical Theory

The classical theory has been criticized on the following grounds:

1. **Narrow View of Organisation:**
   - This theory is limited by its narrow concentration on the "anatomy of formal organisation".
   - This theory ignored the human relations aspect.
   - This theory neglected the interplay of individual personality, informal groups and inter-organizational conflicts in the formal organisation.
   - This theory is focussed on 'organisation without people'.
2. **Assumption of Closed System:**
   - Under this theory, the organisation is viewed as a closed system; *i.e.*, having no interaction with the environment which is totally unrealistic.
   - Whereas, a modern organisation is an open system which has continuous interaction with the environment through the exchange of inputs, outputs and various types of information.
3. **Assumption about Human Behaviour:**
   - As this theory is focussed on 'organisation without people', the human beings were merely treated as any other factors of production and were supposed to obey their superiors.
   - The classical writers ignored the social, psychological and motivational aspects of human behaviour.
4. **Economic Rewards as Main Motivators:**
   - This theory assumed that people at work can only be motivated solely through economic rewards may be money or financial incentive which is wrong.
   - Several researches in human behaviour have contradicted this assumption. Non-monetary factors like job enrichment, better status, etc. can also motivate the workers.
5. **Lack of Empirical Verification:**
   - The classical principles are mostly based on the personal experiences and limited observations of the practitioners. They are not based on empirical research.
   - They lack precision and comprehensive framework for analysis. That is why, it is not clear whether these principles are action recommendations or simply definitions.
6. **Lack of Universality of Principles:**
   - This theory claimed that their principles have universal application. This suggest that the same principles can be applied in different organisations, different management levels and different functions of the same organisation.

- However, the empirical researches, suggest that none of the principles have such characteristics. Moreover, there are many principles which contradict with other principles. For example, the principle of specialization is quite in conflict with the principle of unity of command.

7. **Excessive Emphasis on Rules and Regulations:**
   - Weber's 'ideal' bureaucracy, a major constituent of classical theory, suggested strict adherence to rules and regulations and the scope for individual initiative is thus limited.
   - The result is red-tapism in the organisation. The main objective becomes the observation of rules and regulations while the real objectives for which these rules and regulations are formed are forgotten.

As the classical thought can be studied under three streams. Let us discuss it one by one.

(*i*) **Scientific Management Approach:** The industrial revolution in England gave an immense impetus for the scientific mechanisation approach. It brought about such an extraordinary mechanization of industry that necessitated the development of new management principle and practices. Bringing group of people together for the purpose of working in the factory posed problem for the factory owner. The establishment of formal organisation structure, formal lines of authority, factory systems and procedures had to be undertaken for coordinated effort. In order to deal with these problems, a management movement comes known as "Scientific Management".

Frederick Winslow Taylor (1865–1915) was the first person to recognise and emphasise the need for adopting a scientific approach to the task management. He launched a new movement in 1910 which is known as 'Scientific Management' that is why, Taylor is regarded as the **father of scientific management**.

The introduction of the concept of standard time, standard output, standard cost, standardization of production process, change in the attitude of management and workers to bring about the mutuality of the interests are the important landmarks of Scientific Management. This approach was supported and developed by Henery L Gantt, Frank Gilbreth, Lillian Gilbreth, Harrington Emerson etc.

According to Taylor, "Scientific Management is the substitution of exact scientific investigations and knowledge for the old individual judgement or opinion in all matters relating to the work done in the shop." It implies the application of science to the management of a business concern. It aims at replacement of traditional techniques by scientific techniques.

Principles of Scientific Management:

The Scientific Management is based on five principles which are as follows shown in Fig. 1.19:

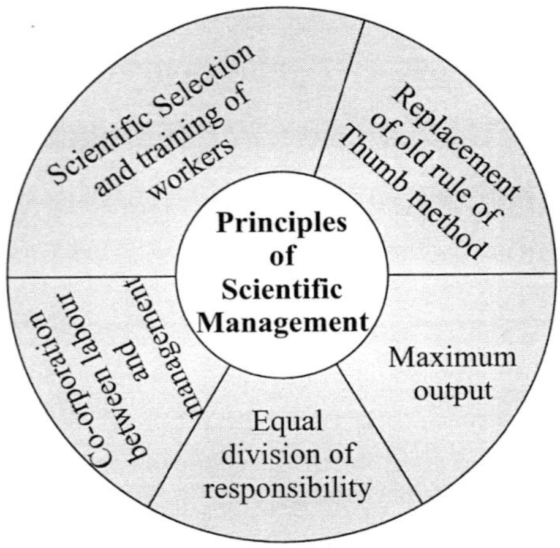

Fig 1.19

1. **Replacement of old rule of thumb method:** Scientific investigation should be used for taking managerial decisions instead of basing decision on opinion, intuition or rule of thumb.
2. **Scientific selection and training of workers:** Management is responsible for their scientific education and training, thus, every organisation should follow a scientific system of selection. In which, the selected workers are to be trained to avoid wrong methods of work.
3. **Coorporation between labour and management:** There should be co-operation between the management and workers. This requires change of mental attitude of the workers and management towards each other. Taylor called it "mental revolution".
4. **Maximum output:** The management and the workers should try to achieve maximum output in place of restricted output. This will be beneficial to both the parties. Maximum output will also be in the interest of the society.
5. **Equal division of responsibility:** The responsibility must be equally divided between the managers and the workers. The management should be responsible for planning and organising the work where as the workers should be responsible for the execution of work as per instructions of the management.

Mental Revolution: The basic idea behind the principles of scientific management is to change the mental attitude of the workers and the management towards each other. Taylor called it "*mental revolution.*"

The mental revolution consists of three aspects:

(*i*) all out efforts for increase in production,

(*ii*) creation of the spirit of mutual trust and confidence, and

(*iii*) inculcating and developing the scientific attitude towards problems.

Taylor suggested that management should try to find the best methods of doing various jobs and introduced standardised materials, tools and equipments so that wastages are reduced. As well as, the workers should be discipline, loyal and sincere in fulfilling the task assigned to them. They should not indulge in wastage of resources. Both the management and the workers should trust each other and cooperate in achieving maximum production.

Thus, Taylor stood for creating a mental revolution on the part of management and workers. Taylor's thinking was confined to management at the shop level, however, he demonstrated the possibility and significance of the scientific analysis of the various aspects of management. To sum up, he laid emphasis on the following:

1. Science, not rule of thumb.
2. Harmony in group action, rather than discord.
3. Maximum output in place of restricted output.
4. Scientific selection, training and placement of the workers.
5. Development of all workers to the fullest extent possible for their own and organisations highest prosperity.

- **Techniques of Scientific Management:** To put the philosophy of scientific management into practice, Taylor and his associates suggested the following techniques:
  1. **Scientific Task Setting:** It is essential to set the standard task which an average worker should do during a working day, Taylor called it "a fair day's work."
  2. **Work Study:** Work study implies an organised, objective, systematic, analytical and critical assessment of the efficiency of various operations in an enterprise. It include the following techniques:
     (*i*) ***Method study*:** It is the study which is conducted to know the best method of doing a particular job. It helps in reducing distance travelled by materials and brings improvement in handling, transporting, inspection and storage of raw materials and goods.
     (*ii*) ***Motion study*:** It is the study of the movement of an operator or a machine. Its purpose is to eliminate useless motions and find out the best method of doing a particular job.
     (*iii*) ***Time study or work measurement*:** Time study is the technique of observing and recording the time required to do each element of an industrial operation. Through this study, the precise time required for each element of a man's work is determined. It helps in fixing the standard time required to do a particular job.
     (*iv*) ***Fatigue study*:** Fatigue, physical or mental, has an adverse effect on worker's health and his efficiency. This study helps in reducing fatigue among the workers.
  3. **Planning the Task:** Taylor emphasised the need for planning the work. He advocated that planning function should be separated from the executive function. Workers should not be asked to choose their own methods and decide what they have to do. The detailed planning should be done by the planning department. The planning department should prepare detailed instructions for the workers as to the type, quality and quantity of the products to be produced.

4. **Standardisation:** Taylor advocated the standardisation of tools and equipment, cost system and several other items. Efforts should be made to provide standardised working environment and methods of production to the workers.
5. **Scientific Selection and Training:** The management should design scientific selection procedure so that the right men are selected for the right job. Workers should be specifically trained for the jobs they are appointed so that they can perform their job is effectively.
6. **Differential Piece-Wage Plan:** This plan was suggested by Taylor to attract highly efficient workers. Under this plan, there are two piece work rates, one is lower and another is higher. The standard of efficiency is determined either in terms of time or output based on time and motion study. If a worker finishes work within standard time or he produces more than standard output within the standard time, he will be given higher piece rate. On the other hand, if a worker is below the standard, he shall be given lower piece rate.
7. **Specialisation:** Taylor advocated 'functional foremanship' to introduce specialisation. He recommended eight foremen in all to control the various aspects of production. He suggested four foremen in the planning department, namely, route clerk, instruction card clerk, time and cost clerk and shop disciplinarian. The four foremen recommended for getting the required performance from the workers include gang boss, speed boss, repair boss and inspector.

Benefits of Scientific Management: Taylor's Scientific management was associated with many benefits to the industry. Scientific management leads to the following benefits:

1. Traditional rule of thumb method is being replaced by scientific techniques for each element of a man's work.
2. Proper selection and training of the workers.
3. Establishment of good relationship between the workers and the management.
4. Achievement of equal division of responsibilities between the workers and the management.
5. Standardisation of tools, equipment, materials and work methods.
6. Detailed instructions and constant guidance of the workers.
7. Incentive wages to the workers for higher production.
8. Elimination of wastes and rationalisation of system of control.
9. Better utilization of various resources.
10. Satisfaction of the needs of the customers by providing higher quality products at lower price.

- **Criticism of Scientific Management:** Taylor's Scientific Management was criticized not only by the workers and the managers but also by the psychologists and the general public. The main grounds of criticism are as follows:

1. The use of the word 'scientific' before 'Management' was objected because what is actually meant by scientific management is nothing but "a scientific approach to management."
2. It was argued that the principles of scientific management as given by Taylor were confined mostly to production management. He ignored certain other essential aspects of management like finance, marketing, accounting, and personnel.
3. Taylor advocated the concept of functional foremanship to bring about specialisation in the organisation. But this is not feasible in practice as it violates the principle of unity of command.
4. Scientific management undermined the human factor in industry. It resulted in monotony of job, loss of initiative, overspeeding workers, wage reductions, job insecurity etc.
5. Trade unionists regarded the principles of scientific management as the means to exploit the labour because the wages of the workers were not increased in direct proportion to productivity increases.

Many of the above objections were later remedied by the other contributors to scientific management like Henri L Gantt, Frank Gilbreth, Lillian Gilbreth and Harrington Emerson. Many of the recommendations of Taylor are still being applied by the modern business undertaking. In short, in can be said that Taylor was the pioneer in introducing scientific reasoning to the discipline of management.

(*ii*) **Management Process School (Administrative Management or Functional Approach):** Management process school is also called the 'traditional' or 'universalist' school as it believes that management principles are applicable to all kinds of group activities.

The advocates of this school perceive management as a process getting things done through and with people operating in organised groups. This approach seeks to analyses the process, to establish a conceptual framework for it, to identify principles underlying it and to build a theory of management as a universal process, regardless of the type of the enterprise, or the level in a given enterprise, although recognising obviously that the environment of management differs widely between enterprises and levels. It looks upon management theory as a way of organising experience so that practices can be improved through research, empirical testing of principles and teaching of fundamental involved in the management process.

Henri Fayol is regarded as the father of this school. Henri Fayol defined management in terms of certain functions and then laid down fourteen principles of management which according to him have universal applicability.

Fayol's Theory of Management

Fayol began by classifying all operations in business organisations under the following six categories as shown in Fig. 1.20.

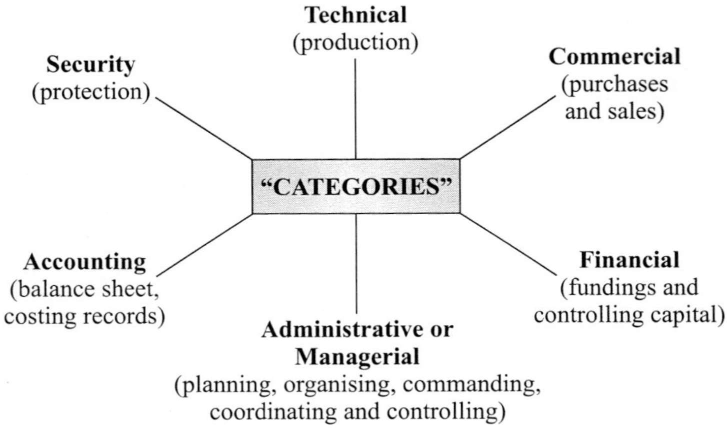

Fig. 1.20

Fayol pointed out that the managerial activity deserved more attention. In his view, management is the process composed of five elements of functions: **planning, organising, commanding, coordination and control**.

Fayol observed:

(*i*) to plan means to study the future and arrange the plan of operations

(*ii*) to organise means to build up the material and human organisation of the business

(*iii*) to command means to make the staff do their work

(*iv*) to co-ordinate means to unite all activities and

(*v*) to control means to see that everything is done in accordance with the standards that have been laid down and the instructions given.

- **Principles of Management**

Fayol completed his theory by stating that to be effective, management should be based on the following fourteen principles as shown in Fig. 1.21.

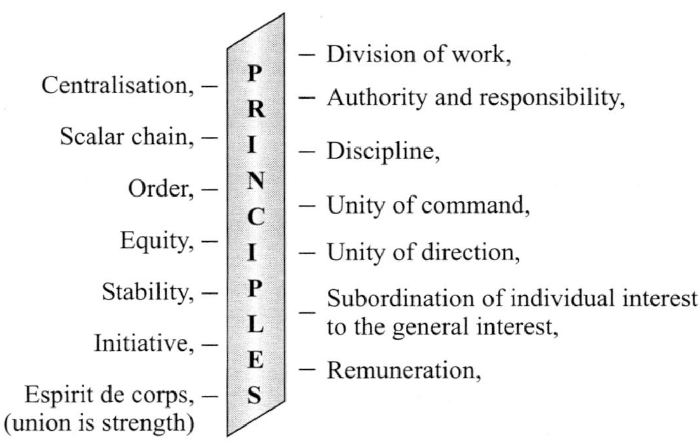

Fig. 1.21

It is significant to point out that 'Fayol' used the word "principles" only for convenience. He did not consider them immutable laws. However, he thought that the principles would be useful to all types of group activities. Thus, he was a universalist in this sense. His theory of management completely revolutionised the thinking of managers as throughout his treatise, there exists an understanding of the universality of the principles. Some of these principles were amplified later on by the writers like Sheldon, Urwick, Mooney and Barnard.

- **Comparison of Taylor and Fayol**

The difference between the contribution of Taylor and Fayol are discussed below:

1. Taylor approach was a kind of efficiency movement. But, Fayol had wider perspective. He tried to evolve principles which could be applied to administration of different spheres.
2. Taylor emphasised on the first line supervision in production area whereas Fayol concentrated on the higher levels of the organisational setup.
3. Taylor used the expression scientific management whereas Fayol had general theory of management. The objective of Taylor was to improve the productivity of workers and eliminate all types of wastes. Fayol developed principles of management having universal applicability.
4. Taylor is known as the father of scientific management. Fayol is called the father of management process or functional management.
5. Taylor paid greater attention to the standardisation of work and tools. This main concern was to increase the efficiency of the workers and managers at shop levels. But Fayol's management theory is from the point of view of the top managers, and his principles are applicable at all levels.

Table 1.7: F.W. Taylor vs Henri Fayol

Point of difference	*F.W. Taylor*	*Henri Fayol*
1. Concern	Taylors main concern were tasks, workers and supervisors.	Fayol's work was concerned more with the efficiency of top level management.
2. Emphasis	Taylor laid great emphasis on standardisation of work and tools. The principles emunicated by him were more applicable at shop level.	Fayol laid emphasis on the principle of general management and the function of managers.
3. Expression	Taylor used the expression scientific management.	Fayol expressed his idea as general theory of administration.
4. Focus	Taylor tried to improve the productivity of the workers, eliminate all kinds of waste.	Fayol tried to develop principle to ensure of better and management of the whole enterprise.
5. Level/ Perspective	Taylor started his study and approach from lowest level in the organisation.	Fayol started his study and approach from the highest level in the organisation.

(Contd.)

6. Major contribution	Tayler's main contribution was development of scientific techniques and scientific principles.	Fayol's main contribution was development of fourteen principles of general management.
7. Personality	Taylor was a scientist and become famous as father of scientific management.	Fayol's had a personality of researcher and practitioner. He become famous as a father of general management.
8. Unity of command	Taylor didn't follow the principles on eight bosses functional foreman to each employee in their respective areas.	Fayol strictly followed the principle, *i.e.,* only one boss for an employee.
9. Application	Applicable to industrial organisation only.	Applicable to all kinds of organisation.

(*iii*) **Bureaucracy Approach: Max Weber** was the one who contributed the theory of bureaucracy to the management thought. He used this word "bureaucracy" to refer to a specific kind of administrative organisation whose characteristics are stated later. Max Weber's primary contribution to management is his theory of authority structure and his description of organisations based on the nature of authority relations within them.

Essentially, it was Weber's contention that there are three types of legitimate authority which are as follows:

1. **Rational-legal authority:** Obedience is owed to a legally established position or rank within the hierarchy of a business, military unit, government and so on.
2. **Traditional authority:** People obey a person because he belongs to certain class or occupies a position traditionally recognized as possessing authority, such as a royal family.
3. **Charismatic authority:** Obedience is based on the follower's belief that a person has some special power or appeal.

Weber's theory of bureaucracy recognizes rational-legal authority as the most important type in organisations. Under traditional authority, leaders are not chosen for their competence and charismatic authority is too emotional and irrational.

- **Characteristics of Bureaucracy:** A bureaucratic organisation displays the following characteristics:

  (*i*) ***Division of Work*:** There is high degree of division of work at both the operative and administrative levels. This leads to specialisation of work.

  (*ii*) ***Hierarchy of Positions*:** There is a hierarchy of authority in the organisation. Each lower position is under the control of the higher one. Thus, there is unity of command. The bureaucratic structure is hierarchical in nature. It is like a hierarchy in which quantity of authority increases as one moves up the ladder in the organisation.

  (*iii*) ***Rules and Regulations*:** The rules, regulations and procedures are laid down by the top administration. Their benefits are as follows:

  - They standardize operations and decisions.

- They serve as receptacles of past learning.
- They ensure equality of treatment to everybody.

(*iv*) ***Impersonal Conduct*:** There is impersonality of relationships among the organisational members. The decisions are entirely guided by rules and regulations and are totally impersonal. The essence of bureaucracy is totally depersonalisation that is there is no room for emotions and sentiments in this type of structure.

(*v*) ***Staffing*:** The personnel are employed by a contractual relationship between the employee and employer. The tenure of service is governed by the rules and regulations of the organisation. The employees get a salary every month which is based on the job they handle and also the length of service.

(*vi*) ***Technical competence*:** The bureaucrats are neither elected nor inherited but they are appointed through selection and the basis of selection is their technical competence. Promotions in bureaucracies are also based on technical qualifications and performance.

(*vii*) ***Official records*:** The administration of a bureaucratic organisation is supported by an efficient system of record keeping. The decisions and activities of the organisation are formally recorded and preserved safely for future reference. This is made possible by extensive filing system. Written documentation and formal filing root out the possibility of loopholes in the oral communication among the organisational members. The official records are considered as the encyclopedia of various activities performed by the organisation. The filing system makes the organisation. The filing system makes the organisation independent of individuals. The official records serve as the memory of the organisation.

- **Merits/Benefits of Bureaucracy:** Weber's ideal bureaucracy has been designed to bring rationality and predictability of behaviour in organisations. Because of its characteristics of division of work, impersonality of relationships, primary of rules, employment and promotion on the basis of technical competence, etc., it also contributes to the efficiency in operations. Bureaucracy is an administrative device that can help in achieving the following advantages:

(*i*) ***Competence*:** The word "competence" means capacity or ability of a person. There is proper delegation of authority in the organisation. People are given task according to their competence.

(*ii*) ***Rules and Regulations*:** All actions are taken carefully because of rules and regulations. There is consistency of actions.

(*iii*) ***Rationality*:** The behaviour of the employees is rationale. They take decisions as per laws, rules and regulations. They don't go by their whims, emotions or prejudices.

(*iv*) ***Predictability*:** The behaviour of the employees is predictable. It is known how they will react under different situations as guidelines are already there in writing.

(*v*) ***Efficiency*:** Bureaucracy leads to efficiency in the organisation. There is division of work leading to specialisation which resulted in efficiency.

(*vi*) ***Impartiality***: Officials are guided by the policies, rules and regulations rather than their personal whims and faces. They are not supposed to shower personal favours on anybody.

- **Demerits/Criticism of Bureaucracy:** Bureaucracy is not free of flaws. It may lead to certain undesirable consequences as discussed below:

  (*i*) ***Rigidity in Operations***: Rules and regulations in a bureaucracy are often rigid and inflexible. Strict compliance with rules and regulations discourages initiative and creativity. It may also provide a cover to avoid responsibility for failures. This structure is not effective under turbulent or dynamic environment. It can't undergo the change required by the fast changing environment.

  (*ii*) ***Delay and Red Tapism***: The rules may be followed in letter and not in spirit. Thus, the rules may become a source of inefficiency leading to delay in operations. The rules may be misused by the person concerned with the implementation of rules. Red tapism and technicalities may follow as a result.

  (*iii*) ***Goal Displacement***: Goal displacement may take place in a bureaucratic organisation. The bureaucrats may give priority to rules and regulations or to the secondary goals and forget about the primary goals. In other words, means become the "ends" and the ends or goals become the "means" leading to goal displacement.

  (*iv*) ***Ineffective Communication***: The bureaucratic structure is fall consisting of several layers of executives. Thus, the communication form the top level to the lowest level will take a very long time.

  (*v*) ***Compartmentalization of Activities***: Jobs are divided into watertight categories which restricts people from performing tasks that they are capable of performing. It encourages preservation of jobs even when they have become redundant.

  (*vi*) ***Empire Building***: Officials in bureaucracy tend to use their position and resources to perpetuate self-interests or the interests of their sub-units. Every departmental head tries to increase the number of his subordinates as a symbol of power and prestige.

  (*vii*) ***Lack of Personal Touch***: Bureaucracy is based on impersonal relationships. It doesn't allow inter-personal relations between employees and informal groups in the organisation.

1. Social System Approach

Chester I Barnard developed the concept of social system. He viewed organisation as a social system that is composed of people who work in cooperation. An organisation comes into existence when:

- there are a number of persons in communication with each other, and
- they are willing to cooperate for a common purpose.

Barnard also recognised informal organisation representing social interactions which generally do not have a consciously coordinated joint purpose. The executives should encourage informal organisations to serve as a means of communication and group cohesiveness.

Barnard's treatment of management differed considerably from that of Taylor and Fayol. Taylor concentrated on improving the task efficiency of the individual. Fayol, on the other hand, moving to the totality concept of management directed his analysis to the operational side, *i.e.,* principles and functions of management. But Barnard started with the individual, moved to cooperative organised endeavour and ended with executive functions.

The broad features of the social system approach are as follows:

- Organisation is treated as a social system. That is why, this approach resembles the human relations approach.
- Relationships exist among the external and internal environment of the organisation.
- Cooperation among group members is necessary for the achievement of organisational goals.
- For effective management, efforts should be made for establishing harmony between the goals of the organisation and those of the various groups within.

Barnard identified the following functions of an executive which are as follows:

(*a*) The maintenance of organisational communication.

(*b*) Securing essential services from individuals in the organisation, and

(*c*) Formulating and defining the purpose.

By performing all these functions, the executives can achieve good human relations in the organisation. Barnard also developed a new concept of authority known as "acceptance authority".

- **Appraisal of Social System School:** As it is well known that Chester I. Barnard is regarded as the Founding Father of Social System School. He studied the inter-relationships within the organisation. His definition of formal organisation is regarded to be a major contribution in the field of management. The main focus is to study the different aspects of social system. For the adherents of this school, organisation is essentially a socio-cultural system composed of groups of people who work in cooperation. For achieving the goals of the organisation, a cooperative system of management can be developed only by understanding the social behaviour of groups and individuals. That is, an organisation is affected by the socio-cultural environment and different types of pressures.

2. Modifications

(*i*) **Neo-Classical Theory (Human Relations Approach):** The classical writers including Weber, Taylor and Fayol neglected the human relations aspect. Where as, the neo-classicists focussed on the human aspect of industry. The classical theory was modified by emphasizing the fact that the organisation is a social system and the human factor is the most important element within it. They conducted some experiments known as Hawthorne Experiments and investigated informal groupings, informal relationships, patterns of communication, patterns of informal leadership, etc. This led to the development of Human Relation Approach.

Elten Mayo is recognized as the Father of Human Relations School. Other prominent contributors to this school are Roethlisberger, Dickson, Dewey, Lewin etc.

This approach is concerned with the recognition of the importance of human element in organisations. It revealed the importance of social and psychological factors in determining workers productivity and satisfaction. The neo-classical approach put stress on inter-personal relations and informal groups at the work place.

The human rationalists argued that it is impossible to achieve organisational objective without the cooperation and willing of people and such cooperation cannot be automatically secured or ordered, it has to be achieved.

The neo-classical approach advocated people-oriented organisation structure which will integrate both formal and informal organisations.

- **Principles of human relation approach:** The basic tenets of neo-classical theory or human relations approach are as under:
  (*a*) The business organisation is a social system.
  (*b*) The individuals behaviour is dominated by the informal group of which he is a member.
  (*c*) An individual employee cann't be motivated by economic incentive (money) alone. His social and psychological needs must be satisfied to improve the level of motivation.
  (*d*) It is ultimately cooperative attitude in an organisation and not the mere command which yields result.
  (*e*) Management must aim at developing social, leadership as well as technical skills. It must take interest in the welfare of workers.
  ((*f*) Morale and productivity go hand in hand in an organisation.
- **Factors Affecting Human Relations:** Organisation should be viewed as a social system. Human relation in the organization are determined by the individual, work-group, leader and work environment as shown in Fig. 1.22.

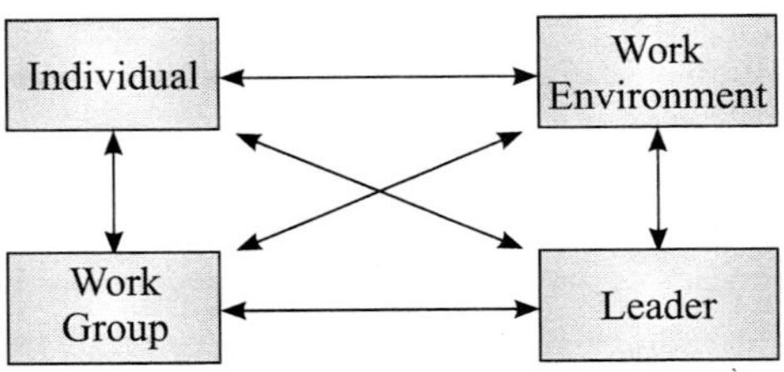

Fig. 1.22 Human relations in action

(*i*) ***Individual*:** The individual is an important element of the organisation and each individual is unique. Individual's behaviour is affected by his feelings, sentiments, values and attitudes. Motivation of employees should give due consideration to their economic, social and psychological needs. Thus, motivation is a complex process.

(*ii*) ***Work-Group*:** The work group is the centre of focus of human relations approach. It has important role in determining the attitude and performance of individual workers. The Hawthorne studies showed that the informal group exert tremendous influence over the behaviour pattern of workers.

(*iii*) ***Work Environment*:** Human relationist advocated the creation of a positive work environment where organisational goals are achieved through satisfaction of employees. In general, when employee's needs are satisfied, the work environment is termed positive and when employee needs are not satisfied, the work environment is termed negative.

(*iv*) ***Leader*:** The human relationists gave great importance to leadership. The leader must ensure full and effective utilisation of all the organizational resources to achieve the organizational goals. He must be able to adjust to various personalities and situations. He must behave in a way that generates respect. As the Hawthorne studies showed, a supervisor can contribute significantly in increasing productivity by providing a free, happy and pleasant work environment.

The essence of the human relation philosophy is to cultivate and develop an environment where the employees as individuals and in groups would wish to contribute their best to the organisational goals.

- **Contributions of Human Relation Approach:** (Features of Neo-classical Approach)

The features of neo-classical approach are as follows:

1. **Social System:** The organisation is a social system which is composed of numerous interacting parts. The social system defines individual roles and establishes norm that may differ from those of the formal organization. The workers follows a social norm which is determined by the co-workers, which defines the proper amount of work, rather than try to achieve the target management thinks they can achieve, even though this would have helped them to earn as much as they physically can.
2. **Social Environment:** The social environment on the job affects the workers and is also affected by them. Management is not the only variable.
3. **Informal Organisation:** The informal organisation doesn't exists within the frame work of formal organization and it affects and is affected by the formal organisation.
4. **Group Dynamics:** The group plays an important role in determining the attitudes and performance of individual workers. At the workplace, the workers often don't act or react as individuals but as members of groups. A person who resists pressure to change his behaviour as an individual often changes it quite readily if the group of which he is a member changes its behaviour.
5. **Informal Leader:** There is an emergence of informal leadership as against formal leadership and the informal leader sets and enforces group norms. He helps the workers to function as a social group and the formal leader is rendered ineffective unless he conforms to the norms of the group of which he is supposed to be incharge.
6. **Two-way Communication:** Two-way communication is the process of interacting or sharing of the information between the sender and receiver and feedback is also given that the information is received. Two-way communication is necessary because it carries necessary information down for the proper functioning of the organisation and transmits upward the feelings and sentiments of people who work in the organisation.
7. **Non-economic Reward:** As it is well known that money is one of the motivators, but not the role motivator of human behaviour. Man is diversely motivated and socio-psychological factors act as important motivators.

- **Criticism of Human Relation Approach:** The neo-classical approach has been criticized on the following ground:
  (*i*) ***Lack of Scientific Validity*:** The human relationists drew conclusions which are based on clinical insight rather than on scientific evidence. The groups chosen for study were not representative in character. The findings based upon temporary groups don't apply to groups that have continuing relationship with one another. Moreover, the experiments focussed on operative employees only.
  (*ii*) ***Limited Focus on Work*:** The human relation approach put all the emphasis on interpersonal relations and on the informal group. It tends to overemphasize the psychological aspects at the cost of the structural and technical aspects.
  (*iii*) ***Over-emphasis on Group*:** The human relations approach over-emphasis the group and group decision-making. But in practice, groups may create problems and collective decision-making may not be possible.
  (*iv*) ***Over-streching of Human Relations*:** It is assumed that all organisational problems are amenable to solutions through human relations. This assumption does not hold good in practice.
  (*v*) ***Over-stress on Socio-physiological Factors*:** The human relations approach undermines the role of economic incentives in motivation and gives excessive stress on social and psychological factors. If the wages are too-low, the employees will feel dissatisfied despite good inter-personal relations at the workplace. Thus, it may be said that the human relations approach seeks to exploit the sentiments of employees for the benefit of the organization.
  (*vi*) ***Negative View of Conflict*:** It views conflict between the goals of the organization and those of individuals as destructive. The positive aspects of conflicts such as overcoming weaknesses and generation of innovative ideas are ignored.

Table 1.8 Scientific Management vs Human Relation Approach

Scientific Management	*Human Relation Approach*
1. It was propounded by F.W. Taylor.	1. It was propounded by Elten Mayo.
2. It originated from the experience of Taylor and others in dealing with the problems of factories.	2. It originated from the Hawthorne Experiments conducted by the psychologists and sociologists.
3. It focussed on the study of the productivity problems of industry.	3. It focussed on the study of individual, his needs and behaviour.
4. It suggested an engineering approach to management problems. It applied scientific management.	4. It suggested human relations as a method of achieving higher productivity. It discarded the engineering approach.
5. The main concepts are scientific task setting, scientific selection and training of people and mental revolution.	5. The main concepts are job satisfaction, motivation and employee morale.
6. It concentrated on improvements in productivity and efficiency.	6. It concentrated on improvements in inter-personal relations and human relations in industry.
7. It is a part of classical theory of organisation.	7. It represents neo-classical theory of organisations.

(*i*) **Behavioural Sciences Approach:** Under this approach, the knowledge drawn from behavioural sciences, namely, psychology, sociology and anthropology, is applied to explain and predict human behaviour. It focuses on human behaviour in organisations and seeks to promote verifiable propositions for scientific understanding of human behaviour in organisations. It lays emphasis on the study of motivation, leadership, communication, group dynamics, participative management etc.

Propositions made by the behavioural scientists are as follows:

(*a*) An organisation is a socio-technical system.

(*b*) Individuals differ in attitudes, perceptions and value system. That is why, they behave differently to different stimuli under different conditions.

(*c*) People working in the organisation have different needs and goals which may differ from the organisational goals. Attempts should be made to achieve fusion between organisational and human needs.

(*d*) A wide range of factors influence inter-personal and group behaviour of people in organisations.

Features of Behavioural Science Approach

The broad features of behavioural science approach are as follows:

1. Behavioural science is an inter-disciplinary approach and integrates the knowledge drawn from different disciplines (psychology, sociology, anthropology, economics, political science and history) for the study of human behaviour.
2. It is an applied science with an objective to apply various researches to solve the organizational problems.
3. It is a normative science which not only suggests cause and effect relationship but also prescribes ways and means to solve organisational problems and effect results.
4. It focuses attention on people from humanistic point of view. It accepts the value of an individual as a thinking, feeling and living organism and his needs and motivations play important role in determining his behaviour in the organization.
5. It is goal-oriented. It recognizes goal conflicts in the organisation and suggests reconciliation of goals of the individuals and the organisation for better organisational climate and greater organizational effectiveness.
6. It adopts system approach which takes into account all the factors affecting organisational behaviour.

Appraisal of Behavioural Science Approach

The study of human behaviour is of great significance in management. Since an individual is a product of social system, his behaviour is not determined by organisational forces alone, but many forces like perception, attitudes, habits and socio-cultural environment also shape his behaviour. Therefore, in understanding human behaviour in the organisation, all these factors must be taken into account. This approach suggests how the knowledge of human behaviour can be used in making people more effective in the organisation.

Behaviourists have enriched management theory through their contributions in the areas of group dynamics, motivation, communication and leadership. The behavioural science have not achieved the precision of the physical sciences. Often the complexities of the human factor and the organisational setting make exact prediction impossible. It should also be noted that the findings of behavioural science research are tentative and require further investigation. They shouldn't be treated as applicable to all situations. Behavioural guidelines can be helpful and profitable, but care must be taken not to accept them as being completely valid and applicable to all situations.

Table 1.9 Human Relation Approach vs Behavioural Sciences Approach

Human Relations Approach	*Behavioural Sciences Approach*
1. It laid emphasis on the individual, his needs and behaviour.	1. It stressed upon groups and group behaviour.
2. It focussed inter-personal relationships.	2. It focussed on group relationships.
3. It was based on Hawthorne experiments and so its scope is limited.	3. It refined the human relation approach and has a wide scope. It is a much more systematic study of human behaviour in organisations.
4. It laid emphasis on informal group, motivation, job satisfaction and morale.	4. The behaviouralists studied group dynamics, informal organisation, leadership, motivation and participative management.
5. It considered conflicts in the organisation as negative and so suggested their settlement.	5. It advocated that conflict is inevitable in modern organisation and it may have positive aspects also.
6. It emphasized working conditions, inter-personal relations, employee oriented leadership and communication system for increasing job satisfaction and morale of the employees.	6. It emphasized the role of job redesign, group dynamics, motivation, leadership and organisational change.
7. It was pioneered by Elton Mayo and his associates.	7. It is based on the work of authors like Maslow, Likert, Keith, Davis, Blake and Mouton, etc.

(*iii*) **Quantitative Approach:** Quantitative approach is also known as, **"Mathematical", "Operations" Research or "Management Science" approach.** The basic feature of quantitative managment thought is the use of mixed teams of scientists from several disciplines. This school uses scientific tools for providing a quantitative base for managerial decisions. The techniques commonly used for managerial decision-making include linear programming, Critical Path Method (CPM), Programme Evaluation Review Technique (PERT), Games Theory, Queuing Theory, Break-Even Analysis etc.

The basic assumptions of quantitative approach are as follows:

1. Management is essentially decision-making and an organisation is a decision-making unit.
2. Organisational efficiency depends upon the quality of managerial decisions.

3. The problem is expressed in the form of a quantitative or mathematical model containing mathematical symbols and relationships.
4. The different variables in management can be quantified and expressed in the form of an equation.
5. Development of decision making may require the skills of many disciplines such as engineering, mathematics, economics, statistics, physical sciences, behavioural sciences, and accountancy.
6. The mathematical models enable the managers to discover significant relationships between variables that could be controlled.

The quantitative approach uses mathematical formulae for finding solutions to the problems that were previously unsolved. The abiding belief of the quantitative approach is that if the management is a logical process, it may be expressed in terms of mathematical symbols and relationships.

Operations research (management science) came into existence during the First World War and gained popularity during the Second World War. But in the 1970s, the quantitative approach to management turned away from emphasis on narrow operations research techniques to the broader perspective of decision techniques and models building. It also incorporated computerised information systems and operations management. The latest emphasis of the quantitative approach marked a move towards a more broadbased management.

Appraisal of Quantitative Approach

The quantitative approach has provided many useful tools for rational decision-making. The mathematical formulation enables practising managers to discover significant relationships that they could control.

This approach provides a rational base for making decisions with precision and perfection. It has been widely used in planning and control activities in industrial organisations. The quantitative approach has led to electronic data processing, management information system, network analysis, linear programming and other techniques.

Quantitative approach suffers from the following limitations:

1. It's main focus is on decision-making and it ignores other functions of management.
2. It doesn't consider the human element in organisation.
3. Quality of decision is based on quality of data. If the data is incomplete or faulty, optimum solutions can't be obtained.
4. This approach assume that all the variables are quantifiable and measurable which may not be fully true in all cases.
5. The mathematical model used may fail to represent the real situation because of which decisions take may not be practicable.

Table 1.10 Scientific Management vs Quantitative Approach

Scientific Management Approach	*Quantitative Approach*
1. Scientific management is concerned with analysis of operating problems at the stop floor level.	1. It is concerned with finding solutions to problems at all levels of the organisation.
2. The conceptual base of scientific management was provided by F.W. Taylor and his asociates.	2. This approach was popularized by W.C. Churchman and his associates.
3. It focussed on improving efficiency of workers and machines by finding one best way of doing things.	3. It focussed on finding right answers to managerial problems.
4. It advocates use of scientific methods in managing.	4. It advocates use of mathematical and statistical techniques for solving management problems.
5. The main techniques of scientific management are time and motion studies.	5. The main technique of quantitative approach is operations research.
6. It lays stress on experiment and research for improving efficiency.	6. It lays stress on developing econometric models for taking managerial decisions.
7. It follows closed-system perspective of the of the organization and deals with its problems in a piecemeal manner.	7. It follows open-system view of the organisation and presents an integrated approach to its problems. It has multidisciplinary orientation in finding solutions to organisational problems.

3. Modern Approach

(*i*) **System Approach:** In the 1960s, a new approach to management appeared which attempted to unify the earlier schools of thought, which is commonly referred to as 'System Approach'. The system approach is based on the generalization that an organisation is a system and its components are inter-related and inter-dependent. According to Cleland and King, "A system is composed of related and dependent elements which, when in interaction, form a unitary whole. It is simply combination of things or parts, forming a complex whole. It's important feature is that it is composed of a hierarchy of sub-systems. The world as a whole can be considered to be a system in which various national economics are sub-systems. That is, each national economy is composed of its various industries, each industry is composed of firms, a firm can be considered a system composed of sub-systems such as production, marketing, finance, accounting and so on". Thus, each system may comprise several sub-system and, in turn, each sub-system may be further composed of sub-systems.

Some of the **characteristics** of an organisation as a system are as follows:

1. A system is goal-oriented.
2. A system consists of several sub-systems which are interdependent and inter-related.
3. A system is engaged in processing or transformation of inputs into outputs.
4. An organisation is an open and dynamic system. It has continuous interface with the external environment as it gets input from the environment and

also supplies its output to the environment. It is sensitive to its environment such as government policies, competition in the market, change in tastes of people, etc.

5. A system has a boundary which separates it from other systems.

Open System Concept: A system may be closed or open. A closed system is self-sufficient and doesn't recognise external environment. Physical and mechanical systems are closed system. A closed system concentrates completely on internal relationships, *i.e.*, interaction between sub-systems only. Because of lack of interaction with environment, it is unable to monitor changes occurring in the external environment. On the other hand, an open system has active interface with the environment through the input process as shown in Fig. 1.23. It can respond to the changes in the environment through the feedback mechanism. That is why, modern authors consider organisation as an open system.

An open system obtain inputs, such as raw materials, labour, capital, technology and information, from the environment. Operations are performed upon the inputs and combined with the managerial process to produce desirable output which are supplied to the environment (*i.e.,* customers). Through a feedback process, the environment's evaluation of the output becomes part of the input for further organisational activity. If the output is satisfied with the output, business operations continue. If it is not, changes are initiated within the business system so that the requirements of the customers are fully met. This is how an open system responds to the forces of change in the environment.

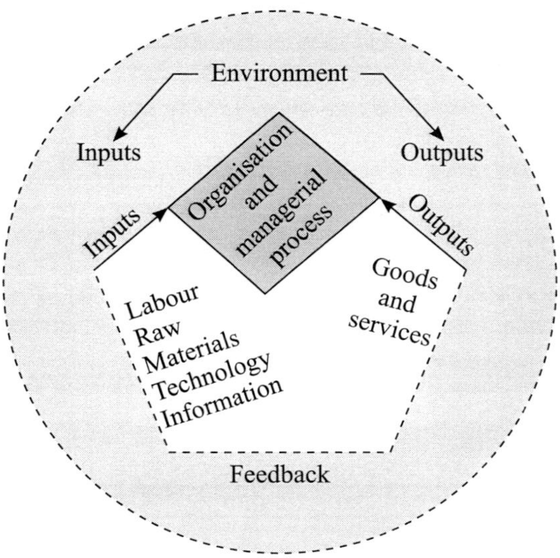

Fig. 1.23 Open system view of organisation

Features of System Approach/Modern Approach: The modern/system approach is far more superior to classical and neo-classical approach because of the following features:

1. **Open System View of Organisation:** The classical theory treated organisation as a closed system. But modern theory considers organisation

as an open system which continuously interact with the environment. It gets various resources from the environment and transforms them into output desired by the environment. Thus, due weightage is to be given to the environmental factors which are affecting the management of an organisation.

2. **Adaptive System:** Organisation is an open system, its survival and growth in a dynamic environment demands an adaptive system which can continuously adjust to changing environment. Management tends to bring changes in the sub-systems of the organisation to cope up with the challenges of environmental forces. Since there is a provision of feedback mechanism, management can evaluate its performance and take corrective actions.

   The basic role of management is considered in terms of its adaptability to environment.

3. **Interdependent Sub-systems:** An organisation is a system consisting of several sub-systems. For example, in business enterprise, production, sales and other departments are sub-system. All these sub-systems are functionally interacting and interdependent. They are tied together as whole through goals, authority flows, resources flows and so on.

4. **Whole Organisation:** This approach provides a unified focus to organisational efforts. It gives managers a way of looking at the organization as a whole that is greater than the sum of its parts. The stress is laid on integration of various sub-systems of the organisation to ensure overall effectiveness of the system.

5. **Synergy:** The output of the system is always more than the combined output of its parts. This is called the law of synergy. (For example, 1 + 1 = 11). The parts of the system become more productive when they interact with each other then when they act in isolation.

6. **Multi-disciplinary:** Modern theory of management is enriched by contributions from various disciplines like psychology, sociology, economics, anthropology, mathematics, operations research and so on.

Contributions of Systems Approach: The systems approach is based on the General System Theory and is of great significance to the management of modern organisation as discussed below:

1. The System Approach examines interrelationships and interdependency among organisational parts and seeks to answer question arising from these relationships. It concentrates on the working of various sub-systems so as to ensure the survival and growth of the system.

2. This approaches an attempt to design an overall theory of management. Interdependency and interrelationships between various sub-systems of the organisations are adequately emphasised. A pressure for change in one sub-system generally has a direct or indirect influence on the other sub-systems also. Thus, this approach acknowledges environmental influences which were ignored by the classical theory. It emphasizes adaptability (ability to respond to change) as well as efficiency. Hence, rigid rules and regulations cannot deal with uncertain and uncontrollable events.

3. The systems approach represents a balanced thinking on organisation and management. It stresses that managers should avoid analysing problems in isolation and rather develop the ability for integrated thinking. It provides a unified focus to organisational efforts.
4. It stresses the dynamic, multi-dimensional, and adaptive nature of organizations. It provides a strong conceptual framework for meaningful analysis and understanding of organisations. It recognizes the interaction and interdependence among the different variables of the environment.

Criticism of System Approach: The system approach is criticized as being too abstract and vague. It cannot easily be applied to practical problems. It doesn't offer specific tools and techniques for the practising manager. Moreover, this approach doesn't recognise difference in systems. It fails to specify the nature of interaction and interdependencies between an organisation and its external environment. Some of the deficiencies are as follows:

1. **Lack of Unification:** The system approach cannot be considered a unified theory of organisation. A "unified theory" is one which can be applied to all types of organisations and present their comprehensive analysis so that various people who want to study organisations from different angles can derive knowledge. That is what systems approach was expected to do. However, the systems approach failed to do so.
2. **Abstract Analysis:** The systems theory is too abstract to be of much use to the practising managers. It indicates that various parts of the organisation are interrelated and this interrelationships is dynamic. But it has failed to spell out the precise relationship between various sub-systems.
3. **Limited View of Organisation-Environment Interface:** This approach has failed to specify the nature of interactions and interdependencies between an organisation and its external environment.
4. **Limited Application:** The system approach has limited application that is it doesn't provide action framework applicable to all types of organisations. For example, modern structural designs, such as matrix organisation, cybernetic control and communication system are applicable to big organisation only. These concepts can't be profitably applied to smaller organisations.

Thus, the system theory has not lived upto the expectations it raised at the beginning. It promised to provide an adequate and comprehensive explanation of organisation, but this promise doesn't seem to be fulfilled.

(*ii*) **Contingency Approach:** Let us take a review of the earlier schools of management, as it, helps to place the current approach to management in perspective. The performance results of the management process school's universalist assumptions were generally disappointing.

- The behavioural approach to management was incomplete.
- Certain quantitative techniques worked in some situations and not in others.
- The quantitative people couldn't solve behavioural problems and behavioural people couldn't overcome operations problems adaptable to quantitative solutions.

- Many authors believe that a system based theory could solve this dilemma. But this approach is also as yet incomplete.

The latest approach to management is known as '**contingency**' or '**situational**' approach. The contingency approach is not new. Pigors and Myers propagated this approach in the area of personnel management as early as in 1950. However, the work of Joan Woodward in the 1950s marked the beginning of the contingency approach to organisation and management. Other contributors include Tom Burns, G.W. Stalker, Paul Lawrence, Jay Lorsch and James Thompson. They analysed the relationship between the structure of the organisation and the environment. Thus, contingency approach incorporates external environment and attempts to bridge the theory-practice gap. It does so in the systems framework. In other words, contingency approach also regards organisation as an open and dynamic system which has continuous interaction with the environment.

Basic Theme of Contingency Approach: According to Kast and Rosenzweig, "The contingency view seeks to understand the inter-relationships within and among sub-systems as well as between the organisation and its environment and to define patterns of relationships or configurations of variables. Contingency views are ultimately directed toward suggesting organisational designs and managerial actions most appropriate for specific situations".

The contingency approach to management is based upon the major premise that there is no one best way to handle any of the management problems.

The application of management principles and practices should be contingent upon the existing circumstances. Functional, behavioural, quantitative and systems tools of management should be applied situationally. There are three major parts of the overall conceptual framework for contingency management:

(*i*) environment;

(*ii*) management concepts, principles and techniques; and

(*iii*) contingent relationship between the first two.

A general framework for contingent management has been shown in Fig. 1.24.

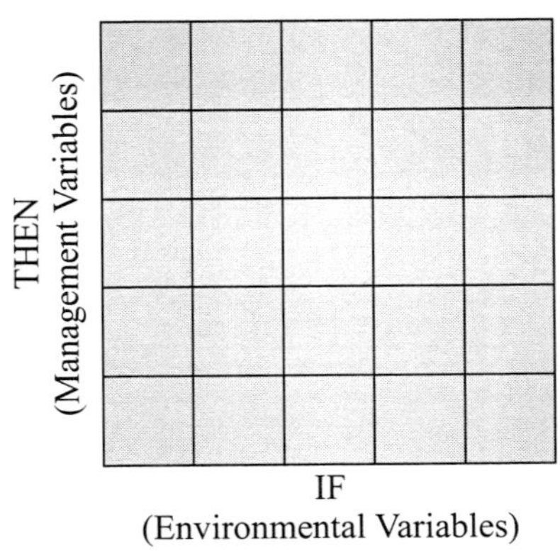

Fig. 1.24 Conceptual framework of contingency model

However, it is an abstract depiction of the contingency model. The environment (If) is an independent variable whereas management (Then) is a dependent variable.

The independent "If's" are along the horizontal axis whereas the dependent "Then's" are along the vertical axis. The goal of the contingency management is to fill in as many cells of the matrix as possible. For example, in the bottom left-hand cell of the matrix, a relevant environmental condition would be identified (one or a combination of the "if's") and then be contingently related to an appropriate management variable (one or more of the "then's") for the most effective goal attainment possible.

The contingency theory stresses that there is no one best style of leadership which will suit every situation. The effectiveness of a particular leadership style vary from situation to situation. For example, participative leadership may be more effective in an organisation which are employing professional personnel in a high technology operation in an atmosphere of non-materialistic orientation and free expression. Where as, authoritarian leadership would be more effective in an organisation which employs unskilled personnel on routine tasks in social values oriented towards materialism and obedience to authority.

Practical Utility of Contingency Approach: The utility of contingency approach will be clear from the following features:

(*i*) Contingency approach advocates that there is no single best way of managing applicable in all situations. The best solution is the one that is responsive to the peculiarities of the given situation. Significant differences exist between different situations. Therefore, management should deal with different situations in different ways. In other words, the effectiveness of any technique is contingent on the given situations. The conditions and complexities of the situation determine which approach is applicable and effective.

(*ii*) Contingency approach rejects universality of management concept and it appeals to common sense but it is much more than common sense. It requires the ability to analyses and diagnose a managerial situation correctly. Use of contingency approach is not possible without the ability to match the management knowledge and skills to the demands of the given situation.

(*iii*) Management should match or 'fit' its approach to the requirements of the particular situation. To be effective, management policies and practices must respond to environmental changes. The organisation structure, the leadership style, and the control system should be designed to fit the particular situation. Thus, management is entirely situational.

(*iv*) This approach is action–oriented as it is directed towards the application of systems concepts and the knowledge gained from other approaches. However, the choice of a particular approach or managerial style would depend upon the type of situation faced.

Evaluation of Contingency Approach: Contingency approach guides the managers to be adaptive to environmental variables, in other words, the managers should develop situational and practical selectivity. Adoption of this approach can be useful in the formulation of strategies, design of effective organisations, planning information systems, establishing communication and control systems, shaping motivational and leadership approaches, resolving conflicts, managing changes, etc.

This approach highlights the multivariate nature of organisations and explains how organisations operate under different conditions with its help, managers can design structures which are highly appropriate to the respective situations. If an organisation is operating in a stable environment, it can have a mechanistic structure characterised by high degree of differentiation, centralization of authority, rigid hierarchical relationships, rules and regulations, etc. But if the environment is dynamic, organic structure would be more appropriate. Organic structure is characterized by decentralised decision-making, collaborative relationship, open communication, scope for innovative decision-making, etc.

Contingency approach suggests the managers to consider environmental contingencies while choosing their styles and techniques. It discounts preconceived notions and liberates the managers from dogmas and rigid behaviour. The management principles and techniques should be tailored to meet the particular situation. Readymade solutions for all the situations are not available. Contingency approach tells the managers to be pragmatic and open-minded.

Contingency approach is an improvement over the systems approach. It not only examines the relationship between sub-systems of the organisation, but also the relationship between the organisation and its environment. The contingency approach seems to hold a great deal of promise for the future development of management theory and practice. The other approaches to management can be all incorporated into the contingency framework.

However, the contingency approach suffers from two *limitations*:

1. Firstly, it doesnot recognise the influence of management concepts and techniques on environment.
2. Secondly, literature on contingency management is yet not adequate.

Table 1.11 Systems Approach vs Contingency Approach

Systems Approach	*Contingency Approach*
1. It lays emphasis on the interdependencies and interactions among systems and sub-systems.	1. It identifies the nature of interdependencies and the impact of environment on organisational design and managerial style.
2. It's main focus is on the internal environment and sub-systems of the organisation.	2. It's main focus is on the external environment of the organisation.
3. It treats all organisations alike. Size of the organisation, and its socio-cultural setting are not considered.	3. Each organisation is to be studied as a unique entity.
4. It studies organisation at the philosophical level.	4. It follows an action-oriented approach and so is pragmatic. It is based on empirical studies.
5. It simply lays down that the organisation interacts with the environment.	5. The impact of environment on the organisation structure and managerial style is the major concern of contingency approach.
6. It provides a theoretical model of understanding the organisation and its sub-systems.	6. It stands for down-to-earth action-oriented approach to organisational problems.
7. Suggests deterministic solutions of management problems	7. It suggests probabilistic and pragmatic solutions of management problems.
8. It doesnot comment on the validity of the classical principles of management.	8. It rejects the blind application of the classical principles of management.

(*iii*) **Operational Approach to Management:** Koontz, O'Donnell and Weihrich, have advocated the operational approach to management. In doing so, they have attempted to draw together the pertinent knowledge of management by relating it to managerial job, *i.e.*, what managers do. This approach recognises that there is a central core of knowledge about managing which exists in management such as line and staff, patterns of departmentation, span of management, managerial appraisal and various managerial control techniques. Many other pertinent elements of knowledge are derived from other fields such as application of system theory, motivation and leadership, decision-making, group behaviour and cooperative systems, communication and mathematical analysis and practices.

Operational approach to management regards management as a universally applicable body of knowledge that can be brought to bear at all levels of managing and in all types of enterprises. At the same time, the approach recognizes that the actual problems managers face and the environments in which they operate may vary between enterprises and levels, and it also recognizes that application of science by a perceptive practitioner must take this into account in designing practical problem solutions.

The operational approach to management is based largely on the following fundamental beliefs that:

1. Management is an operational process initially best dissected by analysing the management functions.
2. If the knowledge of management is to be presented effectively, clear concepts are necessary.
3. Experience with managing in a variety of situations can furnish grounds for distillation of basic truths—theory and principles–which have a clarifying and predictive value in understanding and improving practice.
4. Principles of management can become the focal points for useful research both to ascertain their validity and to improve their applicability.
5. Managing is an art (like medicine and engineering) that should rely on underlying science–concepts, theory and principles and techniques.
6. While the total culture and the physical and biological universe variously affect the manager's environment, as they do in every other field of science and art, management science and theory need not encompass all knowledge in order to serve as a useful foundation of management practice.

1.18 ORGANISATION AS A SYSTEM

Does your organisation have an organisation chart? Most do. Typical "organisation charts" include titles and names in box, arranged in order of hierarchy, connected by lines that show who reports to whom. Organisation charts are useful because they define the individuals and departments that make up the organisation. But if an organization, chart is the only diagram that explains on organisation, it signals an opportunity to do some important re-thinking. An organisation chart doesn't show the interdependent nature of the components that make up the organisation, nor how the work actually gets done. For that, you need to view **the organisation as a system** rather than collection of individuals or department working side-by-side.

The following important aspects of this "systems view" of the organisation:

- The need in society is the primary focus and provides the aim for efforts to improve.
- Suppliers and customers are closely connected to the system.
- The feedback loops shows the system for improvement.
- Customer research and planning are prerequisite for improvement.
- Improvement results from design or redesign of some aspect of the system.
- Everyone in the organisation should participate in improvement.

The success of the system should be measured across multiple dimensions (a family of measures, say the authors starting with a set of ten or fewer), including financial and operational data, as well as staff and customer feedback. Leaders must make sure everyone understands the measure and how they relate to each other.

It's easy for people working in a system to forget how their work relates to and affects those around them or more often those who work in a distant part of the system. The leader's job is to reinforce on a daily basis the systematic nature of the organisation, the interdependence of each unit within it, and the importance of working as a team. Teamwork, leadership, cooperation and positive thinking is the major tool of the organisation system.

1.18.1 What is a System?

Very simply, a system is a collection of parts (or subsystem) integrated integrated to accomplish an overall goal (a system of people is an organisation). System have input, processes, outputs and outcomes, with ongoing feedback among these various parts. If one part of the system is removed, the nature of the system is changed.

System range from very simple to very complex. There are numerous types of systems. For *e.g.*, there are biological systems (the heart, etc.), mechanical systems (thermostat, etc.), and human/mechanical systems (riding a bicycle, etc.), sociological systems (predator, prey, etc.) and social systems (groups, supply and demand, etc.).

Complex systems, such as social systems are comprised of numerous subsystem as well. These subsystems are arranged in hierarchies, and integrated to accomplish the overall goal of the overall system. Each subsystem has its own boundaries of sorts, and includes verious inputs, processes, outputs and outcomes geared to accomplish in overall goal for the subsystem.

> A pile of sand is not a system. If one removes a sand particle, you have still got a pile of sand. However, a functioning car is a system. Remove the carburetor and you have no longer got a working car.

1.18.2 Why is it Important to Look at Organization as System?

The effect of this system theory in management is that writers, educators, consultants etc. are helping managers to look at organisation from a broader perspective. System theory has brought a new perspective for managers to interrupt patterns and events in their organisation. In the past, managers typically took one part and focused on that. Then they moved all attention to another part. The problem was that an organisation could, example have wonderful departments that operates well by themselves but don't integrate well together. Consequently, the organisation suffers as a whole. Now, more managers are recognizing the various parts of the organisation, and in particular, the interrelations of the parts, example the coordination of central offices with other departments, engineering with manufacturing, supervisors with workers, etc. Managers now focus more attention on matters of ongoing organisation and feedback. Managers now diagnose problems not by examining what appear to be separate pieces of the organisation, but by recognizing larger patterns of interactions. Manager maintain prepective by Focusing on the outcomes they want from their organisations. Now managers focus on structures that promote behaviours that determine events—rather than reacting to events as was always done in the part.

1.18.3 System Organisation Theory

The system organisation theory is of recent origin having developed in early sixties. It has a conceptual, analytical base and places a great reliance on impirical research data. It considers *organisation as a system*. A system defined as an organised or complex whole; an assemblage or combination of things or part forming a complex unitary whole. Parts of a system are know as *sub-systems*. The various sub-systems are inter-related and they are arranged according to some scheme so that whole is more than the sum of parts. This is done to ensure the efficient functioning of the systems as a whole. A system has a boundary which maintains proper relationship between the system and its environment.

Systems are broadly of two of two types—(1) *open system* and (2) *closed system*.

An open system interacts with its environment whereas a closed system has no interaction with the environment. All living systems are open system but all non-living systems are closed systems. Organisation being an open system continuously interacts with its environment. Therefore, to understand organisation, its boundary must be identified. Organisation in interactions with its environment can be understood as input-output models. Inputs are information, energy and materials which organisation takes from its environment. It transforms them with the help of men and machines and supplies the outputs to the environment.

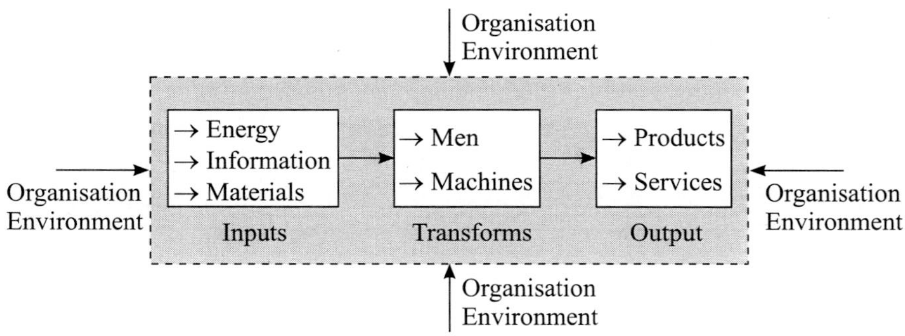

Reaction of environment to the outputs is called feedback mechanism with which the organization can evolute and correct itself. Organisation as a system has multiple objectives. It comprises various subsystem. *e.g.*, technical sub-system, social sub-system, power sub-system. etc. These sub-systems are inter dependent and there is a need for interlinking them through some processes.

Thus, systems organisation theory involves study of an organization by identifying (*a*) its strategic parts, (*b*) the nature of their material dependency, (*c*) the processes in the system which link the parts together and make for their adjustments to one another, and (*d*) the goals sought to be accomplished by the system.

1.18.4 Parts of Organisation System

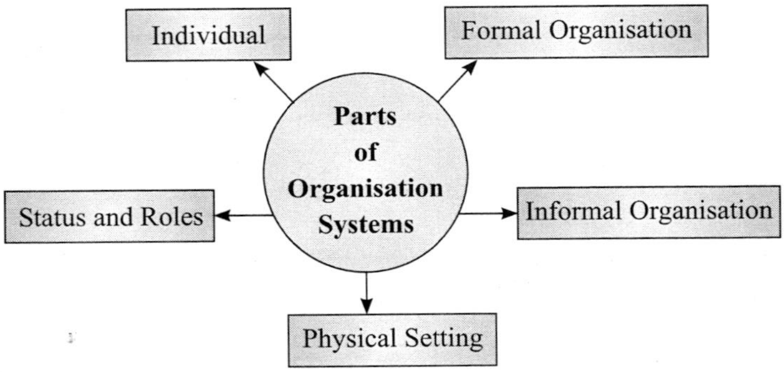

The main parts of the organisation systems are given below:

1 **Individual:** Individual and his personality structure is the basic part of the system. The motives and attitudes of an individual decide what he expects to achieve by participating in the organisation system.

2. **Formal Organisation:** It is inter-related pattern of jobs designed to regulate the actions of individuals and other resources in the organisation. It requires the individual to perform his job and the individual demands fulfilment of his expectations by performing the job. There is often an incongruency between the goals of the organisation and those of its members.

3. **Informal Organisation:** There is significant interaction between the individual and the informal group to which he belongs. The informal group demands the individual to conform to the behaviour patterns laid down by it. The individual in turn seeks to accomplish his goals by associating with the informal group. Both these sets of expectations interact resulting in modifying the behaviour of each other.

4. **Status and Roles:** Every organisation has prescribed pattern of roles the individuals are expected to play. These roles determine their status. When the demands on an individual made by the formal and informal organization are contradictory, a role conflict arises. A fusion between the two roles becomes necessary. The fusion process is a force which acts to wield divergent elements together in order to preserve the organisational integrity.

5. **Physical Setting:** The physical surroundings in which a job is performed is an important component. Inter action present in the complex man-machine system need to be carefully examined. The human engineer cannot approach the problem in a purely technical fashion. The psychological, social and Physiological conditions of organisation members should be considered so as to fit machines to men.

1.18.5 Linking Processes

The various parts of system are interconnected. The processes through which they are interconnected are as follows:

(*i*) **Communication:** If is an effective mechanism that links all the organisational segments together. It involves receiving messages from the external environment and exchanging such information among the individuals. It is also a control and coordination mechanism for linking the decision centres in the systems.

(*ii*) **Decision Making:** Decision–Making is another process for linking various parts in the organisation. According to March and Simon, the decision to produce depends upon the interaction between individuals and demands of the organisation. On the other hand, decision to participate depends on the demands and rewards of the organisation.

(*iii*) **Balance:** According to W.G. Scott, balance refers to an equilibrating mechanism whereby various parts of the system are maintained in harmoniously structured relationship with each other. Balance helps to ensure integrity in the face of rapidly changing environment.

CASE STUDY

DABBAWALLAH Case Study

Dabbawallah, is a person in Mumbai (Bombay), India, whose job is carrying and delivering freshly made food from home in lunch boxes to office workers. The word "Dabbawala" is literally translated as "one who carries a box"; "Dabba" means a box (usually an cylindrical aluminium container), while "wala" is a term of reference to the preceding word (literally translated, the closest meaning would be "tiffin-man"). Though the profession seems to be simple, it is actually a highly specialized trade that is over a century old and has become integral to Mumbai's culture. The dabbawala originated when India was under British rule: many British people who came to the colony didn't like the local food, so a service was set up to bring lunch to these people in their workplace straight from their home. Nowadays, Indian business men are the main customers for the dabbawalas, and the services provided are cooking as well as delivery.

Mumbai is a densely populated city with a huge flow of traffic. Because of this, lengthy commutes to workplaces are common, with many workers traveling by train. Instead of going home for lunch or paying for a meal in a café, many office workers have a cooked meal sent by a caterer who delivers it to them as well, essentially cooking and delivering the meal in lunch boxes and then having the lunch boxes collected and re-sent the next day. This is usually done for a monthly fee. The meal is cooked in the morning and sent in lunch boxes carried by dabbawalas, who have a complex association and hierarchy across the city. A collecting dabbawala, usually on bicycle, collects dabbas from homes or, more often, from the dabba makers (who actually cook the food). The dabbas have some sort of distingushing mark on them, such as a color or symbol (most dabbawalas are illiterate).

The dabbawala then takes them to a designated sorting place, where he and other collecting dabbawalas sort (and sometimes bundle) the lunch boxes into groups. The grouped boxes are put in the coaches of trains, with markings to identify the destination of the box (usually there is a designated car for the boxes). The markings include the rail station to unload the boxes and the building address where the box has to be delivered. At each station, boxes are handed over to a local dabbawala, who delivers them. The empty boxes, after lunch, are again collected and sent back to the respective houses.

Everyone who works within this system is treated as an equal. Regardless of what function a dabbawala, everyone gets paid about 4,000 rupees (which equals around 50 British pounds), not a lot of money when one considers that they also return the lunch's packaging to the home.

More than 175,000–200,000 lunches get moved every day by an estimated 4,500–5,000 dabbawalas, all with an extremely small nominal fee and with utmost punctuality. According to a recent survey, there is only one mistake in every 6,000 deliveries. In fact, the American business magazine Forbes gave a Six Sigma performance rating for the precision of dabbawalas. This rating indicates a 99.999999 accuracy percentage of correctness, meaning one error in every six million transactions—an astonishing (and perhaps unbelievable) degree of exactness.

The service is uninterrupted even on the days of extreme weather, such as Mumbai's characteristic monsoons. The local dabbawalas at the receiving and the sending ends are known to the customers personally, so that there is no question of lack of trust. Also, they are well accustomed to the local areas they cater to, which allows them to access any destination with ease. Occasionally, people communicate between home and work by putting messages on chits inside the boxes. Of course, this was before the telecommunications revolution.

The main reason for their popularity could be the Indian people's aversion to fast food joints and love of home made food. Some of them have delivered lectures in top B-schools of India. The BBC has produced a documentary on Dabbawalas, and Prince Charles, during his visit to India, made a point to visit them (he had to fit in with their schedule, since their timing was too precise to permit any flexibility). Dabbawalas have been practising this art for more than a century. Tiffin is an old-fashioned English word for a light lunch, and sometimes for the box it is carried in. Dabbawalas are sometimes called tiffin-wallas.

In Hong Kong, from the time of the mass influx of people from China during 1945–1950s, until people got richer in the 1970s, the same tradition of tiffin carriers existed as in Mumbai. Typically tiffin boxes were coloured with flowers and other devices for the factory workers lunch.

Practical Implementation

7.00 am After being given an address, Simon collects a Dabba to transport across Mumbai. This Dabba contains a typical Indian lunch-curry, vegetables, and a type of bread called Roti.

8.00 am The Dabbawallahs often carry the Dabbas on their heads, Simon gives it a go with the one he's been charged with! He walks to the Dabbawallahs' meeting point, for the start of the Dabba's journey.

9.00 am This is how it's usually done! It's estimated that a Dabbawallah carries around 40 Dabbas on his head at one time. That's like carrying 40 packed-lunch boxes full of food, on your head!

10.00 am Simon helps the Dabbawallahs lift a crate of Dabbas on to a train. He realises just how heavy it is and can't imagine how the Dabbawallahs can balance it on their heads.

11.00 am After a really cramped train journey, the Dabbawallahs shift their loads on to a cart. Each cart holds around 200 Dabbas. These carts are then pushed across the city ready for Dabba delivery!

12.30 pm Mumbai is well known for being a manic city, vehicles whizz round the city's streets, tooting horns, and coming very close to the Dabbawallah's cart. Simon finds it quite a hair-raising experience.

1.00 pm After what seems like a decade of pushing a loaded cart through the hot streets of Mumbai, Simon reaches his destination and hands over the Dabba he's been looking after.

2.00 pm The food is gobbled up by the Dabba's recipient, while Simon grabs a quick tea-break outside in the now, blistering heat. Once the food is finsihed, the Dabba is handed back to Simon...

2.45 pm Simon must now return the Dabba to where he picked it up this morning. He must repeat the process in reverse, there are no shortcuts for a Dabbawallah, or for a Blue Peter presenter in training.

EXERCISES

Short Questions Answers

1. Explain management as a science.
2. Describe the functions of management.
3. Explain management as a process.
4. Differentiate between management and administration.
5. Briefly explain the features of management as a profession.
6. Management is perfect science Ellaborate.
7. Describe the nature and characteristics of management.
8. Explain the significance and limitations of the systems approach to management.
9. Distinguish between behavioural approach and human relations approach to management.
10. Distinguish between the following:
    (*a*) Contingency Approach and Systems Approach.
    (*b*) Scientific Management and Quantitative Approach.

Long Questions Answers

1. What do you understand by management? Explain in brief the main functions of management.
2. Briefly explain the Henri Fayol principles of management.
3. Explain the difference between management and administration.

4. Describe the evolution of management?
5. Why is management regarded both as a science and as an art? Explain.
6. "To manage is to forecast and plan to organise to command to coordinate and to control". Discuss this statement.
7. Write in detailed note on the professionalisation of management in India while explaining the concept of profession.
8. Who is manager? What are the qualities of a successful manager?
9. How are the functions of management inter-related? Discuss.
10. Explain the various functions of management. How are the functions of planning and control interdependent?
11. Discuss the contribution of systems approach to the study of management.
12. What is contingency approach to management? Critically evaluate the contingency approach?
13. Compare and contrast various theories of management which approach would you suggest and why?
14. "Fayol is considered as the father of modern management theory". Discuss.
15. "Elton Mayo is known as the father of Human Relations School". Comment.
16. Distinguish between Human Relations and Scientific Management Approach to Management.
17. Outline the major contributions of either Elton Mayo or F.W. Taylor to the modern management theory.
18. Write short note on:
    (*i*) The Human Relations School.
    (*ii*) The Mathematical School.
    (*iii*) The Contingency Approach.

■■■

CHAPTER

2

Management Information System

In today's competitive world of business systems, management Information systems have strategic purpose in business management. The role of MIS is now limited to business operations management. The new role of MIS is to support business management in strategic analysis, strategy formulation, and strategy implementation to achieve short-term and long-term business goals. Design, and architecture of MIS is to run customer centric customer driven business with efficient use of *e*-business technologies.

Computer, networks, communication, internet and web technologies together have radically changed the business management process and have an impact on the role of the manager. The objective of this part is to understand the cultural and operational changes in business management due to the application of various technologies.

Today the world is of computer science and engineering. So, the companies is urging the needs of the people to whom they can provide them large number of jobs.

This reflects great importance for us to know and understand information system. That helps us in managing the organization. Let's understand in brief what information is and how it is different from data, nature and structure of organisation.

2.1 MEANING OF INFORMATION SYSTEM

Before going further lets understand the terms which make us better to know more in depth about the information system.

What is Data?

Data is a raw material with which we start any works and our project too. Collecting these raw material (data) include lots of money. So before collecting any data one should focus that it should combined the data with necessary and sufficient data. To make it useful we need machine and if useless it is processed to create INFORMTION.

What is Information?

INFORMATION is processed data, which used by managers to take actions soon to run the organization efficiently with achieving the objective maximum profit with minimum wastage of output. The data processed by machines gives information.

So, all together data and information conclude that first data is collected, later after collecting data it is transmitted into information with the help of machine. Make system which is called INFORMATION SYSTEM.

Types of Information System

1. **Strategic:** Strategic information are long term planning and involve direction to be done for future target to be achieved. These are unstructured.
2. **Tactical Information:** These tactical information state that these are for short term planning in order to fulfill day to day decision which help in increasing the profit with less wastage and help in upgrading the performance of the employer and employees.
3. **Operational Information:** Operational information need for day-to-day operations of an organization. **Example**: *Daily sales, Billing.*
4. **Statutory Information:** This information is basically needed by the law which is sent to the government authorities when required. Keeping in mind about the image of the organization in the mind of customer, employer working in the company, government etc. **Example**: *Sales tax return.*

2.2 CONCEPT OF MANAGEMENT INFORMATION SYSTEM

IT industry is facing new technologies everyday and not many people are able to cope with such rapid changes. Costly mistakes result while making critical decisions in the acquisition of computer systems and the implementation. The problem is even more pressing for small to medium size companies were resources are not quite easily available.

The concept of MIS has emerged substantially over the years. In the 50's and 60's the management saw the potential of computers to process large amounts of data speedily and accurately.

The departments that were involved with such activities were known as **Electronic Data Processing** (EDP) departments. The focus of EDP was record keeping *e.g.*, Accounting data. Then in the 70's there was a shift from data to information. The focus was not an data but on the analysis of organization data.

There was a shift in the philosophy. Such a concept came to be widely known as 'Management Information System'. In the 70's the top management relied on the staff of EDP and MIS to supply the necessary information. The 80's saw the **Personal Computer (PC)** revolution. The personal computer to the desk-top metaphor changed the picture completely.

The biggest pay-off for such direct use was the "what-if" analysis capability. This led to the emergence of **Decision Support System (DSS).** The information and decision hungry managers of 80's saw a huge potential in the expert systems as a result of spectacular growth in the **Artificial intelligence** area. Combined with DSS philosophy the expert systems could supply a superior class of managerial information support, known as **Knowledge Based Systems (KBS).** *The EDP targeted the operational level of management. The MIS/DSS/KBS target the middle management.

Attempt were made to provide information to top management as well, known as Executive Information System, (EIS).

1950s to 21st Centuary (Emergence of MIs)

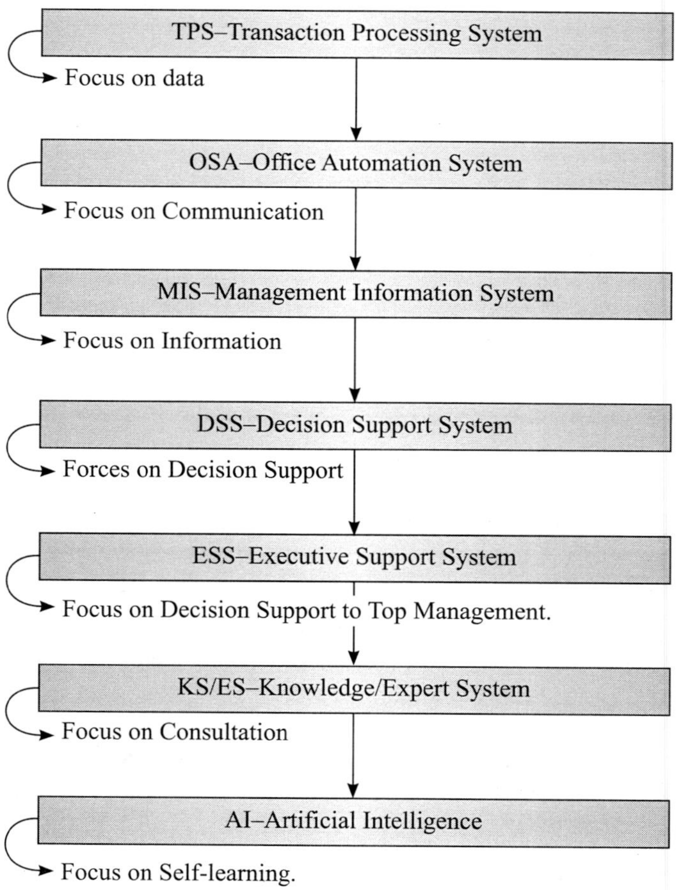

Fig. 2.1

If we talk about the concept of management information system (MIS) the **data is transferred/input** in hardware, software, networking, humans and communications through various means and is referred as IT resources. Then this information is **combined with** corporate mission and **is used** for business strategies, business process and management objective to develop functional strategies **for achieving** organization goals.

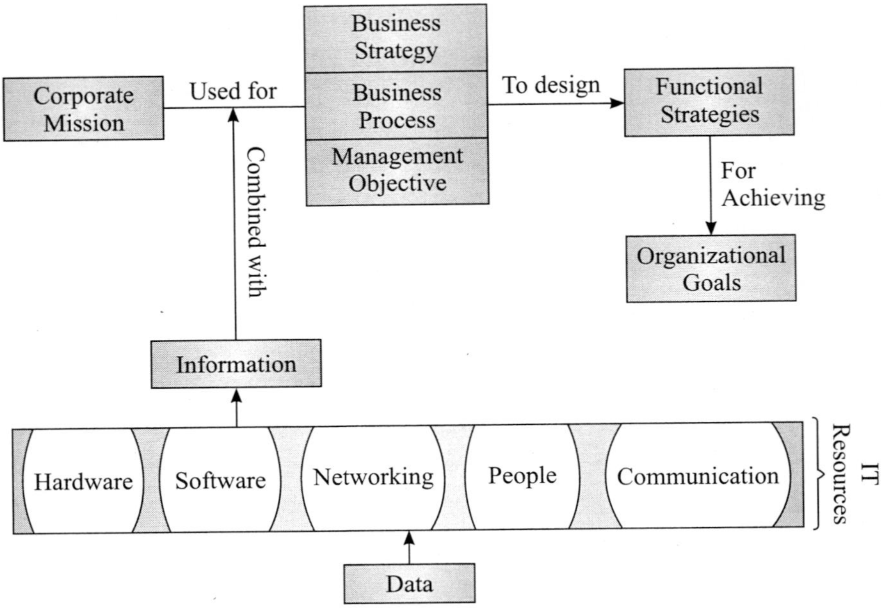

Fig. 2.2

2.3 DEFINITION OF MIS

The Management Information System (MIS) is a concept of the last decade or two. It has been understood and described in a number of ways. It is also popularly known as information system, the information and decision system, the computer based information system.

The MIS has more than one definition, some of which are given below:

1. The MIS is defined as a system which provides information support for decision-making in the organisation.
2. The MIS is defined as an integrated system of man and machine for providing the information to support the operations, the management and the decision-making function in the organisation.
3. The MIS is defined as a system based on the database of the organisation evolved for the purpose of providing information to the people in the organisation.
4. The MIS is defined as a computer-based information system.

Though there are a number of definitions, all of them converge on one single point, *i.e.,* the MIS is a system to support the decision-making function in the organisation. The difference lies in defining the elements of MIS. However, in today's world, the MIS is a computerised business processing system generating information for the people in the organisation to meet the information needs for decision-making to achieve the corporate objectives of the organisation.

In an organisation, small or big, a major position of the time goes in data collection, processing, documenting and communicating it to the people. Hence a major portion of the overheads goes into this kind of unproductive works in the organisation.

Every individual in an organisation is continuously looking for some information which is needed to perform his/her works. Hence, the information is people-oriented and it varies with the nature of the people in the organisation.

The development and management of information technology tools assists executives and the general workforce in performing any tasks related to the processing of information. MIS and business systems are especially useful in the collection of business data and production of reports to be used as tools for decision-making.

MIS is a computer system, usually based as a mainframe or minicomputer and designed to provide management personnel with up-to-date information on an organization performance, *e.g.*, inventory and sales. These systems output information in a form that is usable by managers at all levels of the organization: strategic, tactical and operational. The term system in MIS implies order, arrangement and purpose. The information can be used for various purposes,

- strategic planning
- delivering increased productivity.
- reducing service cycles.
- reducing product development cycles.
- reducing marketing life cycles.
- increasing the understanding of customer's needs.
- facilitating business and process re-engineering.

2.4 IMPORTANCE OF MIS

Management Information System is a integrated set of component or enities that interact to achieve a particulars function, objective or goal. Therefore it is a computer based system that provides information for decisions making on planning, organizing and controlling the operation of the sub-system of the firm and provides a synergistic organization in the process.

The component of an information system includes. A hardware which is used for input/output process and storage of data, software are used to process data and also to instruct the hard-ware components, data bases which is the location in the system where all the organization data will be automated and procedures which is a set of documents that explain the structure of the management information system.

There are various driving factors of management information system for example:

Technological revolutions in all sectors make modern managers to need to have access to large amount of selective information for the complex tasks and decisions. There are huge amount of information available to today's manager and this had therefore meant that managers are increasingly relying on management information system to access the exploding information. Management information services helps manager to access relevant, accurate, up-to-date information which is the more sure way of making accurate decisions. It also helps in automation and incorporation of research and management science techniques into the overall management information system for example–probability theory.

The management information services are capable of taking advantage of the computational ability of the company like processing, storage capacity among others.

Based on this relevancy, management information system should be installed and upgraded in various organizations since today's managers need them to access information for managerial decision-making and also management functions.

MIS is an important resource needed to develop other resources. An MIS is a special purpose systems useful for management in an organization. Managers need relevant information, which is information that increases their knowledge and reduces uncertainty. Thus it is usable by the manager for its intended purpose.

1. It is essential tool for managers in planning and decision-making. It is a well coordinated information system, a database, to provide management with needed information to plan and make decisions.
2. It is useful for planning. As every organization makes long-term and short-term plans with the help of information like sales and production, capital investments, stocks, etc. management can easily plan.
3. MIS is always management oriented and keeps in view every level of management and gets the desired information.
4. MIS provides updated results of various departments to management.
5. MIS is always management oriented and keeps in view every level of management and gets the desired information.

2.5 OBJECTIVES AND GOALS OF MIS

Mission

The mission of the management information system (MIS) is to provide data processing support and enhance an electronic infrastructure to optimize delivery of services to the user departments. In an organization goals and objectives are inter-related to each other.

Goal 1: Improve efficiency in all aspects of services provided by MIS.

Objective 1: Establish an Information Technology (IT) Advisory Board to develop priorities leading to accomplishment of country goals and objectives.

Objective 2: Interconnect all data functions:

(*i*) Converge satellite data centres at the MIS data center.

(*ii*) Centralize Geographical Information System (GIS).

(*iii*) Acquire the high-speed fiber connections between all offices required to efficiently pass large data packets across geographic boundaries.

(*iv*) Centralize criminal justice name and address data an one database.

Objective 3: Reduce number of hardware platforms and centralize administration of backups, storage management, security functions, and scale upward as increased capacity is required.

(*i*) Upgrade mainframe to Multiprise server for internet and internal uses.

(*ii*) Licence Virtual Machine (VM) operating system to run multiple platforms indepently.

Objective 4: Standardize and implement automation tools to contain technique services support costs.

(*i*) Promote use of Local Areas Networks (LANs) and networks computers to replace stand alone PC's to speed device configuration and software distribution, secure data backup functions, and maximize data sharing in unit works groups.

Objective 5: Implement and Maintain Disaster Recovery plan for country data. Equip alternate data center. Conduct regular data recovery tests.

Goal 2: Plan business application systems that meet business needs and country goals.

Objective 1: Ensure application systems meet business needs.

(*i*) Assess strengths and weaknesses of all systems on an ongoing basis.

(*ii*) Submit project plans to IT board review process.

Objective 2: Plan for comprehensive data presentation systems which include data, text, images, voice, and multimedia systems.

Goal 3: Deliver business applications quickly.

Objective 1: Organization should buy applications tools if they are cost-effective.

Objective 2: Select best platform for application considering functionality and supportability.

Objective 3: Deliver 3-month work unity segments to keep users and developer aligned.

Objective 4: Use industry-standard methodologies for development. Avoid specialized tools requiring non-standard training.

Objective 5: Form patnerships with users to best accomplish project completion and satisfaction.

(*i*) Promote use of cross-functional participation on projects.

(*ii*) Establish clear, written goals and role guidelines for each project member. Communicate to define clear service level expectation for each project.

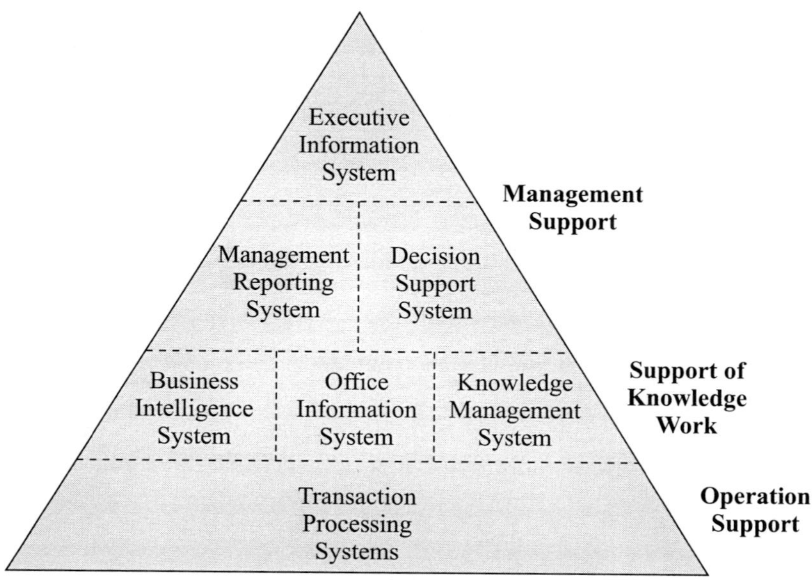

Fig. 2.3

2.6 CHARACTERISTICS OF MIS

- Management information system support structured and semistructured Decision at the operational and management control levels. They are also useful for planning purpose of senior management staff.
- Management information, systems are generally reporting and control oriented. They are designed to report on existing operation and therefore to help provide day-to-day control of operations.
- MIS rely on existing corporate data and data flows.
- MIS have little analytical capability.
- MIS generally aid in decision-making using past and present data.
- MIS are relatively inflexible.
- MIS have internal rather than an external orientation.
- Information requirements are known and stable.
- MIS require a lengthy analysis and design process.
- Provides reports with fixed and standard format: Hard copy and soft copy reports.
- Uses internal data stored in computer system.
- Requires formal requests from users.

Characteristics of an effective management information system:

(*i*) Has senior management support and sponsor.
(*ii*) Has gone through business risk analysis.
(*iii*) Has good project management controls.
(*iv*) Has good user interaction/ feedback controls.
(*v*) Has good security management controls.
(*vi*) Has good management reporting system.
(*vii*) Has good return on investment.
(*viii*) Is delivered on time.
(*ix*) Delivers all the functional requirements and liked by the users.

2.7 FUNCTIONS OF MIS

Following are the seven major functions of MIS:

(*i*) Facilitate the management of the business.
(*ii*) Provide management with one adequate decision support system by providing information that is timely, accurate, consistent, complete, and relevant.
(*iii*) Deliver complex material throughout the institution.
(*iv*) Support the organization's strategic goals and direction.
(*v*) Reduce expenses related to labor-intensive manual activities.
(*vi*) Enhance communication among employees.
(*vii*) Makes work effective and efficient.

MIS supplies decision-makers with facts, supports and enhance the overall decision-making process and enhances job performance throughout an institution. At the most senior levels, MIS provides the data and information to help the board and management make strategic decisions. At other levels, MIS allows management to monitor the institution's activities and distribute information to other employees, customers, and members of management.

The impact of MIS on the functions is in its management. With a good MIS support, the management of marketing, finance, production and personnel becomes more efficient. The tracking and monitoring of the functional targets becomes easy. The functional managers are informed about the progress, achievements and shortfalls in the activity and the targets. The manager is kept alert by providing certain information indicating and probable trends in the various aspects of business. This helps in forecasting and long-term perspective planning. The manager's attention is brought to a situation which is exceptional in nature, inducing him to take an action or a decision in the matter. A disciplined information reporting system creates a structured database and a knowledge base for all the people in the organisation. The information is available in such a form that it can be used straight away or by blending and analysis, saving the manager's valuable time.

2.8 MIS: A SUPPORT TO THE MANAGEMENT

The management process is executed trough a variety of decisions taken at each step of planning, organising, staffing, directing, coordinating and control, MIS aids decision-making. If the management is able to spell out the devisions required to be taken, the MIS can be designed suitably. The decisions required to be taken in these steps are tabulated in Table 2.1.

Table 2.1 Decisions in Management

Steps in Management	*Decision*
Planning	A selection from various alternatives—strategies, resources, methods, etc.
Organising	A selection of a combination out of several combinations of the goals, people, resources, method and authority.
Staffing	Providing a proper manpower complement.
Directing	Choosing a method from the various methods of directing the efforts in the organisation.
Coordinating	Choice of the tools and the techniques for coordinating the efforts for optimum results.
Controlling	A selection of the expectional conditions and providing the decision guidance to deal with them.

The objective of the MIS is to provide information for a decision support in the process of management. It should help in such a way that the business goals are achieved in the most efficient manner. Since the decision making is not restricted to a particular level, the MIS is expected to support all the levels of the management in conducting the business operations. Unless the MIS becomes a management aid, it is not useful to the organisation.

Types of Management Information System

A management information system (MIS) is a computer-based system that provides the information necessary to manage an organization effectively. An MIS should be designed to enhance communication among employees, provide an objective system for recording information and support the organization's strategic goals and direction. There are four types of MIS that will be introduced in ascending order of sophistication.

1. Transaction Processing Systems: These systems are designed to handle a large volume of routine, recurring transactions. They were first introduced in the 1960s with the advent of mainframe computers. Transaction processing systems are used widely today. Banks use them to record deposits and payments into accounts. Supermarkets use them to record sales and track inventory. Most managers use these systems to deal with tasks such as payroll, customer billing and payments to suppliers.

2. Operations Information Systems: These systems were introduced after transaction processing systems. An operations information system gathers comprehensive data, organizes it and summarizes it in a form that is useful for managers. Most of these systems access data from a transaction processing system and organize it into a form usable by managers. Managers use operations information systems to obtain sales, inventory, accounting and other performance-related information.

3. Decision Support Systems (DSS): A DSS is an interactive computer system that can be used by managers without help from computer specialists. A DSS provides managers with the necessary information to make intelligent decisions.

A DSS has three fundamental components:

(*a*) *Database management system (DBMS):* Stores large amounts of data relevant to problems the DSS has been designed to tackle.

(*b*) *Model-based management system (MBMS):* Transforms data from the DBMS into information that is useful in decision making.

(*c*) *Dialog generation and management system (DGMS):* Provides a user-friendly interface between the system and the managers who do not have extensive computer training.

4. Expert Systems and Artificial Intelligence: These systems use human knowledge captured in a computer to solve problems that ordinarily need human expertise. Mimicking human expertise and intelligence requires that the computer (1) recognize, formulate and solve a problem; (2) explain solutions and (3) learn from experience. These systems explain the logic of their advice to the user; hence, in addition to solving problems they can also serve as a teacher. They use flexible thinking processes and can accommodate new knowledge.

2.9 MANAGEMENT EFFECTIVENESS AND MIS

'Negandhi Estafen' provides a good model for the analysis of management effectiveness, which generates an enterprise effectiveness in achieving the goals and objectives. The models puts a lot of emphasis on the management philosophy and the environment factors on which the effectiveness is dependent. The environment factors provide the opportunities to survive and grow with certain constraints, while the management philosophy sets the guidelines for deciding the management practices of run the enterprise.

MANAGEMENT AND MIS

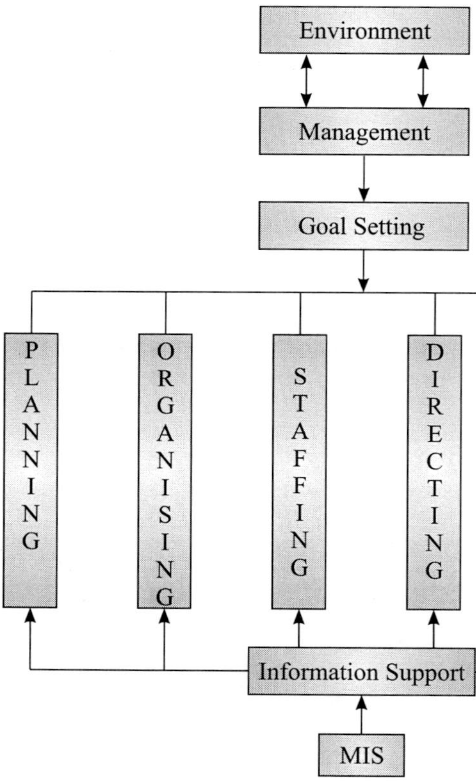

Fig. 2.4 MIS support to management process

While the environment factors are difficult to control, it is left for the management to change its philosophy towards the various players in the business, viz. the employees, the consumers, the suppliers, the government, the community and the shareholders. Basically, it is a change in attitude towards these players. For example, how to look at the employees? If the attitude is to treat them as business partners, you will empower them and create a sense of belonging to the organisation. Such an attitude will have an impact on the management practices, where the employees will play a decisive critical role. It will affect the organisation structure by reducing its size and the reporting levels.

If the attitude towards the consumer is changed to fulfill the expectations giving rise to a higher satisfaction, then the management practices in the product design, manufacturing and marketing will undergo a sufficient change. The product life cycle will undergo a significant change. The product life cycle will then be short, and more features and functions will be added to the product fulfilling not only the functional needs but also the service needs of the consumer.

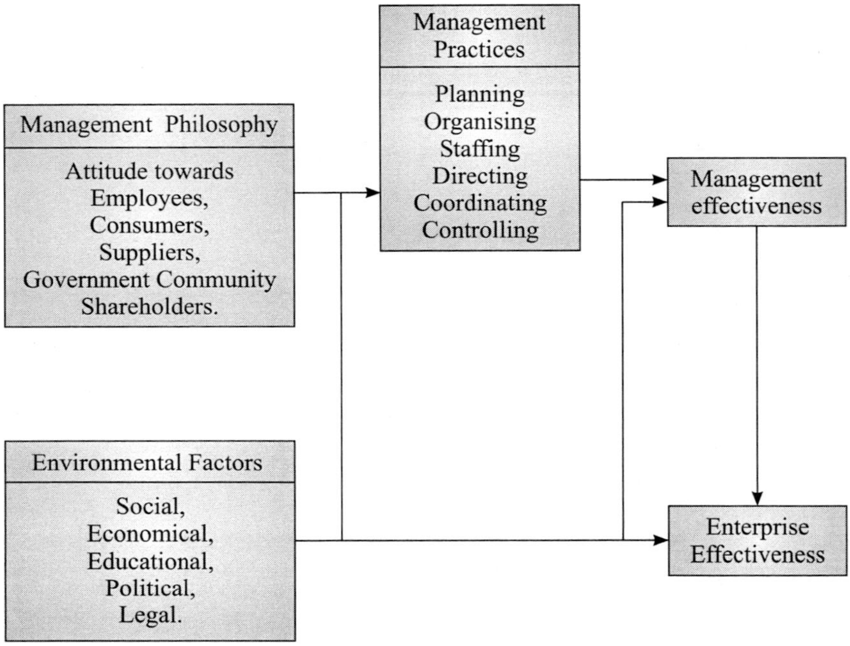

Fig. 2.5 Negandhi Estafen Model for Analysis of Management Effectiveness

The management practices, therefore, emerge out of the managements philosophy and the environment, in which it operates. The management effectiveness, would largely depend on both these factors. The MIS design would, therefore, be different depending upon the management practice followed by several organisations in the same industry. Such design improves the management effectiveness leading to an improvement in the enterprise effectiveness.

2.10 MIS: ORGANISATION

MIS design should be designed, viewing the organization and should give due weightage to the human side of the organisation and its culture. The task and technology are the physical aspects of the organisation which can be ascertained very easily. But culture and people are very difficult to assess from the design point of view. The structure of the five sub-systems should be considered while designing the MIS. MIS design should give reports in line with the organization structure. This means that the main decision makers and the power centers must be recognized in the MIS. Let us discuss these aspects of the organisation, structure and their implications.

In a tall hierarchy with a high degree of centralization, the MIS should give control information to the higher management where decision-making in concentrated.

If the system is structured on the functional basis where the functional head is a key decision maker and all the functions have equally important role to play, then the MIS will have a functional design with the information support to the functional head. Further, in such a set-up, an integrated MIS would be necessary, reporting the corporate status of the business to the top management.

If the organisation works on a standarized system where rules, policies, systems and procedures have been laid down, then these become part of the MIS. The processing routines in the MIS incorporate these features as an integral part. This is safe as it has already been approved by the management of the organisation. Along with the information, if the decision-making responsibilities are also clearly defined and allocated, then the MIS can produce information reports by processing the data and summarizing the results in line with the decision maker's position in the structure.

In the basic model of the organization is modified as a product or a project organization system, then the MIS should focus on the management of planning and control of the multiple functions. Besides these functions, he has to know the trouble spots and shows the interconnection with the other functions. It must summaries all information relating to the span of control of product or project manager. The MIS should be able to cater to the view of the product or the project manager and also to the top management.

In the organisation culture provides sufficient incentive for efficiency and results, the MIS should support this culture by providing such information which will aid the promotion of efficiency. If the culture encourages delegation of power and authority, then the MIS should incorporate the decision making rules in the system.

The organization system is an open system and MIS should be so designed that it highlights the critical business, operational, technological and environmental changes to the concerned level in the organization, so that the action can be taken to correct the situation. The principle of the feed forward control should be extensively used as a design feature to provide a prior warning to the decision maker.

Since the organisation system has a dynamic role to play to meet the changing needs of a business, the MIS becomes a common support system for playing the dynamic role. When an organization is moving through the business phases of introduction, growth, maturity and decline. MIS should provide an information support, relevant to that phase of the business cycle. This means the designer of the MIS should foresee such requirements and make the design flexible enough to support such requirements.

The organizational learning helps to tone up the behaviour of the organization. The MIS should support the learning mechanism by identifying the cause and effect in a given situation. It should keep the records of action and provide help to analyze the best action in a given situation. It should be help to build various decision models for use by the managers. The information support should be such that the group of enterprising manager should be able to improve their capabilities to perform batter.

The design of the MIS, in isolation from organizational factors, is destined to fall as it just not fit into the structure. Since organization systems in the same business differ for various reasons such as the leadership style, the management style, culture and group of people as a body and so on, it is difficult to evolve a standard model of the MIS for a business and/or an industry.

MIS plays a very important role in creating organization behaviour which in turn sets the goals for achievement. Technology and people decide the organization structure and style of the management. Figure 2.6 explains the impact and relationship of MIS on the organization behaviour.

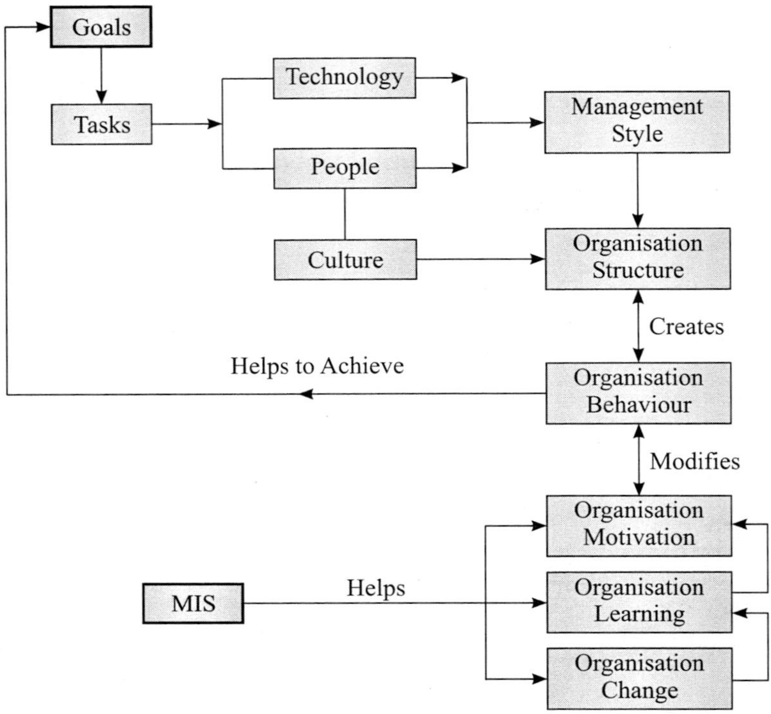

Fig. 2.6 Organisational Behaviour and Management Information System

2.11 MIS: BUSINESS PLANNING

Business environment is prone to changes and this factor makes business planning very complex. Some factors such as the market forces, technological changes, complex diversity of business and competition have a significant impact on any business prospects. MIS is designed to assess and monitor these factors. The MIS design is supposed to provide some insight into these factors enabling the management to evolve some strategy to deal with them. Since these factors are a part of the environment, MIS design is required to keep a watch on environment factors and provide information to the management for a strategy formulation. Strategy formulation is a complex task based on the strength and weakness of the organization and the mission and goals it wishes to achieve. Strategy formulation is the responsibility of the top management relies on the MIS for information.

There are various business strategies such as overall company growth, product, market, financing and so on. MIS should provide the relevant information that would help the management in deciding the type of strategies the business needs. Every business may not require all the strategies all the time. The type of strategy is directly related to the current status of business and the goals it wishes to achieve. The MIS is supposed to provide current information on the status of the business vis-a-vis the goals. MIS is supposed to give a status with regard to whether the business is on a growth path or is stagnant or is likely to decline, and the reasons thereof. If the status of the business shows a declining trend, the strategy should be of growth. If business is losing in a particular market segment, then the strategy should be a market or a product strategy.

The continuous assessment of business progress in terms of sales, market, quality, profit and its direction becomes the major role of MIS. It should further aid the top management in strategy formulation at each stage of business. The business does not survive on a single strategy but it requires a mix of strategy operating at different levels of the management. For example, when a business is on the growth path, it would require a mix of price, product and market strategies. If a business is showing a decline, it would need a mix of price-discount, sales promotion and advertising strategies.

The MIS is supposed to evaluate the strategies in terms of the impact they have on business and provide an optimum mix. The MIS is supposed to provide a strategy-payoff matrix for such an evaluation.

In business planning, MIS should provide support to top management for focusing its attention on decision-making and action. In business management the focus shifts from one aspect to another. In the introductory phase, the focus would be on a product design and manufacturing. When the business matures and requires and requires to sustain or to consolidate, the focus would be on the past sales services and support. The MIS should provide early warning to change the focus of the management from one aspect to the other.

Evolving the strategies is not the only task the top management has to perform. It also has to provide the necessary resources to implement the strategies. The assessment of resource need, and its selection becomes a major decision for the top management. The MIS should provide information on resources, costs, quality and availability, for deciding the cost effective resource mix.

When the strategies are being implemented, it is necessary that the management gets a continuous feedback on its effectiveness in relation to the objective which they are supposed to achieve. MIS is supposed to give a critical feedback on the strategy performance. According to the nature of the feedback, the management may or may not make a change in the strategy mix, the focus and the resource allocation.

MIS has certain other characteristics for the top management. It contains forecasting models to probe into the future—the business model for evaluation of the strategy performance by simulation business conditions. It contains functional models such as the model for a new product launching budgeting, scheduling and the models using PERT/CPM technique for planning.

MIS for the top management relies heavily on databases which are external to the organization. The management also relies heavily on the internal data which is evolved out of transaction processing. Management uses the standards, the norms, the ratios and the yardsticks while planning and controlling the business activities.

They are also used for designing strategies and their mix. The MIS is supposed to provide correct, precise and unbiased standards to the top management for planning.

We can summaries the role of the MIS in the top management function as follows. MIS supports by way of information, to.

(*i*) decide the goals and objectives,

(*ii*) determine the correct status of the future business and projects,

(*iii*) provide the correct focus for the attention and action of the management,

(*iv*) evolve, decide and determine the mix of the strategies,

(*v*) evaluate the performance and give a critical feedback on the strategic failures,

(*vi*) provide cost-benefit evaluation to decide on the choice of resources, the mobilization of resources, and the mix of resources,

(*vii*) generate the standards, the norms, the ratios and the yardsticks for measurement and control.

2.12 MANAGERIAL DECISION-MAKING AND MIS

2.12.1 Decision-Making Concepts

The word 'decision' is derived from the Latin root decido, meaning to cut off. The concept of decision, therefore, is settlement, a fixed intention bringing to a conclusive result, a judgement, and a resolution. A decision is the choice out of several options made by the decision maker to achieve some objective in a given situation.

Business decisions are those, which are made in the process of conducting business to achieve its objectives in a given environment. In concept, whether we are talking about business decision or any other decision, we assume that the decision maker is a rational person.

The major characteristics of the business decision-making are:

(*a*) Sequential in nature.

(*b*) Exceedingly complex due to risks and trade offs.

(*c*) Influenced by personal values.

(*d*) Made in institutional settings and business environment.

The business decision-making is sequential in nature. In business, the Decisions are not isolated events. Each of them has a relation to some other division or situation. The decision may appear as a "snap" decision but it is made only after a long chain of developments and a series of related earlier decisions.

The decision making process is a complex process in the higher hierarchy of management. The complexity is the result of many factors, such as inter-relationship among the experts or decision makers, a job responsibility, a question of feasibility, the codes of morals and ethics, and a probable impact on business.

The personal values of the decision maker play a major role in decision-making. A decision otherwise being sound on the business principle and economic rationality may be rejected on the basis of the personal values, which are defeated if such a decision is implemented. The culture, the discipline and the individual's commitment to goals will decide the process and success of the decision.

Whatever may be the situation, of one analyses the factors underlying the decision-making process, it would be observed that there are common characteristics in each of them. There is a definite method of arriving at a decision; and it can be put in the form of decision process model.

The decision-making process requires creativity, imagination and deep understanding of human behaviour. The process covers a number of tangible and intangible factors affecting the decision process. It also requires a foresight to predict the post-decision implications and a willingness to face these implications. All decisions solve a 'problem' but over a period of time they give rise to a number of other 'problems'.

Rational Decision Making

A rational decision is the one which, effectively and efficiently, ensures the achievement of the goal for which the decision is made. If it is raining, it is rational to look for a cover so that you do not get wet. If you are in business and want to make profit, then you must produce goods and sell them at a price higher than cost of production.

In reality, there is no right or wrong decision but a rational or an irrational decision. The quality of decision-making is to be judged on the rationality and not necessarily on the result it produces.

The rationality of the decision made is not the same in every situation. It will vary with the organisation, the situation and the individual's view of the business situation. The rationality, therefore, is a multi-dimensional concept. For example, the business situational decisions in a private organisation and a public sector undertaking differ under the head of rationality. The reason for this difference in rationality is the different objectives of the decision makers. Any business decision if review by a shareholder, a consumer, an employee, a supplier and a social scientist, will result in a different criticism due to difference individual's rationality. This is because each one of them will view the situation in different contents and the motive with in different objectives, since, wether a decision is right or wrong depends upon a specific rational view.

The question which further arises: It a decision "rational"? If it terns out to be wrong in terms of the results it produces, can we cast doubts on the rationality?

Simon Herbert A* differentiates among the types of rationality. A decision, in a given situation is:

- Objectively rational if it maximises the value of the objective.
- Subjectively rational if it maximises the attainment of value within limitation of the knowledge and awareness of the subject.
- Consciously rational to the extent the forces of the decision-making is deliberate and a conscious one.
- Organisationally rational to the degree of the orientation towards the organisation.
- Personally rational to the extent it achieves an individual's personal goals.

In other words, so long as the decision maker can explain with logic and reason, the objectivity and the circumstances in which the decision is made, it can be termed as a rational decision. Whether the rationality applied is appropriate or not could be point for debate.

Gross Bertram M** suggests three dimensions of rationality.

- First, the degree of satisfaction of human interest.
- Second, the degree of feasibility in achieving the objectives.
- Third, a consistency in decision-making.

If a decision maker shows consistent behaviour in the process of decision-making, then one can say that he meets the test of the rationality.

The Problems in Making Rational Decisions

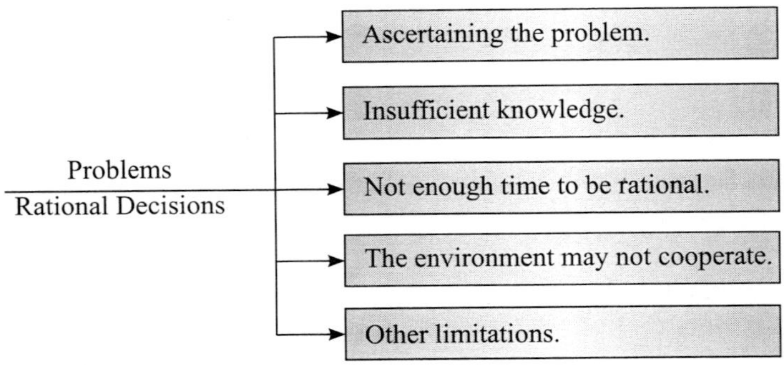

Fig. 2.7

(*a*) **Ascertaining the Problem:** As Peter Drucker points out, "the most common source of mistakes in the management decisions is the emphasis on finding the right answers rather than the right questions". The main task is to define the right problem in clear terms. The management may define the problem as the "sales are declining". Actually, the decline of sales is symptomatic, the real problem may be somewhere else. For example, the problem may be the poor quality of the product and you may be thinking of improving the quality of advertising.

(*b*) **Insufficient Knowledge:** For perfect rationality, total information leading to complete knowledge in necessary. An important function of a manager is to determine whether the dividing line is reached between insufficient knowledge and the enough information to make a decision.

(*c*) **Note enough time to be Rational:** The decision maker is under pressure to make decisions. If time is limited, he may make a hasty decisions which may not satisfy the test of rationality of the decision.

(*d*) **The Environment may not Cooperate:** Sometimes, the timing of the decision is such that one is forced to make a decision but the environment is not conducive for it. The decision may fail the test of rationality as the environmental factors considered in the decision-making turn out to be untrue.

For example, in a product pricing the factor of oil and petroleum product price is considered as stable. But the post-decision environment proves the consideration to be wrong.

(*e*) **Other limitations:** Other limitations are the need for a compromise among the different position, misjudging the motives and values of people, poor communications, misappraisal of uncentainties and risks, and inability to handle the available knowledge and human behaviour.

Now do we then ensure rationality? It is ensured, if the process of decision making is carried out systematically, whereby all the aspects of the decision-making discussed above are taken care of Herbert Simon said that a decision maker follows the process of decision-making disregarding the decision of the type of decision and the motive behind the decision. This process is followed consciously or without knowing it. We can put this process in the Decision-Making Model.

2.12.2 Decision Making Process

Decision-making is a process of selection from a set of alternative courses of action which is thought to fulfil the objectives of the decision problem more satisfactorily than other.

- According to McFarland, "A decision in an act of choice wherein an executive forms a conclusion about what must be done in a given situation. A decision represents a course of behaviour chosen from a number of possible alternatives". A decision involves the act of choice and the alternative chosen out of the available alternatives. The process concerned with searching and evaluating alternatives to a problem and selecting the best alternative is known as decision-making.
- The core of decision-making process is described by Herbert Simon in a model. He describes the model in three phases as shows in Fig. 2.8, *viz.* (*a*) Intelligence; (*b*) Design; and (*c*) Choice. MIS follows this model in its development stage.

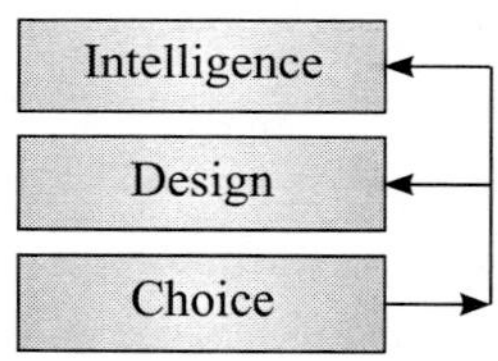

Fig. 2.8 Herbert Simon Model

Intelligence

Raw data collected, processed and examined. Identifies a problem calling for a decision.

Design

Inventing, developing and analysing the different decision alternatives and testing the feasibility of implementation. Assess the value of the decision outcome.

Choice

Select one alternative as a decision, based an the selection criteria.

In the intelligence phase, the MIS collects the data. The data is scanned, examined, checked and edited. Further, the data is soured and merged with other data and computations are made, summarised and presented. In this process, the attention of the manager is drawn to all problem situations by highlighting the significant differences between the actual and the expected, the budgeted or the targeted.

In the design phase, the manager develops a model of the problem situation on which he can generate and test the different decisions to facilitate its implementation.

If the model developed is useful in generating the decision alternatives, he then further moves into phase of selection called a choice.

In the phase of choice, the manager evolves a selection criterion such as maximum profit, least cost, minimum waste, least time taken, and highest utility. The criterian is applied to the various decision alternatives and the one which satisfies the most is selected.

In these three phases, if the manager fails to reach a decision, he starts the process all over again from the intelligence phase were additional data and information is collected, the decision-making model is refined, the selection criteria is changed and a decision is arrived at.

The MIS achieves this is an efficient manner without repeated use of the Simon Model again and again. An ideal MIS is supposed to make a decision is arrived at.

An example of the Simon Model would illustrate further its use in the MIS. For example, a manager finds on collection and through the analysis of the data that the manufacturing plant is under-utilised and the products which are being sold are not contributing to the profits as desired. The problem identified, therefore, is to find a product mix for the plant, whereby the plant is fully utilised within the raw material and the market constraints, and the profit is maximised. The manager having identified this as the problem of optimisation, now examines the use of linear programming (LP) model. The model is used to evolve various decision alternatives. However, selection is made first on the basis of feasibility, and then on the basis of maximum profit.

The product mix so given is examined by the management committee. It is observed that the market constraints were not realistic in some cases, and the present plant capacity can be enhanced to improve the profit. The same model is used again to test the revised position. Therefore, additional data is collected and an analysis is made to find out whether the average 20 percent utilisation of the capacity can be increased. A market research for some products is made and it is found that some constraints need to be removed and reduced. Based on the revised data, LP model is used, and a better optimum solution obtained.

Types of Decision

The type of decisions are based on the degree of knowledge about the outcomes or the events yet to take place. If the manager has full and precise knowledge of the event or outcome which is to occur, then the decision making is not a problem. If the manager has full knowledge, then it is a situation of certainity. If he has partial knowledge or a probabilistic knowledge, then it is decision-making under risks. If the manager does not have any knowledge whatsoever, then it is decision-making under uncertainty.

A good MIS tries to convert a decision-making situation under uncertainity to the situation under risk and further to certainity. Decision-making in the operations management, is a situation of certainity. This is mainly because the manager in this field, has fairly good knowledge about the events which are to take place, has full knowledge of environment, and has a predetermined decision alternatives for choice or for selection.

Decision-making at the middle management level is of the risks type. This is because of the difficult in forecasting an event with hundred percent accuracy and the limited scope of generating the decision alternatives.

At the top-management level, it is a situation of total uncertainty on account of insufficient knowledge of the external environment and the difficulty in forecasting business growth on a long-term basis.

A good MIS design gives adequate support to all the three levels of management.

Nature of Decisions

Decision-making is a complex situation. To resolve the complexity, the decisions are classified as programmed and non-programmed decisions.

Programmed Decision: If a decision can be based on a rule, method or even guidelines; it is called the programmed decision. If the stock level of an item is 200 numbers, then the decision to raise a purchase requisition for 400 numbers, is a programmed-decision-making situation. The decision maker here is told to make a decision based on the instructions or on the rule of ordering a quantity of 400 items when its stock level reaches 200.

If such rules can be developed wherever possible, then the MIS itself be designed to make a decision and even execute. The system in such cases plays the role of a decision maker based on a given rule or a method, since the programmed decisions are made through MIS, the effectiveness of the rule can be analysed and the rule can be reviewed and modified from time to time for an improvement. The programmed decision-making can be delegated to a lower level in the management cadre.

Non-Programmed Decision: A decision which cannot be made by using a rule or a model is the non-programmed decision. Such decisions are infrequent but the stakes are usually larger. Therefore, they cannot be delegated to the lower level. The MIS in the non-programmed-decision situation, can help to some extent, in identifying the problem, giving the relevant information to handle the specific decision-making situation. The MIS, in other words, can develop support systems in the non-programmed-decision-making situations.

2.12.3 Methods for Decision-Making Alternatives

There are several methods to help the manager decide among the alternatives. The methods basically are search processes to select the best alternative upon satisfying certain goals.

There are methods for selection of decision alternatives with the goals in view are:

(*a*) Optimisation Techniques.

(*b*) Payoff Analysis.

(*c*) Decision Tree Analysis.

All the operational research models use optimisation techniques, to decide on the decision alternatives. When a decision making situation can be expressed, in terms of decision versus the proable event and its Pay-off value, then it is possible to construct a martin of the decision versus the events described by a value for each combination. The manager can then apply the criteria such as the maximum expected value, the maximum profit and the maximum loss or the minimum regrets.

The method of decision tree can be adopted, if the decision-making situation can be described as a chain of decisions. The process of the decision-making is sequential and a chain of decision achieves the end objective.

The use of both the Pay-off matrix and the decision tree requires a probabilities knowledge of the occurrence of events. In many situations this knowledge is not available and the MIS has to provide the information support in this endeavor.

1. Optimisation Techniques: Optimisation techniques contains various methods like Linear Programming, Integer Programming, Dynamic Programming, Queuing Models, Inventory Models, Capital Budgeting Models and so on. These methods are used in cases where decision-making situation is closed, deterministic and requires to optimize the use of resources under conditions of constraints. To handle these situations, software packages are available. These methods are termed as operational research (OR) techniques. All the OR methods attempt to balance the two aspects of business under conditions of constraint.* In the Inventory Model, the cost of holding inventory versus the cost of procuring the inventory is balanced under the contraint of capital and meeting the demand requirement.* In the Queuing Model, the cost of waiting time of the customer versus the cost of an idle time of the facility is balanced under the constraint of investment in the facility and the permissible waiting time for the customer.* In the capital budgeting model, the return on investment is maximised under the capital constraint versus the utility of the investment. The MIS supports the formulation of a model, and then using it for the decision-making.

2. The Payoff Analysis: The Payoff Analysis is used to made the decision, when all the alternatives and their outcomes are not known with certainity. **The Payoff matrix** is constructed where the rows show the alternatives and the columns shows the conditions or the states of nature with the probability of occurrence. The intersection of column and the row shows the value of an outcome resulting out the alternative and the state of the nature. A typical Payoff matrix in pricing decision is shown in the given Table 2.2.

Table 2.2: Payoff I

Your decision alternative	*Competitor's Decision: Probability*	*No change*	*Increase*	*Decrease*	*Expected gain*
		0.50	*0.20*	*0.30*	
No change in the price		4	5	8	5.40
Increase the price		6	4	3	4.70
Decrease the price		10	12	4	8.60

For example, if the decision chosen is 'no change in the price and the competition' also does not change the price, then your gain is '4'. The decision is taken by choosing that decision alternative which has the maximum expected value of outcome. Since, the expected value in case of the third alternative is the highest, the decision would be to decrease the since.

The concept of utility relates to the money value considered by the decision maker. Utility is measured in terms of utile. Money has a value of a different degree to different decision makers depending upon the amount, and also the manner in which it is received. If rupee one is equal to one utile, the Rs. 100 million is not 100 million utiles but could be much more. The utile values will be different if the money is received in one lot as against in parts in several years. The utility function is different for different decision makers. The utile value of utility has an influence an the risks taking ability of the decision maker. A well placed manager with a sound business will tend to gamable or take more risk, than a manager not so well placed in the business. In such decision-making situations, the monetary values of the outcomes are replaced by the value of utiles, suitable to the decision maker's utility function. In our example of pricing, if we replace the values by utiles, the matrix would be as given in Table 2.3.

Table 2.3: Payoff II

Decision	*Competitor's Choice: Probability*	*No change*	*Increase*	*Decrease*	*Expected utility*
		0.50	*0.20*	*0.30*	
No change in the price		4	50	200	72.00
Increase in Price		200	4	400	220.80
Decrease in Price		100	20	4	54.12

Since the highest value of utility is 220.80 utiles, the decision would be to increase the price.

3. Decision Tree Analysis: When a decision maker must make a sequence of decisions, the decision tree analysis is useful in selecting the set of decisions. Decision tree analysis is one of the device for representing a diagrammatic presentation of sequential and multi-dimensional aspects of a particular decision problem for systematic analysis and evaluation is 'decision tree'—whereby the decision problem, alternative courses of action, states of nature and likely outcomes of alternatives are diagramatically or graphically depicted as if they are branches and sub-branches of 'Horizontal tree'.

As the name implies, a decision tree consists of network of nodes, branches, probability estimates and payoffs. Nodes are of two types, decision node (*designated as a square*) and a chance node (*designated as a circle*). Alternatively course of action or strategies originate from a decision node as the main branches (decision branches). At the terminal of each decision branch, there is a chance node whereform chance events emanate in the form of sub-branches (Chance branches). The respective payoffs and the probabilities associated with alternative courses and the chance events are shown alongside the chance branches. At the terminal of the chance branches are shown the expected payoff values of the outcome.

- There is basically two types of decision trees—**deterministic** and **probabilistic**. These can further be divided into single-stage and multi-stage trees. A single-stage deterministic decision tree involves making only one decision under conditions of certainity (no change events). In a multi-stage deterministic decision tree a sequence or chain of decisions are to be made. A problem involving only one decision to be made under conditions of risk or uncertainty (more than one chance event) can be represented in a single stage probabilistic decision tree while the same type of problem calling for a sequence of decisions is depictable in multistage probabilistic decision tree.

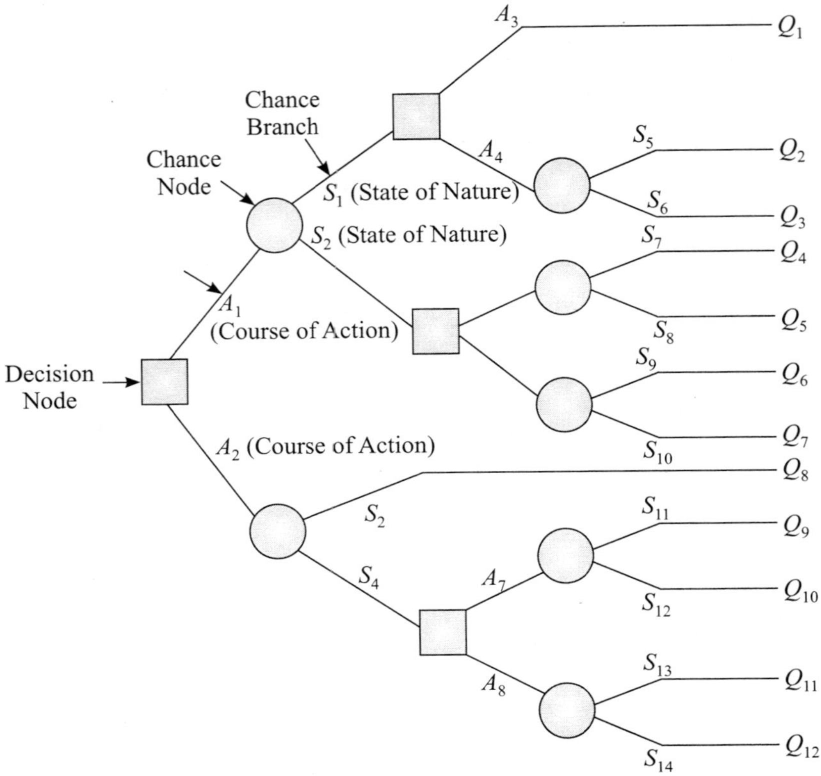

Fig. 2.9 Portray of a Decision Tree

2.12.4 MIS and Decision-Making

It is necessary to understand the concepts of decision-making as they are relevant to the design of the MIS. The Simon Model provides a conceptual design of the MIS and decision-making, wherein the designer has to design the system in such a way that the problem is identified in precise terms. That means the data gathered for data analysis should be such that it provides diagnostics and also provides a path to bring the problem to surface.

In the design phase of the model, the designer is to ensure that the system provides models for decision-making. These models should provide for the generation of decision alternatives, test them and pave away for the selection of one of them. In a choice phase, the designer must help to select the criteria to select one alternative amongst the many.

The concept of programmed decision-making is the finest tool available to the MIS designer, whereby he can transfer decision-making from a decision in drawing a decision tree, one must follow certain basic rules and conventions as stated below:

1. Identify all decisions (and their alternatives) to be made and the order in which they must be made.
2. Identify the chance events or state of nature that might occur as a result of a each decision alternative.

3. Develop a tree diagram showing the sequence of decisions and chance events. The tree is constructed starting from left and moving towards right. The square box '□' denotes a decision point which the available courses of actions are considered. The circle 'O' represents the chance node or event, the various states of nature or outcomes emanate from this chance event.
4. Estimate probabilities that possible events or states of nature will occur as a result of the decision alternatives.
5. Obtain outcomes (usually expressed in economic terms) of the possible interactions among decision alternatives and events.
6. Calculate the expected value of all possible decision alternatives.
7. Select the decision alternative (or course of action) offering the most attractive expected value maker to the MIS and still retain the responsibility and accountability with the decision maker or the manager. In case of non-programmed decisions, the MIS should provide the decision support systems to handle the variability in the decision-making conditions. The decision support systems provide generalised model of decision-making.

The concept of decision-making systems, such as the closed and the open systems, help the designer in providing a design flexibility. The closed systems are deterministic and rule based, therefore, the design needs to have limited flexibility, while in an open system, the design should be flexible to cope up with the changes required from time to time.

The methods of decision making can be used directly in the MIS provided the method to be applied has been decided. A number of decision-making problems call for optimisation, and OR models are available which can be made a part of the system. The optimisation models the static and dynamic, and both can be used in the MIS. Some of the problems call for a competitive analysis, such as a payoff analysis. In these problems, the MIS can provide the analysis based on the gains, the regrets and the utility. The concepts of the organisational and behaviourial aspects of decision-making provide an insight to the designer to handle the organisational culture and the constraints in the MIS. The concepts of the rationality of a business decision, the risk awareness of the managers and the tendency to avoid an uncertainity, makes the designer conscious about the human limitations, and prompts him to provide support in the MIS to handle these limitations. The reliance on organisational learning, makes the designer aware of the MIS and makes him provide the channels in the MIS to make the learning process more efficient. The relevance of the decision-making concepts is significant in the MIS design. The significance arises out of the complexity of decision-making, the human factors in the decision-making, the organisational and behaviour aspects, and the uncertain environments. The MIS design addressing these significant factors turns out to be the best design.

2.13 MIS: INTERACTION WITH EXTERNAL ENVIRONMENT

The use of Management Information System in the decision-making is the latest approach in technology. There are three main approaches of MIS:

- Decision Support System
- Knowledge Management
- Artificial Intelligence System.

2.13.1 Decision Support System

Decision support systems are an application of Herbert Simon Model. As explained earlier, the model has three phases, viz. Intelligence, Design and Choice. The decision support system basically helps the information system in the intelligence phase where the objective is to identify the problem and then go to the design phase for solution. The choice of selection criterian varies from problem to problem and then go to the design phase for solution. The choice of selection criterian varies from problem to problem. It is, therefore, required to go through these phases again and again till a satisfactory solution is found. In the use of three phase cycle, you may use either enquiry, analysis, and models or accounting systems to come to a rational solution.

The decision support system helps in making a decision and also in its performance evaluation. These systems can be used to validate the decision by performing sensitivity analysis on various parameters of the problem.

In decision making, we know that there are two types of decisions– Programmable and Non-Programmable. The programmable decision, because of its rule base structure, can be computerised, as inputs, processing methodology, analysis and choice of decision making are predetermined. Decision support system, can be built around the rule in case of programmable decision situation. While in non-programmable decisions, the rules are not fixed or predetermined, and requires every time the user to go through the decision-making cycle as indicated in the Herbert simon model.

The decision support system refers to a class of system which support in the process of decision-making and does not always give a decision itself. The nature of the decision is such that the decision makers need a variety of information, when same or similar types of decisions are to be made. These needs are such that wider additional demands on information would be made, the moment a piece of information is received. The calls on the information are continuously made till the decision maker is fully satisfied. The reason for changing the demands is also because the methods of decision making undergo a change from time to time. The quantum and the scope of information also changes depending upon the risk in decision-making. The higher the risk, more information may be sought.

Types of Decision Support Systems

- Status Inquiry System
- Data Analysis System
- Information Analysis System
- Accounting System
- Model Based Systems

1. ***Status Inquiry Systems:*** The number of decisions in the operational management and some at the middle management are such that they are based on one or two aspects of a decision-making situation. It does not call for any elaborate computations, analysis, choice, etc. for decision-making. If the status is

known, the decision is automatic, *i.e.,* the status and solution is unique relation.

2. ***Data Analysis Systems:*** These decision systems are based on comparative analysis, and use of a formula or an algorithm. But, these processes are not structured and therefore, vary. The cashflow analysis, the inventory analysis and the personal inventory systems are examples of the analysis systems. The use of simple data processing tools and business rules are required to develop this system.
3. ***Information Analysis Systems:*** In this system, the data is analysed and information reports are generated. The reports might be having exceptions as a feature. The decision makers use these reports for assessment of the situation for decision-making. The sales analysis, the accounts receivable systems, the market research analysis the MRP systems are examples of this system.
4. ***Accounting Systems:*** These systems are not necessarily for decision-making but they are desirable to keep track of the major aspects of the business or a function. The contents of these systems is more data processing leading to formal reporting, with exceptions, if necessary. These systems account items such as cash, inventory, personnel and so on and relate it to a norm or norms developed by management, for control and decision.
5. ***Model Based System:*** These systems are simulation models or optimisation models for decision-making. These decisions generally, are one time and infrequent and provide general guidelines for operation or management. The product mix decision, the material mix, the job scheduling rules, and the resource or assets or facilities planning systems are the examples.

In order to illustrate these decisions support system, let us take the example of **materials management** function and the variety of the decision and the type of decision systems used therein to support and evaluate the decision.

Decision	***Types of System Required***
Finding and selection of vendor	Inquiry System
Procurement	Performance Analysis System
Pricing	Data Analysis
Selections of vendor based on price quality and performance	Information analysis system
Selection of order quantity	Model based system
Inventory rationalisation	Valuation of inventory and Accounting System.
Management of inventory within various financial and stocking constraints.	Inventory optimisation model.

We can further analyse these systems in terms of the input source, the system, the hardware and the type of user as shown in table.

Some more facts about decision support systems, are:

(*i*) The decision support systems are developed by the users and system analysts jointly.

(*ii*) The decision support system uses the principles of economics, science and engineering, and also the tools and techniques of management.

(*iii*) The data used in the decision support system is drawn from the information systems developed in the company.

(*iv*) The decision support systems are developed in isolation and form an independent system subset of the management information system.

(*v*) The most common use of the decision support system is to test the decision alternatives and also to test the sensitivity of the result to the change in the system and assumptions.

(*vi*) The data and information for the decision support system are used from the internal sources such as the database and the conventional files, and from the external sources.

2.13.2 Knowledge Management

The scope of knowledge management however depends on the perspective the organisation would take. One broader perspective is that KM includes strategic, tactical and operational knowledge management. A limited perspective is 'Knowledge Management' for improving operational efficiency of the organisation.

The emergence of KM is out of needs of managing competitive business better. It is out of managing risks in better way as the stakes are very high. The forces which drive KM are internal as well as internal.

These forces create initiatives in the organisation to build Knowledge management System (KMS). KMS is a logical extension to sophisticated MIS, which uses latest information technology to improve process capabilities, customer service and business performance.

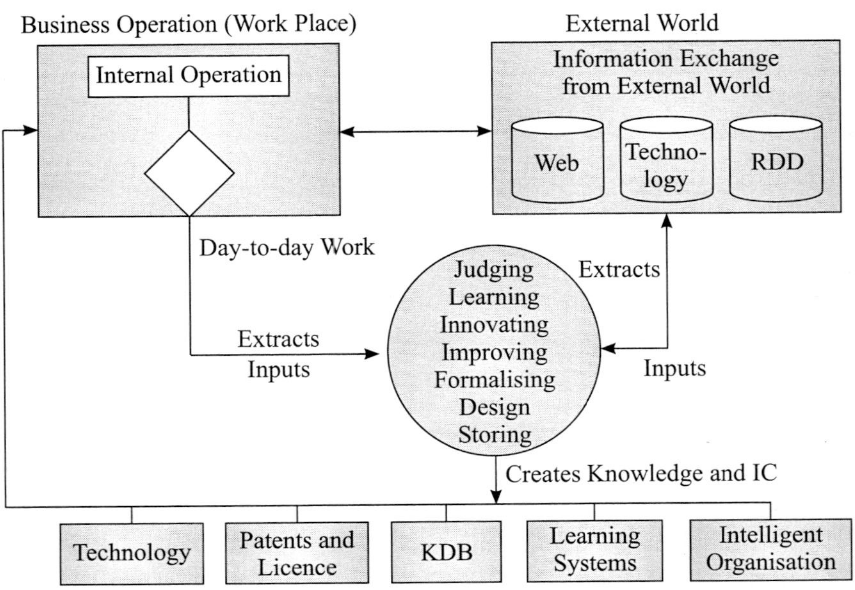

Fig. 2.10 Knowledge Building Model

Knowledge management has following processes.

- Define, capture, manipulate, store and develop.
- Develop information systems for knowledge creation.
- Design applications for improving organisation's effectiveness.
- Create knowledge set, *i.e.,* intellectual capital to increase economic value of the organisation.
- Keep IC continuously on upgrade to use it as a central resource.
- Distribute and share to concerned.

Effective use of knowledge depends on how a user decides to perform the work and personal ability to pick and choose and further manipulate for effective behaviour. Hence, with formal KM in place in the organisation, use of KM still depends on the persons initiative, ability to envision and the problem solving skills.

Figure 2.10 shows some beliefs in terms of true and false about knowledge, knowledge management and system.

KM system is a superset of systems build over present information system architecture. It is not a transaction processing system and it does not have standard system components, which when implemented would create IC.

It would be appropriate to clear some doubts on knowledge and KM for its acceptance in the organisation as an asset.

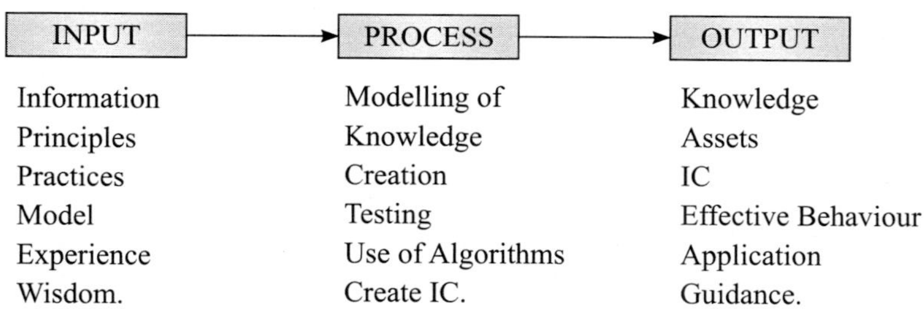

Fig. 11 KM system: A conceptual view

Time has come to handle the knowledge–IC systematically, for use in strategic management of business. The knowledge over a period gets developed in the organisation and it resides in people, information, files, database. It is not always, explicit but tacit in character and content, to know and acquire. To bring knowledge as critical input in the management process, it is necessary to have knowledge management systems. The knowledge management system therefore deals with definition, acquisition, construction, storage, delivery and application of knowledge. KMS handles two types of knowledge. One is tacit and either is explicit.

2.13.3 Artificial Intelligence (AI) Systems

All human beings have intelligence, which they use for problem solving, intelligence when supported by knowledge and reasoning abilities becomes an artificial intelligence. When such an artificial intelligence is packed into a database as a system, then what we have is AI system.

AI systems fall into three basic categories, viz., the expert systems (knowledge based), the natural language (native languages) systems, and the perception system vision, speech, touch.) AI is a software technique applied to the non-numeric data expressed in terms of symbols, statements and patterns. It uses the methods of symbolic processing, social and scientific reasoning and conceptual modelling for solving the problems. The AI systems are finding applications in configurations, design, diagnosis, interpretation, analysis, planning, scheduling, training, testing and forecasting.

The AI systems do not replace people. They liberate experts from solving common/simple problems leaving the experts to solve complex problems. Artificial Intelligence system help to avoid making same mistakes, and to respond quickly and effectively to a new problem situation. The knowledge-based expert system is a special AI system. It has wide applications in business and industry. AI systems thrive on artificial intelligence. AI is a field of science and technology where knowledge from other disciplines such as computer science, biology, psychology, mathematics, statistics and engineering is blended together. The goal of AI is to develop computer functions of features as close to human intelligence, described as reasoning, learning, problem solving, exhibiting creativity, respond quickly, sort quickly anfiguons and incomplete and or erroneous information or situation. AI domain is very large and its application spreads in wide areas of business and Industry. Cognitive science application uses knowledge and human information processing capabilities to produce major application as expert systems. The seaver provides intelligent information sets, knowledge database, rule base and goal driven reaching capability to solve the problem. Expert systems are designed to make humanlike inferences leading to an advice to decision maker.

Robotics, application uses AI, engineering science, and physiology to produce computer intelligence to guide a 'computer driven machine' to perform like human being, having capabilities of perception, touch, manipulation, locomotion, navigation and so on. Robotics applications are found useful in manufacturing, material handling and transportation.

Natural interface application uses AI to build natural, realistic, multi sensory human-computer interface. This interface enables you to build a 'Virtual Reality'. Virtual reality becomes very handy support to test, feel or experience the 'reality' before it is actually created.

CASE STUDY

Information Technology Management at Max Pipes

Max Pipes was a company with turnover close to ₹ 500 crore. The company was actively supplying agricultural products across India. The pro-agro government policies and various discounts offered by government contributed to the company's progress, and made the company a leader in this area. Not only did it lead in the business of pipes and drip irrigation, but it also captured all the allied businesses and had plans to grow further. The managing director, Mr Honda, was not keen on introduction of IT at various levels, though the company had already computerized its accounting and some of its quality assurance processes. Being in the process industry they had to deal with a lot of inventories and processing of raw materials.

There were also different aspects of waste management. They had made some efforts in the past to automate some information processing but did not achieve the desired results. SQC department also made a few automation efforts but the company stopped those, as there was no ROI based on their calculations.

But their policies did not seem to work for long. There were a few companies in the market which had started to claim some benefits through IT. Even Max had noticed the changes in the business environment, where all the organizations and society were gearing up for IT. Honda realized that it was time to look for serious changes in policies and develop and manage IT to achieve business goals. He called a meeting of his senior executives and explained that they had lost, during the last year, about 20 million dollars because of their inability to optimize the production capacity. They could not store raw material beyond a certain limit. In a fluctuating market the demand kept on changing. If they produced more they did not have the space to store it. If they produced less there was always a possibility of losing a contract. It was difficult for an MD to manage the overall requirements of the market and the order inventories. If the inventories were not ordered in advance, the vendors were unable to supply it within the stipulated time. As their orders were discreet they were not eligible for the discounts and that was also affecting their pricing.

The IT head introduced Honda to a new inventory management system and Management Information System (MIS) to keep track of the inventory and departmental needs. He also explained to him the usage of IT for performance management system. Honda was not in favour of IT as in their previous attempt at implementation of IT, they had realized that the ROI (Return On Investment) was very low. Even the management was reluctant to go ahead with IT. But the overall proposal given by the IT head was so exciting that Honda decided to go in for it.

Initially they appointed KK Consultancy for systems study for development of MIS and automation. KK Consultancy studied the system and after four months, came up with their recommendations. Max Pipes decided to implement the recommendations. They deployed the new inventory management system. As per the recommendations of the consultants they also went ahead with automation at various pockets. Later all the systems were connected with fibre optic backbone and other features such as online ordering, inventory-related alarms, and capacity planning were developed.

The company spent over one and a half million dollars on infrastructure development, resource training, and infrastructure related to IT. In the next six months many departments began using the MIS and Honda started getting reports on time. Unfortunately, in the next six months the production went down from 330 metric tons to 300 metric tons. The overall profit of the organization dipped by 25% in those six months.

Honda was very upset with the results and felt that the IT initiatives had contributed to this loss and he almost made up his mind to stop the IT initiatives. He discussed this with Mr Kelvin; the company's IT head, and made it clear that he could not go ahead with IT with no ROI. The IT head was not of the same opinion but he failed to convince Honda.

Kelvin knew that the organization's IT was not to be blamed for the losses. It had more to do with the implementation of IT. Kelvin went to Honda and sought another six months. He told him that since they had already invested a huge amount in infrastructure and training, six more months would not make much of a difference. Besides it takes time for the actual returns to materialize.

Initially, Honda was not willing to give another chance to IT and Kelvin. But he thought it was worth trying as huge investments had already been made. He gave the green signal to Kelvin but made it clear that it would be the last chance; in case of a failure he would go back to his old methods.

For the next one month Kelvin studied the IT implementation process afresh. He studied all the gaps and made the following observations:

- Unavailability of relevant information
- Too much of data
- Training
- Reluctance to use the system
- Underestimating human role in the new environment
- Special efforts to prove IT wrong
- Improper usage
- Not using the relevant information
- Using the information blindly

Kelvin then started implementing IT closely. He started from the top. He gave very detailed training to Honda on how to use information. He also developed small tools to generate alerts. The implementation went on smoothly. He even used forecasting systems for predicting demand and placed tentative orders for raw materials. Every supervisor was trained to use various tools and obtain information.

At the end of six months there was improvement in production, which came back to normal. Honda said he could not see the impact of IT in improving the performance. Kelvin was of the opinion that the measure used for IT performance measurement was not right. Kelvin presented an IT performance matrix to Honda.

He said that IT had a role in even a 300 million dollar order that they got for next year. It had played a key role in completing all orders on time without delay. Better reputation in the market this quarter was a result of IT. In spite of the poor market conditions the company's sales were good because of correct IT strategies adopted by the company. He showed that the relationship between IT and business was balanced and their Business Sales Cycle was also balanced. Kelvin explained the advantages of these initiatives for the business in the next quarters.

The overall picture was so bright that Honda approved more budget for IT and Kelvin was appointed as the new CIO of the company. In the next one year the company doubled its profits. There was always a thrust on proper implementation and measurement of IT.

2.14 WHAT IS THE ROLE OF MANAGEMENT INFORMATION SYSTEM (MIS) IN SUSTAINING EFFICIENCY AND EFFECTIVENESS OF AN ORGANIZATION?

A management information system (MIS) is a system that provides information needed to manage organizations effectively. Management information systems are regarded to be a subset of the overall internal controls procedures in a business, which cover the application of people, documents, technologies, and procedures used by management accountants to solve business problems such as costing a product, service or a business-wide strategy. Management information systems are distinct from regular information systems in that they are used to analyze other information systems applied in operational activities in the organization. Academically, the term is commonly used to refer to the group of information management methods tied to the automation or support of human decision making, *e.g.* Decision Support Systems, Expert systems, and Executive information systems.

An 'MIS' is a planned system of the collection, processing, storage and dissemination of data in the form of information needed to carry out the management functions. In a way, it is a documented report of the activities that were planned and executed. According to Philip Kotler "A marketing information system consists of people, equipment, and procedures to gather, sort, analyze, evaluate, and distribute needed, timely, and accurate information to marketing decision makers."

2.14.1 Applications of MIS

With computers being as present everywhere as they are today, there's hardly any large business that does not rely extensively on their IT systems. However, there are several specific fields in which MIS has become invaluable.

1. Strategy Support: While computers cannot create business strategies by themselves they can assist management in understanding the effects of their strategies, and help enable effective decision-making. MIS systems can be used to transform data into information useful for decision making. Computers can provide financial statements and performance reports to assist in the planning, monitoring and implementation of strategy. MIS systems provide a valuable function in that they can collate into coherent reports unmanageable volumes of data that would otherwise be broadly useless to decision makers. By studying these reports decision-makers can identify patterns and trends that would have remained unseen if the raw data were consulted manually. MIS systems can also use these raw data to run simulations - hypothetical scenarios that answer a range of "what if" questions regarding alterations in strategy. For example, MIS systems can provide predictions about the effect on sales that an alteration in price would have on a product. These Decision Support Systems (DSS) enable more informed decision making within an enterprise than would be possible without MIS systems.

2. Data Processing: Not only do MIS systems allow for the collation of vast amounts of business data, but they also provide a valuable time saving benefit to the workforce. Where in the past business information had to be manually processed for filing and analysis it can now be entered quickly and easily onto a computer by a data processor, allowing for faster decision making and quicker reflexes for the enterprise as a whole.

3. Management by Objectives: While MIS systems are extremely useful in generating statistical reports and data analysis they can also be of use as a Management by Objectives (MBO) tool. MBO is a management process by which managers and subordinates agree upon a series of objectives for the subordinate to attempt to achieve within a set time frame.

Objectives are set using the **SMART** ratio: that is, objectives should be

S - Specific

M - Measurable

A - Agreed

R - Realistic

T - Time-Specific

The aim of these objectives is to provide a set of key performance indicators by which an enterprise can judge the performance of an employee or project. The success of any MBO objective depends upon the continuous tracking of progress. In tracking this performance it can be extremely useful to make use of an MIS system. Since all SMART objectives are by definition measurable they can be tracked through the generation of management reports to be analyzed by decision-makers.

2.14.2 Benefits of MIS

The field of MIS can deliver a great many benefits to enterprises in every industry. Expert organizations such as the Institute of MIS along with peer reviewed journals such as MIS Quarterly continue to find and report new ways to use MIS to achieve business objectives.

1. Core Competencies: Every market leading enterprise will have at least one core competency—that is, a function they perform better than their competition. By building an exceptional management information system into the enterprise it is possible to push out ahead of the competition. MIS systems provide the tools necessary to gain a better understanding of the market as well as a better understanding of the enterprise itself.

2. Enhance Supply Chain Management: Improved reporting of business processes leads inevitably to a more streamlined production process. With better information on the production process comes the ability to improve the management of the supply chain, including everything from the sourcing of materials to the manufacturing and distribution of the finished product.

3. Quick Reflexes: As a corollary to improved supply chain management comes an improved ability to react to changes in the market. Better MIS systems enable an enterprise to react more quickly to their environment, enabling them to push out ahead of the competition and produce a better service and a larger piece of the pie.

4. Significant cost benefits, time savings, productivity gains and process re-engineering opportunities are associated with the use of data warehouse for information processing.

5. Data can easily be accessed and analysed without time consuming manipulation and processing.

6. Decisions can be made more quickly and with confidence that the data are both time-relevant and accurate.

7. Integrated information can be also kept in categories that are meaningful to profitable operation.

CASE STUDY

MIS in Banking Sector (AXIS BANK)

Axis Bank, previously called UTI Bank, was the first of the new private banks to have begun operations in 1994, after the Government of India allowed new private banks to be established. The Bank was promoted jointly by the Administrator of the Specified Undertaking of the Unit Trust of India (UTI-I), Life Insurance Corporation of India (LIC), General Insurance Corporation Ltd., National Insurance Company Ltd., The New India Assurance Company, The Oriental Insurance Corporation and United Insurance Company Ltd. UTI-I holds a special position in the Indian capital markets and has promoted many leading financial institutions in the country. For example, using MIS strategically can help a company to become a market innovator. By providing a unique product or service to meet the needs of customers, a company can raise the cost of market entry for potential competitors and thus gain a competitive advantage. Another strategic use of MIS involves forging electronic linkages to customers and suppliers. This can help companies to lock in business and increase switching costs. Finally, it is possible to use MIS to change the overall basis of competition in an industry. For example, in an industry characterized by price wars, a business with a new means of processing customer data may be able to create unique product features that change the basis of competition to differentiation.

In today's scenario MIS plays a pivotal role in Organizations. Organizations worldwide make extensive use of MIS. It's designed by the top management of an organization, is a tool to assembling & accumulating facts & figures of all the important business processes.

Thus here are some of the major importance/advantages of MIS in organizations:

1. The organization that uses MIS is able to record, process, route & tabulate all important business transactions. As & when need arises the organization is able to incorporate the needed changes & improvements in the area of concern.

2. MIS facilitates informed decision making. It usually represents a number of options from which one can choose the best.

3. The top management analyses whether its resources are being utilized optimally.

4. A two way communication flow is greatly enhanced by the MIS. The management freely tells the job responsibilities to its employees. The employees in return discuss their doubts & grievances.

5. MIS supports the planning & controlling function of managers in the organization. Managers use past/historical data as well as the current data to analyse the performance & hence apply controlling measures.

6. MIS encourages DECENTRALISATION in the organization. Decentralization is possible when there's a system to measure operations at the lower levels.

7. It brings coordination. It facilitates integration of specialized activities by keeping each department aware of the problems & requirements of other departments.

2.15 THE IMPORTANCE OF A GOOD INFORMATION MANAGEMENT SYSTEM

All companies use information to set strategies and accomplish business objectives. Every business, program or system must address well-defined objectives, which will add value, either directly to the bottom line or toward the achievement of the organization's goals and objectives. Good management information objectives usually fall into one of three categories:

- Service (effective and efficient),
- Profit (or cost-avoidance, and
- Social (moral, ethical and legal) responsibility.

A good management information system will only reap the benefits if the companies gain insight to better align strategies and identify critical relationships and gaps along four key company dimensions–**people, process, culture and infrastructure**.

A good information system provides a framework for companies to evaluate themselves relative to these dimensions. By understanding and improving alignment with these critical dimensions, companies can maximize the value and impact of information as a strategic corporate asset to gain competitive advantage.

The following are the most important reasons to have a good management information system:

1. To control the creation and growth of records: Despite decades of using various non-paper storage media, the amount of paper in an office continues to escalate. An effective records information system addresses both creation control (limits the generation of records or copies not required to operate the business) and records retention (a system for destroying useless records or retiring inactive records), thus stabilizing the growth of records in all formats.

2. To reduce operating costs: Record keeping requires administrative dollars for filing equipment, space in offices, and staffing to maintain an organized filing system (or to search for lost records when there is no organized system).

It costs considerably less per linear foot of records to store inactive records in a Data Records Center versus in the office. [Multiply that by 30% to 50% of the records in an office that doesn't have a records management program in place], and there is an opportunity to effect some cost savings in space and equipment, and an opportunity to utilize staff more productively - just by implementing a records management program.

3. To improve efficiency and productivity: Time spent searching for missing or misfiled records are non-productive. A good record management program (*e.g.* a document system) can help any organization upgrade its recordkeeping systems so that information retrieval is enhanced, with corresponding improvements in office efficiency and productivity. A well designed and operated filing system with an effective index can facilitate retrieval and deliver information to users as quickly as they need it.

Moreover, a well managed information system acting as a corporate asset enables organizations to objectively evaluate their use of information and accurately lay out a roadmap for improvements that optimize business returns.

4. To assimilate new records management technologies: A good record management program provides an organization with the capability to assimilate new technologies and take advantage of their many benefits. Investments in new computer systems whether this is financial, business or otherwise, don't solve filing problems unless current manual recordkeeping or bookkeeping systems are analyzed (and occasionally, overhauled) before automation is applied.

5. To ensure regulatory compliance: In terms of recordkeeping requirements, China is a heavily regulated country. These laws can create major compliance problems for businesses and government agencies since they can be difficult to locate, interpret and apply. The only way an organization can be reasonably sure that it is in full compliance with laws and regulations is by operating a good management information system which takes responsibility for regulatory compliance, while working closely with the local authorities. Failure to comply with laws and regulations could result in severe fines, penalties or other legal consequences.

6. To minimize litigation risks: Business organizations implement management information systems and programs in order to reduce the risks associated with litigation and potential penalties. This can be equally true in Government agencies. For example, a consistently applied records management program can reduce the liabilities associated with document disposal by providing for their systematic, routine disposal in the normal course of business.

7. To safeguard vital information: Every organization, public or private, needs a comprehensive program for protecting its vital records and information from catastrophe or disaster, because every organization is vulnerable to loss. Operated as part of a good management information system, vital records programs preserve the integrity and confidentiality of the most important records and safeguard the vital information assets according to a "Plan" to protect the records. This is especially the case for financial information whereby ERP (Enterprise Resource Planning) systems are being deployed in large companies.

8. To support better management decision making: In today's business environment, the manager that has the relevant data first often wins, either by making the decision ahead of the competition, or by making a better, more informed decision. A good management information system can help ensure that managers and executives have the information they need when they need it.

By implementing an enterprise-wide file organization, including indexing and retrieval capability, managers can obtain and assemble pertinent information quickly for current decisions and future business planning purposes. Likewise, implementing a good ERP system to take account of all the business' processes both financial and operational will give an organization more advantages than one who was operating a manual based system.

9. To preserve the corporate memory: An organization's files, records and financial data contain its institutional memory, an irreplaceable asset that is often overlooked. Every business day, you create the records, which could become background data for future management decisions and planning.

10. To foster professionalism in running the business: A business office with files, documents and financial data askew, stacked on top of file cabinets and in boxes everywhere, creates a poor working environment. The perception of customers and the public, and "image" and "morale" of the staff, though hard to quantify in cost-benefit terms, may be among the best reasons to establish a good management information system.

EXERCISES

Questions for Discussion

1. What were the reasons due to which Max Pipes could not achieve any Return on its IT investments during the first two quarters?
2. What were the flaws in IT management during the first six months?
3. What were the flaws in the IT performance measurement?
4. Discuss IT implementation with reference to Max Pipes case.

Short Question Answers

1. What do you understand by data and information?
2. Why there is a need of management information system?
3. Define the types of information system.
4. What importance of MIS is in today's competitive world?
5. Elaborate the role of MIS in society.
6. What are the characteristics of management information system?
7. Define decision support system with its importance.
8. Define distinguishing features of a knowledge management system.

Long Question Answers

1. Briefly explain the management information system with its components.
2. Elaborate the role of information system in business world.
3. Explain the types of management information system with importance of each system.
4. Explain how management information system helps in decision making for organizations?
5. Define decision support system with its use in the company's.
6. "MIS is a combination of human and computer based resources". Explain.
7. "Model Management Subsystem includes financial, statistical, management science or other models". Elaborate.
8. Define knowledge management system and explain its benefits with significance of its components.
9. Explain the types of management information system?
10. What is the role of Management Information System (MIS) in sustaining efficiency and effectiveness of an organization?
11. Explain the applications of MIS?
12. What are the benefits of MIS?
13. Why good management information system is required? Explain its importance?

■■■

CHAPTER 3

Organization Structure and Organization Analysis

Introduction

Organizational Analysis: An organizational analysis is a process by which an organizations systems, capacity, and functionality are assessed in order to increase its efficiency, performance, and output. An organizational analysis aims to understand behavioural relationships, structure and technology with the use of various models and theories. Within the field of organizational analysis, management is typically defined as the process of utilizing numbers resources, (personnel, materials, physical plant, equipment, information, time and money) in order to accomplish organizational goals.

An analysis, or sometimes called restructuring, of an organization may become necessary when either external or internal forces have created a problem or opportunity.

Whereas, organizational analysis, in management science, is the study of the processes that characterize all kinds of organizations, including business firms, government agencies, labour unions, and voluntary association such as sports clubs, charities, and political parties. Any organization is a social unit with three properties:

1. it is a corporate (or group) actor,
2. it claims a special and limited purpose (such as making profits or providing medical care), and
3. its creators intend it to last beyond the accomplishment of a single action, if not indefinitely.

Modern cultures are marked by an increase in the importance, influence, and power of organizations. Consequently, contemporary studies in social science and management have emphasized the analysis of organizations. Yet much of the research is narrowly focused on the properties associated with particular types of organizations, such as hospitals, prisons, government agencies, businesses, schools and churches. While many of its findings are associated with business management, the field of organizational analysis is far more general, that is, it studies the processes that apply to all kinds of the organization. One goal of such enquiry is the identification of more effective management strategies. Some examples of pressures that may lead to the implementation of an organizational analysis are the environment shifts, technology changes, organizations grow, leadership changes.

Organizational Structure: An organizational structure consists of activities such as task allocation, coordination and supervision, which are directed towards the achievement of organizational aims. It can also be considered as the viewing glass or perspective through which individuals see their organization and its environment. An organization can be structured in many different ways, depending on their objectives. The structure of an organization will determine the modes in which it operates and performs. Organizational structure affects the organizational action in two big ways. First, it provides the foundation on which standard operating procedures and routines rest. Second, it determines which individual get to participate in which decision-making processes, and thus to what extent their views shape the organization's actions.

Organizational structure refers to the way in which an organization's activities are divided, grouped and co-ordinated into relationships between managers and employees, managers and managers and employees and employees. Thus, an organization's departments can be formally structured in three major ways:

- by function,
- by product/market,
- or in matrix form,

Whereas, there are six key elements that managers need to address when they design their organizations structure which are shown in Fig. 3.1.

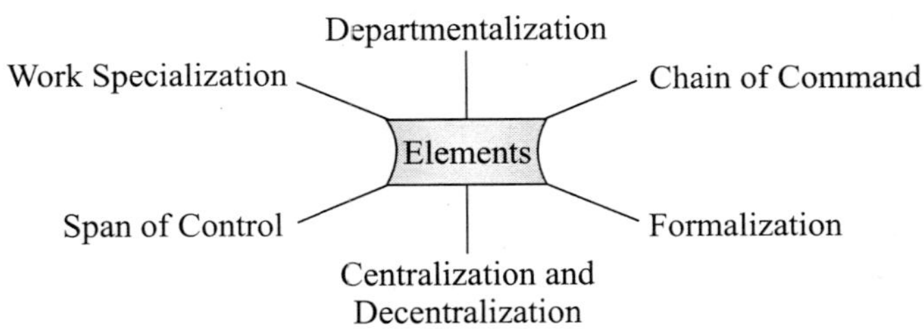

Fig. 3.1

3.1 PLANNING APPROACH TO ORGANIZATIONAL ANALYSIS

Organizational analysis and planning focuses on cultivating and maintaining an efficient workforce through the design and structure of an organisation, as well as the relationships and behaviour of individuals within organizations. Specifically, organizational analysis is concerned with developing models and theories that accurately captures the functioning and development of organizations and that account for the ways in which organizations respond to and bring about changes. Organizational planning, on the other hand, involves designing an organizations structure and dividing up the responsibilities of an organization. The goals of organizational analysis and planning typically have been to determine the best way to view and organize a company in order to manage it successfully and to bring about greater efficiency.

3.1.1 Models of Organizational Analysis

One of the basic techniques of organizational analysis is "modeling" that is developing models of organizations that delineate the way they function and evolve in order to identify the best way of managing each one. Modeling enables managers to determine the crucial variables in particular circumstances so they can experiment with different combinations of variables to achieve their desired results. For example, managers can determine the best combination of technology and organizational structure for their company by using organizational models.

Organizational models typically focus on behavior, structure or technology. In consideration of these variables, four general models of organizational analysis exist: the rational (also called classical model), the natural system (also called the participative model), the sociotechnical, and the cognitive model which is shown in Fig. 3.2.

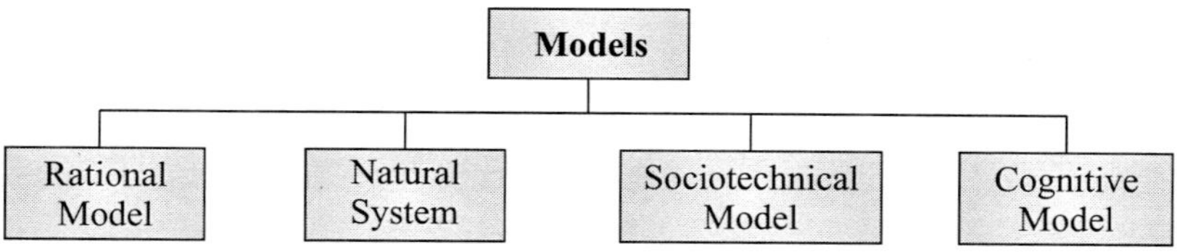

Fig. 3.2

Let us study it one by one.

(*a*) **Rational Model:** A pioneer of the rational model of organization was Frederick Taylor, who was influential near the start of the twentieth century. Taylor's background in engineering prompted his organizational analysis on efficiency. In Taylor's view, there was one best way the most efficient way-to-perform a task. Scientific management sprang from his work, with resultant time and motion studies in which tasks were timed and employee motions were gauged for efficiency. The best way to perform a task, in Taylor's view, was the way that accomplished the task in the least amount of time. He extended this view from employees to management, suggesting that nearly all organizational tasks could become more efficient if scientific principles were applied.

This was at the down of the introduction of the automobile to America. Although many western European nations began manufacturing automobiles before the early 1900s, production efficiency still had a long way to go. Applying scientific management principles helped Ford Motor Company develop the first American, mass-produced automobile. Frederick Taylor, then was correct. Scientific Management did work, but it was not without problems. The main problem was that it ignored the boredom that repetitive tasks created for workers. Workers became simply replaceable parts in the organizational machine.

In addition, the rational model of organization presupposes that decisions about an organization's structure are reached because of the rational assessment of an organization's needs, goals and external influences. And like Taylor's scientific management, this is true in some situations, but is not comprehensive enough to tell the whole story of how needs, goals and external influences affect organizational analysis and planning.

The rational model assumes that deviations from rationality result from errors in judgement and calculation as well as from ignorance. This model treats organization as mechanical groups because it conceives of the organization as having structures of different parts, and all of these parts can be modified and manipulated inorder to improve the efficiency of the entire organization. Furthermore, individual parts of the organization are viewed as modifiable through deliberate effort. Finally, this model sees the long-term development of the organization as modifiable and controllable through planned modification inorder to accomplish definite goals.

The rational model is still pervasive among managers and corresponds to the pyramidal organizational structure, in which top managers are at the apex and employees are at the bottom. Managers possess the authority in this model, defining and assigning tasks to the employees, who are charged with completing the tasks. They must begin by giving employees clear and detailed instructions. After that, managers must evaluate employee performance and distribute rewards and punishments based on the way employees performed their tasks.

Managers assume that worker motivation is directly correlated with economic rewards and punishments meted out by the managers. Motivation, from a rational perspective, simply involves increasing pay or threatening workers with various punishments. Hence, according to this model, managers rely on pay and related forms of compensation to motivate the workers to complete their tasks efficiently in order to achieve company goals. Example of rational model is shown in Fig. 3.3.

The problem with this assumption is that there are many motivators other than money, there can be many ways to perform a given task, and there are many organizational goals that are not rational. The rational model is thus a starting point for thinking about organizational analysis, but certainly not encompassing enough to provide a complete picture.

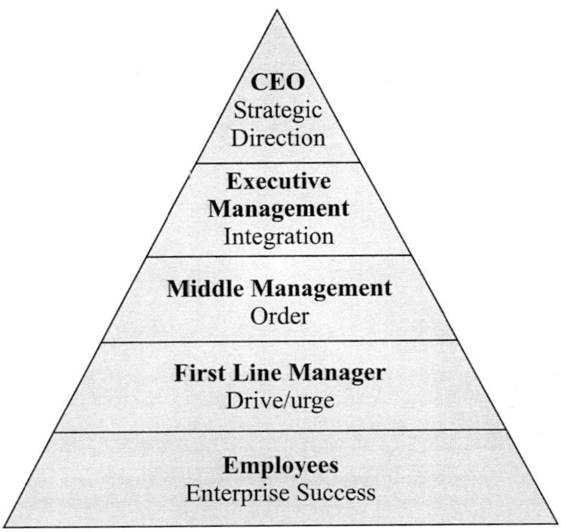

Fig. 3.3 Management Pyramid

(*b*) **Natural System Model:** In contrast to the rational model, the natural system model views organizations holistically, that is, as systems. The natural system model sees an organization as not only striving to accomplish its own goals, but also other important goals. An organizational system is regarded as an institution in its own right that has needs of its own. Hence, according to this model, an organization seeks to maintain a balance of its goals and various needs which may restrict the way it pursues other goals.

Unlike the rational model of organization, the natural system model sees the modification of an organization as unplanned and adaptive reactions to unstable conditions that threaten the balance of the organization as an entire system. The way an organization responds to problem is characterized as a defence mechanism and as being influenced by the common values ensconced in the members of the organization. This model concentrates on threats to an organization's equilibrium, that is on events and activities with the potential of disrupting an organization's balance.

When deviations from organizational plans and goals occur, they are seen not as the product of error or ignorance, but as the result of limitations brought about by an organizations social structure. This model generally is based on the concept of organizations as organisms in which all the parts are interconnected and interdependent. Consequently, changes in one part of an organization are thought to have an impact on other parts of the organization, and so planned modification of the organization is difficult.

In practical terms, the natural system model strives to balance the needs of all the members of the organization as well as other stakeholders, such as customers, shareholders and suppliers. This model holds that organizations function best when members belong to at least one effective work group (department, committee, or staff group), thereby contributing to the goals of organizations. Members who belong to more than one work group help link the different units of the organization together and facilitate communication and the exchange of information throughout the organization.

The natural system model views change as affecting the entire organization, not just individual or individual units. Consequently, managers cannot change just one small part of an organization, rather they must change the whole organization. As a result, planning for change must be comprehensive and systematic. Theoretically, the natural system model helps prevent conflicts in that changes take place only with the involvement of each member of the organization. Therefore, commitment to change is greatly increased and conflict over change is limited. Natural social system is shown in Fig. 3.4.

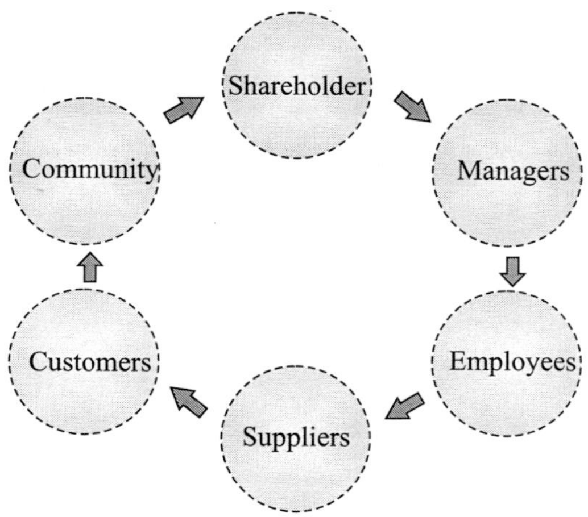

Fig. 3.4

(*c*) **Socio-technical Model:** Because of the limitations of the previous models of organization, theorists have developed other models to capture the essence and functioning of modern organizations. The sociotechnical model does not rely on the mechanical and biological analogies of the rational and natural system models. Instead, the socio-technical model views organizations as having a greater ability to modify their form and structure. Neverthless, like the natural system model, the sociotechnical model sees organizations as evolving. An organization changes when the expectations of its members change as a result of their collaboration with other members and the exchange of information.

This model views organization as systems that interact with their environment. Through the course of this interaction, organizational behaviour is affected by human, social, technological, and organizational inputs. These inputs are all interdependent, thus a change in one causes a change in the others. The basic tenets of the sociotechnical model include the belief that behaviour in organizations can have a number of causes, that organizations are systems, and that informal social systems are different from formal social systems. An organization's main task is accomplished through the process of inputs being converted into outputs. The organization is designed around these tasks. Similarly each unit of the organization is designed around its specific subtask. The sociotechnical model assumes that an organization's effectiveness is determined by its design to perform its main task. Organizations have differentiated, yet integrated, units based on three primary factors: technology (including techniques, skills, and materials), geographic location, and time (work shifts). According to this model, if an organization is effectively designed around its main task and if its units are differentiated and integrated effectively, then the number of conflicts will be reduced. Socio-technical model is shown in Fig. 3.5.

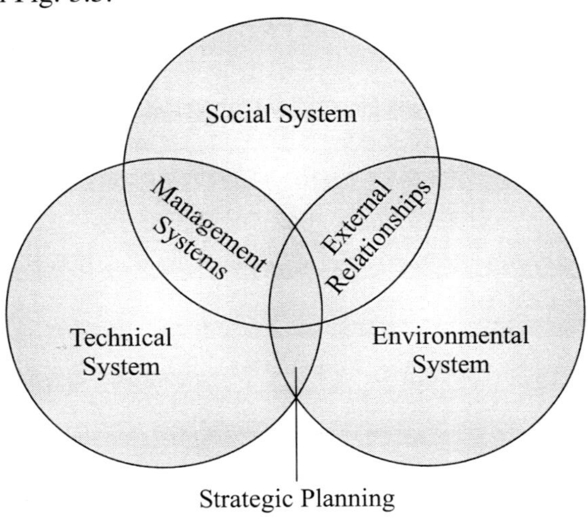

Fig. 3.5

(*d*) **Cognitive Model:** The cognitive model of organization consists of three primary components: cognition, the decision-making or problem-solving process, and an organizational setting. Cognition refers to the information-processing units of an organization and its organizational units. The decision-making or problem-solving component is a series of steps, operations, and procedures that an organizational unit uses to make decisions or solve problem. The organizational setting component is the arrangement of the organization, that is, the way tasks are distributed and the way processes are coordinated.

Although the rational model of organization focuses on clarifying and assigning tasks, it doesnot address the other aspects of organizations. In particular, it provides little in terms of the ways organization solve problems once tasks are clarified and assigned. The cognitive model moves beyond this level of organizational analysis by focussing on the processes through which organizations assign specific activities and time for the activities to be performed.

The cognitive model focuses on the decision-making process of an organization. An organization makes decision in accordance with its objective and based on available information. Since this model views individuals as having the capacity to do only a few things at a time, the organization functions as the combination of these limited capacities and facilitates the overall completion of a number of complex tasks, which are broken into series of subtasks so that the individuals can perform them. These subtasks are the areas of specialization within an organization. Specialization, in turn, brings about the flow of specific information to and from specialized units.

This model provides several key insights into the working of organizations. It conceives of an organization as a process that develops from the interaction of human cognition, organizational structure, and the types of decisions that need to be made. Because of these characteristics, the cognitive model focuses on the development and adaption of organizations in different circumstance. Further more, this model accounts for the way in which specialization affects organizational behaviour and coordination.

In conclusion, because each different organizational model has its advantages and disadvantages, managers must decide which one (or ones) best captures the workings of their company by valuating the assumptions and key processes of each, as well as by determining which one can solve the kinds of problems that needed to be solved.

3.1.2 Types of Organizational Planning Structures

Organizational planning involves designing an organization's structure to maximise efficiency. This includes dividing a company into different units, department and teams like shown in Fig. 3.6.

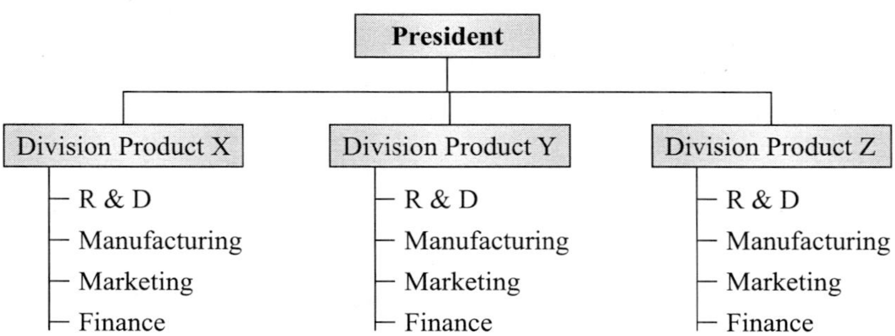

Fig. 3.6

As in this figure, divisional structure divides the employees (in the company) based on the product/customer segment/geographical locations. For example, each division is responsible for certain product and has its own resources such as finance, marketing, warehouse, maintenance etc.

Prior to this division, managers must consider a company's goals and business obstacles as well as alternative company structures. Business goals may include, for example, increasing the flow of information, promoting teamwork, and reducing redundancy. Next, managers must consider the different organizational structures environment. The selection may be made by taking different units based on one of the following six criteria as shown in Fig. 3.7.

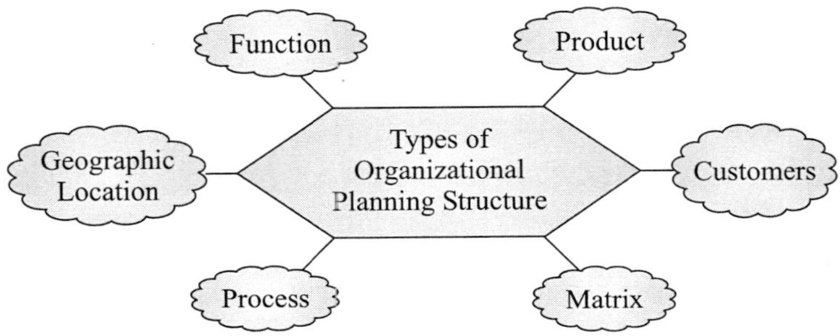

Fig. 3.7

(*a*) **Function:** A manufacturing firm typically includes functional units such as engineering, production, finance, sales and personnel. These different functions are controlled by managers who head each function. In this organizational structure, each function primarily focuses on its core tasks (*e.g.*, production or finance).

(*b*) **Product:** Large, diversified companies may find it advantageous to divide their tasks based on product groups, such as food stuffs, farm chemicals, and pharmaceuticals. This organizational structure has the benefit of enabling each business unit to produce desired results. However, it can lead to high administrative costs and redundancy.

(*c*) **Customers:** Companies may be divided into different unit based on target customers. For example, a book publisher may be organized by retail bookstores, mail-order book stores, online book stores, and school system tier, such as elementary, middle school, high school, and higher education.

(*d*) **Geographic Location:** Many sales and service companies use geographic location as the basis for creating departments within the organization. This organizational structure calls for members of each group to concentrate on particular locations for which they are responsible.

(*e*) **Process:** Some companies are organized by process. Process organization is common in manufacturing and clerical companies. For example, natural gas companies have different units for exploration, production and distribution. This type of organizational structure enables each event to have its own specialists.

(*f*) **Matrix:** Matrix organizational structures include not only general functional units like production, sales and finance, but also product or geographic units. Company executives frequently average the product units directly. The product unit, in turn collaborate with and coordinate the functional units. By adopting the matrix arrangement, companies attempt to reap the benefits of the functional and the product or geographic structures, while bypassing the inefficient and redundant aspects of the product structure. A company with a matrix organizational structure has functional units such as development, production and sales matched in a matrix by product or geographic location.

Company executives and managers must strive to select the organizational structure that best suits their fields of business, that offers the optimal amount of control, specialization, and co-operation and that facilitates key business activities while also taking into consideration concerns for efficiency and effectiveness.

3.2 ORGANIZATIONAL STRUCTURE-DESIGN

As we have discussed before that organization structure defines how job tasks are formally divided, grouped and co-ordinated. That is, the way tasks are divided up, how the work flows, how this flow is coordinated and the forces and mechanisms that allow this coordination to occur.

Organization structure is the process of building a team of highly talented, professional, ambitious and enthusiastic individuals to achieve or set goals and targets. Effective employee management and business expansion are the main reasons for the necessity of a systematic organizational structure and design. Many people face countless difficulties while starting their dream organization. The scope of word 'Organization' is very wide. An organization can be company, a business or even a school or a college. A small business will need a simple organizational structure, while large multinational companies or universities with several colleges under its control would need a large or complex organizational structure.

Whether small or large, the first step in forming an organization is to set up a management team which will supervise the working of the entire organization. The management team includes the chairman/promoter of the organization, board of directors and non-executive chairman. The mission and vision of the organization should be defined by the management.

Let us explain a simple organizational structure with the help of an example. A primary school or start-up private limited company is a good example of a simple organizational structure. Apart from the management, these organizations have fewer employees as compared to the big organizations. The main advantage of a simple organizational structure is the simplicity in caring out the day-to-day activities. In primary school, the school head can personally take care of every student's needs due to less number of students. The boss of a small firm can solve his employee's problems immediately and can personally interact with him for performance related issues. Like large organizations, small companies also have marketing department, personnel department, legal advisors and accounts department. Generally, small organizations have few branches and less annual turnover than the larger ones. The ultimate power is with the head, who has the final say on any matter related to that company.

The "organizational chart" cannot fully capture the organizational structure but gives us a place to begin when studying it. Let us explain the organizational chart with the help of example of some basic company organization structure in Fig. 3.8.

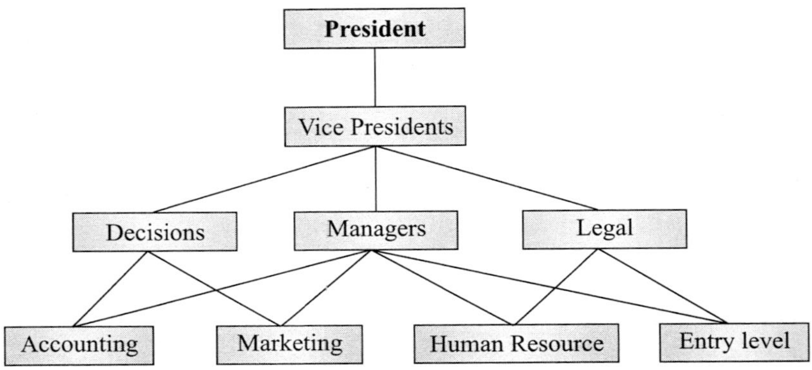

Fig 3.8

Two fundamental requirements of an organizational structure are

(1) **Division of labour** into distinct tasks. Division of labour leads to specialization.

(2) **Coordination** of the labours so that workers able to work in concert to accomplish the goals of the organization.

Coordination occurs through

- informal communication,
- formal hierarchy, and
- standardisation.

Informal communication has high media-richness and mainly includes sharing of the information which is very important in teams as the persons in the team share all the formal information, they also need to have informal communication. Where as, formal hierarchy is direct supervision which is common in large firms. But, the major problems faced is, formal heirarchy is costly, also and less popular with young staff. Standardisation, on the other hand, is basically the process of developing and implementing technical standards which consists of formal instructions. It knows it goals or output clearly.

3.2.1 Key Elements of Organizational Structure

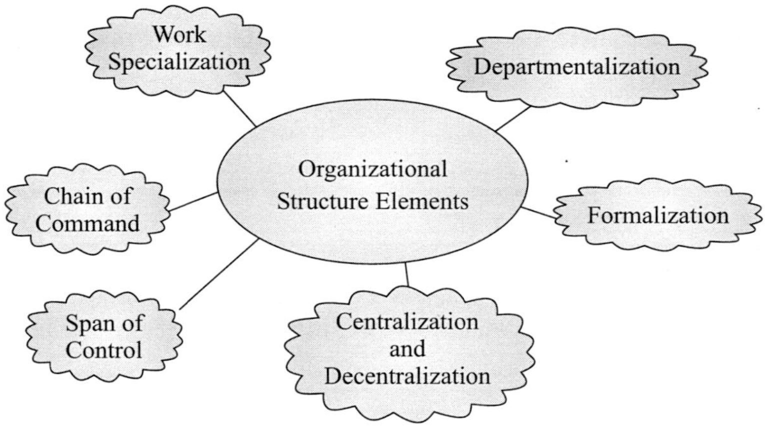

Fig. 3.9

Figure 3.9 shows the six key elements of organizational structure.

1. Work specialization: Work specialization describes the degree to which tasks in an organization are divided into separate jobs. The main idea of this design is that an entire job is not done by one individual. It is broken down into steps, and a different person completes each step. Individual employees specialize in doing part of an activity rather than the entire activity. It involves repitive performance of a few skills.

It can also be viewed as a means to make the most efficient use of employee's skills. Some task require highly developed skills whereas others can be performed by the untrained.

Division of Labour is the subdivision of work into separate jobs assigned to different people. It potentially increases work efficiency. It makes efficient use of employee's skills. It increases employee's skills through repetition. Specialized training is more efficient.

2. Departmentalization: It is the basis by which jobs are grouped together. For instance, every organization has its own specific way of classifying and grouping work activities. After reviewing the plans, usually the first step in the organizing process is departmentalization. Once jobs have been classified through work specialization, they are grouped so those common tasks can be coordinated. Detpartmentalization is the basis on which work or individuals are grouped into manageable units.

There are five traditional methods for grouping work activities:

(*a*) ***Departmentalization by function***: It organizes by the functions to be performed. The functions reflect the nature of the business. The advantage of this type of grouping is obtaining efficiencies from consolidating similar specialities and people with common skills, knowledge and orientations together in common units. As shown in Fig. 3.10, it groups jobs by functions performed. It can be used in all kinds of organizations, it depends on the goals each of them wants to achieve. Example of functional departmentalization is shown in this figure.

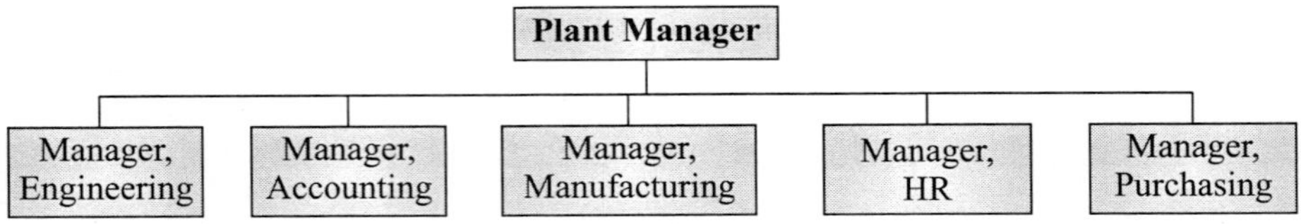

Fig. 3.10

Different aspects on this type of departmentalization:

Positive Aspects	*Negative Aspects*
• Efficiencies from putting together similar specialities and people with common skills, knowledge and orientations. • Coordination within functional area. • In-depth specialization.	• Poor communication across functional areas. • Limited view of organizational goals.

(*b*) ***Departmentalization by product*:** Assembles all functions needed to make and market a particular product are placed under one executive. For instance, major departmental stores are structural around product groups such as home accessories, appliances, women's clothing, men's clothing, and children's clothing. It groups job by-product line. Each manager is responsible of an area within the organization depending of his/her specialization.

Different aspects on this type of departmentalization:

Positive Aspects	*Negative Aspects*
• Allows specialization in particular products and services. • Managers can become experts in their industry. • Closer to customers.	• Duplication of functions. • Limited view of organizational goals.

(*c*) ***Departmentalization by geographical regions*:** Groups on the basis of territory or geography. For example, Merck a major pharmaceutical company, has its domestic sales departmentalized by regions such as Northeast, Southeast, Midwest, Southwest and Northwest. Let us explain it with the help of an example which is shown in Fig. 3.11 in which Sales Vice president have four regions under it with four sales director of western, southern, mid western and eastern region.

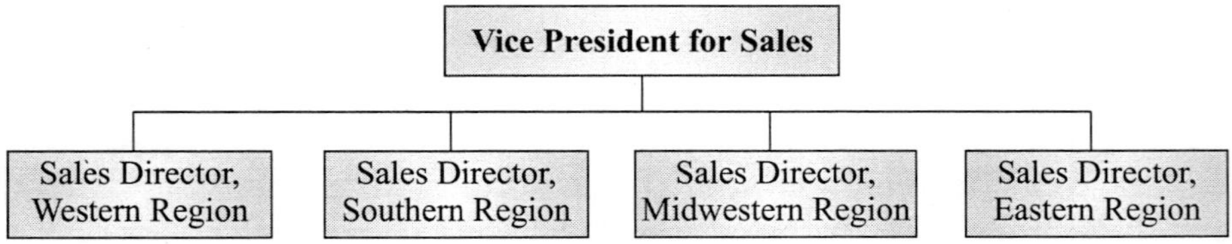

Fig. 3.11

Different aspects on this type of departmentalization:

Positive Aspects	*Negative Aspects*
• More effective and efficient handling of specific regional issues that arise. • Serve needs of unique geographic markets	• Duplication of functions. • Can feel isolated from other organizational areas.

(*d*) ***Departmentalization by process*:** Groups jobs on the basis of product or customer flow. Each process requires particular skills and offers a basis for homogeneous categorizing of work activities. A patient preparing for an operation would first engage in preliminary diagnostic tests, then go through the admitting process, undergo a procedure in surgery, receive post operative care, be discharged and perhaps receive out-patient attention. These services are each administered by different departments.

Different aspects on this type of departmentalization:

Positive Aspects	*Negative Aspects*
• More efficient flow of work activities.	• Can only be used with certain types of products.

(*e*) ***Departmentalization by customer*:** Groups jobs on the basis of a common set of needs or problems of specific customers. For instance, a plumbing firm may group its work according to whether it is serving private sector, public sector, government, or not-for-profit organizations. A current departmentalization trend is to structure work according to customer, using cross-functional teams. This group is chosen from different functions to work together across various departments to interdependently create new product or services. For example, a cross functional team consisting of managers from accounting, finance and marketing is created to preface a technology plane. Figure 3.12 shows example of customer departmentalization

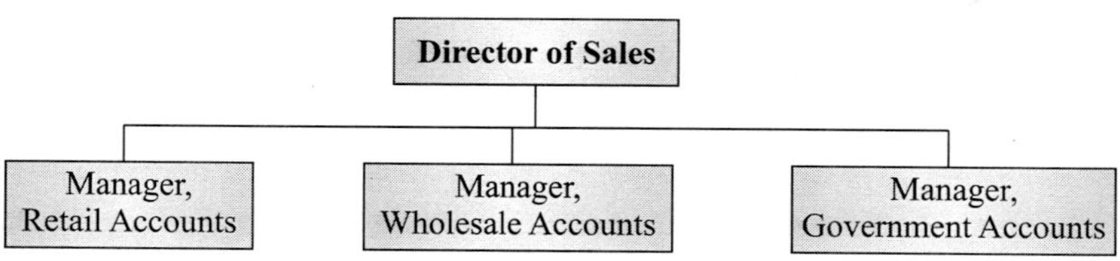

Fig. 3.12

Different aspects on this type of departmentalization:

Positive Aspects	*Negative Aspects*
• Customer's needs and problems can be met by specialists.	• Duplication of functions • Limited view of organizational goals.

3. Chain of Command: Chain of command is the unbroken line of authority that connects each level of management with the next level. The chain of command helps organization to function smoothly by making two things clear:

- who is responsible for each task,
- and who has the authority to make official decisions.

In organizations, employees are assigned.

1. Responsibility for their jobs; they are obligated to perform those duties and to achieve goals and objectives associated with their positions.
2. Accountability for reporting their results to supervisors or team members and justifying outcomes that fall below expectations.

Managers ensure tasks are accomplished by exercising authority. Authority is the power to make decisions, issue orders, carry out actions, and allocate resources.

Delegation is the assignment of work and the transfer of authority and responsibility to do that work.

Possible chain-of-command systems:

1. *Line organization*: The simplest and most common chain-of-command system. Everyone knows who is accountable to whom. Know which tasks and decisions each person is responsible for. It does have a disadvantage—the technical complexity of a firm's activities may require specialized knowledge that individual managers don't have and can't easily acquire.
2. *Line and staff organization*: Combines specialization with management control. Managers are supplemented by functional groupings of people known as staff. These staff employees provide advice and specialized services but they are not in the line organization's chain of command.

   - **Power and Authority:** Organizational structure is a means of facilitating the achievement of organizational objectives. Such structures are not static, but dynamic. They reorganize in response to changing condition that occur in the environment, new technology, or organizational growth. Organization structure are dependent upon the employees whose activities they guide. Supervisors rely upon power and authority to ensure that employees get things done.
   - **Authority:** The organizational structure provides the framework for the formal distribution of authority. **Formalization** is the degree to which tasks are standardized and rules and regulations govern employee behaviour. It influences the amount of discretion an employee has over his or her job. In an organization with high degrees of formalization, job descriptions and policies provide clear direction. Where formalization is low, employees have a great deal of freedom in deciding how they conduct their work. Within the same organization, different departments may have different degrees of formalization. For example, in a hospital doctors have freedom in selecting treatments, drugs and methods for treating patients. However, the hospital physical plant staff has a strict schedule for cleaning buildings, moving lawns and maintaining the facilities.

     Authority is the legitimate power of a supervisor to direct subordinates to take action within the scope of the supervisor's position. Formal authority in the organization can be traced all the way back to the U.S. constitutional right to over property. The owner of the organization has the authority to make decisions. For example, entrepreneurial forms have an informal arrangement of employees and centralization of decision-making authority, the owner.

- **Forms of Authority:** Three forms of authority are line authority, staff authority and team authority.
  - (*i*) Line Authority is direct supervisory authority from superior to subordinate. Authority flows in a direct chain of command from the top of the company to the bottom.
  - (*ii*) Chain of command is an unbroken line of reporting relationships that extends through the entire organization that defines the formal decision-making structure. It helps employees know to whom they are accountable and whom to go to with a problem.
  - (*iii*) Line Departments are directly linked to the production and sales of specific products. Supervisors in line departments, such as marketing and production—give direct orders, evaluate performance, and reward or punish those employees who work for them.
  - (*iv*) Unity of command within the chain states that each person in an organization should take orders from and reports to only one person. This help prevent conflicting demand being placed on employees by more than one boss. However, the trend towards employee empowerment, fueled by advances in technology and changes in design from downsizing and engineering have tempered the importance of being accountable to only one superior.
  - (*v*) Span of control refers to the number of employees that should be placed under the direction of one manager. Spans within effective organizations vary greatly. The actual number depends on the amount of complexity and the level of specialization. In general, a wide span of control is possible with better-trained, more experienced, and committed employees.

4. Span of Control: It is important to a large degree because it determines the number of levels and managers an organization has. Also, determines the number of employees manager can efficiently and effectively manage. The span of control in an organization is defined as the number of employees reporting directly to one supervisor. Traditionally, the span of control has been defined as a number between 4 and 7 subordinates under one manager. The span of control is a concept which developed during the 19th century for the organization of large European armies into smaller and smaller subunits. That is, one could have 7 soldiers in one platoon with 1 leader, 7 platoons with one higher-ranking leader, etc. to form a large army which was relatively easy for one general at the top to manage under conditions of combat. Wider spans of management increase organizational efficiency.

5. Centralization and Decentralization: Centralization is the remaining of power or authority to upper-levels of management. The advantages to centralization involve a closer control of company operations-including policies and practices. By retaining power at the top, those decisions that directly affect a company will be made by the proper governing board.

Decentralization is the degree to which authority is dispersed throughout the lower levels of an organization. One of the main advantages to decentralization is that decisions can be made considerably faster, without having to seek approval from upper levels of authority. Also, the presence of decentralization acts as a motivational tool for employees to work their way through the ranks. Decentralization also cases the workload for busy executives who maynot have time to oversee all aspects of the company

- Reasons to move from centralization to decentralization:
  (*a*) ***Considerable company growth*:** a company may find it advantageous to switch to a decentralization model in order to keep organized. In this manner, a company may establish new divisions or departments to keep track of its growth.
  (*b*) ***Geographic*:** decentralization could involve the dispersing of power from a centralized headquarters to localized branches. This designation of power is most commonly used in the situations where a domestic company has offices in a foreign country.
  (*c*) ***Technological Complexity*:** a company may also choose to decentralize its chain of command in order to better facilitate its technological departments. This maneuver may also indirectly demonstrate a sense of confidence to the technology employees, as it proves a company has confidence in their actions. A positive side effect could be the creation of an environment where ideas can be freely explored and where they may flourish.
  (*d*) ***Time*:** decentralization is sometimes necessary as a time solution. For instance, a company might experience a time limit to finish a project, and the most feasible solution would be to account lower-level managers that are more directly involved in the project to supervise.

More Centralization	***More Decentralization***
1. Environment is stable.	1. Environment is complex uncertain.
2. Lower-level managers are not as capable or experienced at making decisions as upper-level managers.	2. Lower-level managers are capable and experienced at making decisions.
3. Lower-level managers donot want to have say in decisions.	3. Lower-level managers want a voice in decisions.
4. Decisions are significant.	4. Decisions are relatively minor.
5. Organizations is facing a crisis or the risk of company failure.	5. Corporate culture is open to allowing managers to have a say in what happens.
6. Company is large.	6. Company is geographically dispersed.
7. Effective implementation of company strategies depends on managers retaining say over what happens.	7. Effective implementation of company strategies depend on managers having involvement and flexibility to make decisions.

6. Formalization: It refers to the degree to which jobs within the organization are standardized and the extent to which employee behaviour is guided by rules and procedures. As formalization is the degree to which jobs within the organization are standardized, so it is of two levels.

- **High levels**
  - minimum amount of discretion over what, when and how
  - consistent output
  - explicit job decriptions
  - organizational rules
  - clearly defined procedures.
- **Low Levels**
  - unprogrammed behaviours
  - exercise discretion
  - great deal of freedom
  - less standardization
  - consider alternatives.

3.2.2 Designing an Effective Organization Structure

Effective organization design considers five interrelated components:

1. Leadership
2. Decision-making and structure
3. People
4. Work processes and systems
5. Culture

This is shown in Fig. 3.13.

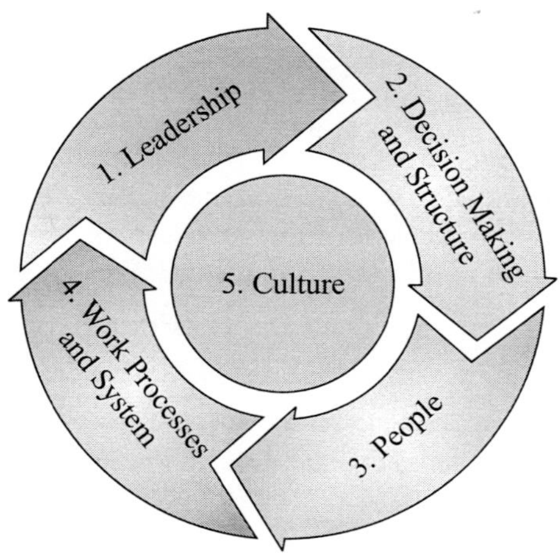

Fig. 3.13

- Clear vision and priorities
- Cohesive leadership team

- Clear roles and accountabilities for decisions
- Organizational structure that supports objects.
- Organizational and individual talent necessary for success.
- Performance measures and incentives aligned to objectives.
- Superior execution of programmatic work processes.
- Effective and efficient support processes and systems.
- 'High performance' values and behaviours
- Capacity to change.

- **Principles of effective Organizational Design**
  1. All five components of the "wheel" (in figure 3.13) is considered: A common misstep is to focus on structure alone (boxes and reporting lines) as the solution.
  2. Five components are aligned to one another: One element that "doesn't fit" can limit the performance of the whole system.
  3. Then, align strategy and organization to one another: Organizational strengths and weaknesses influence the range of feasible strategies; in turn, organizations should evolve with any new strategic direction.
- **Symptoms of an Ineffective Organization:**
  1. *Lack of coordination*: Lack of coordination happens when the work is not finished, or the team is isolated or out of step. Basic root cause of this symptom is leadership, decision-making and structure and work processes and systems.
  2. *Excessive conflict*: Conflict in excess is due to needless friction among internal groups. Root cause is again leadership, decision-making and structure and people (related to poor performance measures or incentives, not lack of talent or skill per se).
  3. *Unclear Roles*: Roles or goals of the organization is not clear which is due to functions which overlap and/or fall through the cracks. Basic root cause is decision-making and structure.
  4. *Gap in skills or misused resources*: Skills are basically a job/or activity that require training and practice. It is due to missing or underutilized skills or resources (ability or skill of the people). Likely root causes are leadership (skill of leader), people and work processes and systems.
  5. *Poor work flow*: When there are lot of disruptions, or when the process is very cumbersome. Likely root cause is work processes and systems.
  6. *Reduced responsiveness*: When there is slow reactions to environmental shifts, the response is reduced. Causes are leadership, decision-making and structure and culture.
  7. *Conflicting communication*: Communication is basically act or means of conveying of information. External stakeholders are confused and are complaining because of the miscommunication. Its root cause is work processes and systems.

8. *Low Staff Morale*: Low staff morale is the lack of confidence or drive among the team members and poor teaming (right person are not selected together in order to make a particular team). Its root cause is leadership and culture.

- **Basic Principles of Effective Structure**

1. *No "right" answer*: There is no silver bullet which means every structure has its strengths and weaknesses, no structure is perfect.
2. *But a better answer*: However, there is likely to be a "better" structure for a strategy and stage of development. Analysis can help to determine alternative structures that will support the strategy.
3. *Making necessary compromises*: As the organization knows its strengths and weaknesses; compromises in structure are often necessary. The final structure is likely to be a "hybrid" of the best options.
4. *Managing tradeoff*: Whatever structure is selected, it is essential that the organization manage its inherent weaknesses or tradeoffs. The "levers" that help manage these tradeoffs are the other four elements of an effective organization (processes, people, leadership and culture)

- **Components of Structure:** Structure has two components: groupings and linking of activities which is shown in Fig. 3.14.

Grouping

- How individuals, jobs, functions or activities are differentiated and aggregated
- Optimizes information flow within the group but typically creates barriers with other groups.

Linking

- Mechanisms of integration used to coordinate and share information across groups.
- Enables leadership to provide guidance and direction across the organization.

An optimal structure balances differentiation (through grouping) with integration (through linking)

Fig. 3.14

We tend to spend 90% of our energy on grouping but very little on linking because:

(1) Grouping decisions are usually the essence of the change in structure (new units or reconfiguration of old units).

(2) Grouping is equated with the new hierarchy.

(3) People's job (especially those that have changed) are the result of new grouping decisions.

(4) People (incorrectly) assume that most organizational problems are caused by having the wrong grouping.

Bridgespan's organization diagnostic data reveal that coordination and linkages are one of the structural problem facing non-profits.

- **Five grouping models to consider**

Type	*Common in ...*
• Functional	• Single-program organizations; most frequently used by our clients
• Geographic	• Multi-site organizations; frequently used by networks
• Program ("Product")	• Multi-service organizations and foundations
• Customer/Market	• Small organizations with narrow customer focus, or large non-profits where programs/customers align
• Matrix	• Large and sophisticated nationwide and/or global organizations

Most organization end up with a "hybrid" structure, combining elements of different models but with one dominant approach.

1. **Functional Model:** Functional model is organized around key functions or departments.

Functional model is appropriate when organization is small and/or has single programmatic focus and no need to manage across a large geographic area. It is shown in Fig. 3.15.

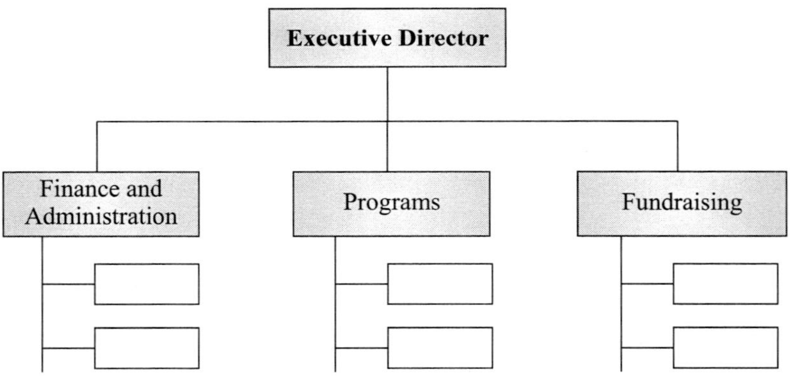

Fig. 3.15

Let us take an overview of pros and cons of functional model and ways to manage cons.

- **Pros**

1. Functional model develops depth of skills in a particular function or department that is most jobs are functional in nature.
2. It promotes functional innovation, scale and lower cost.
3. It is simply, easy to understand by each department their core responsibilities and to hold them accountable.

- **Cons**
  1. If an organization have different priorities and measures, then, functions can work at counter-purposes.
  2. This model focus on function rather than overall organization or beneficiaries.
  3. Processes across functions can break down in which individuals are unclear on their role.
  4. Where as, cross-functional decisions get pushed up for ED resolution.
- **Ways to manage Cons**
  1. Functional model requires clarity of vision and priorities which are translated into measurable departmental and individual goals.
  2. It ensure staff see their role in serving customers through processes.
  3. It ensure key work processes are defined, including roles across departments.
  4. It also make decision-making explicity (*e.g.*, RAPID tool)

2. **Geographic Model:** Geographic model is organized around major geographies like US, Africa, Asia. This model is most appropriate when organization is large with multiple programs, often different across geographies. Whereas, local differences are critical for success like regulation, fundraising, economies. Geographic model is shown in Fig. 3.16.

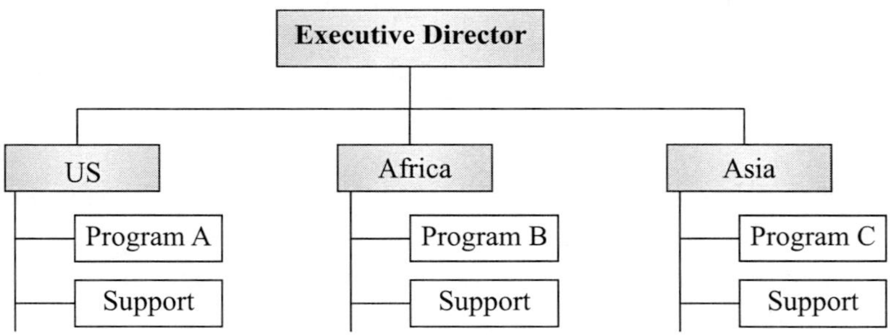

Fig. 3.16

Now, let us discuss pros and cons of this model and ways to manage cons.

- **Pros**
  1. Resources are needed to be succeeded within a geography are available.
  2. This model allows greater customization of programs or services by religion.
  3. It enables clear focus with accountability for results by geography.
  4. It also enables focus on geographic funding sources.
- **Cons**
  1. Geographic model requires strong skills, particularly of geo head.
  2. It leads to functional duplication and potential loss of control.
  3. Work processes and output (*e.g.*, services) may differ across geos.
  4. This model creates confusion about who makes decision.
  5. Organization becomes heterogeneous but not a unified culture.

Ways to manage cons

1. Geographic model is focused on hiring, training and best practice sharing.
2. I.D. functions to manage globally (*e.g.*, finance) create hybrid structure.

3. It determines if any key processes should be done in common way.
4. Decision-making is made explicit (*e.g.* RAPID tool)
5. It develop and reinforce elements of common culture (if important).

3. **Program ("product") Model:** Program or product model is organized around major programs. This model is most appropriate when programs are very different from one another (*e.g.,* different customers, economics, etc.); all these factors are similar across geographies. Resources and skills needed to succeed by program are very different. This model is shown in Fig. 3.17.

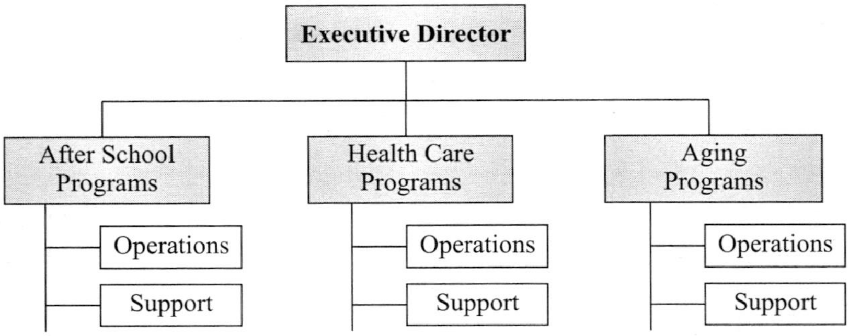

Fig. 3.17

Now, let us discuss pros and cons of program model and ways to manage those cons.

- **Pros**
  1. Product model promotes depth of understanding within a particular program area and also promotes program innovation.
  2. Resources are needed to succeed within a program are available.
  3. This model enables clear focus with accountability for program results.
  4. It also enables focus on funding sources which are often program oriented.
- **Cons**
  1. This model requires strong skills, particularly program heads.
  2. This model leads to functional duplication and potential loss of control.
  3. It is difficult to coordinate common customers across programs.
  4. Here also, organization becomes heterogeneous, not a unified culture.
- **Ways to manage cons**
  1. It is focused on hiring, training, and best practice sharing.
  2. I.D. functions to manage commonly (*e.g.,* finance) create hybrid structure.
  3. Work processes and systems are developed that enables management and tracking of customers.
  4. Elements of common culture are developed and reinforced (if important).

4. **Customer/Market Model:** Customer or market model is organized around customers/clients (beneficiaries) served. This model is appropriate when the customers are very different and have different service requirements (resulting in different programs provided).

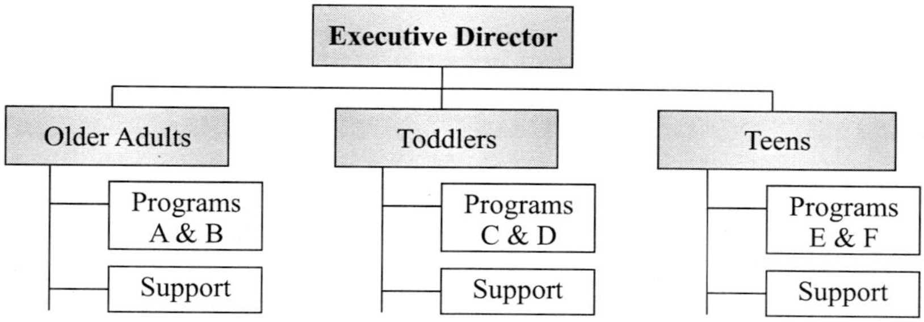

Fig. 3.18

This model is shown in Fig. 3.18. Now, let us take an overview on pros, cons and ways to manage cons.

- **Pros**
  1. Customer focus enables development of programs tailored to a population's needs and able to get "results".
  2. This model enables clear focus with accountability for results by customer group.
  3. This model enables clear focus on most important customers and/or markets.
- **Cons**
  1. This model requires strong skills, particularly customer unit heads.
  2. It leads to functional duplication and potential loss of control.
  3. It results in duplication if clients have similar needs (*e.g.,* health).
  4. Organization becomes heterogeneous and not an unified culture.
- **Ways to manage cons**
  1. This model is focus on hiring, training, and best practice sharing.
  2. I.D. functions to manage commonly (*e.g.,* finance) create hybrid structure.
  3. This model determines if a few program areas should be provided across customer groups.
  4. Elements of common culture is developed and reinforced (if important).

5. **Matrix Model:** Matrix model is organized to manage multiple dimensions, *e.g.,* program and geography. This model is appropriate when: Almost NEVER: very difficult to manage. This model is shown in Fig. 3.19.

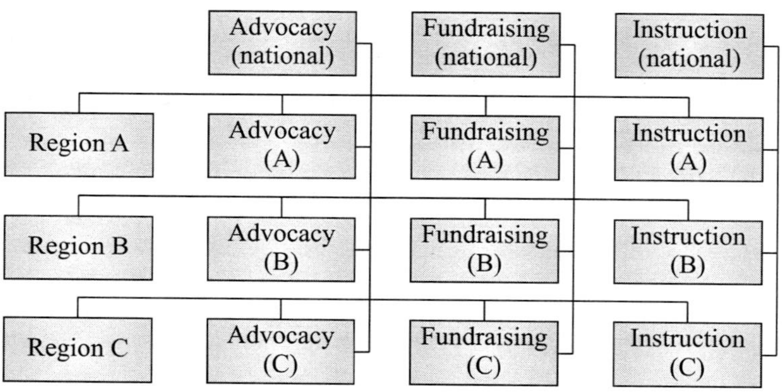

Fig. 3.19

- **Pros**
  1. This model enables organization to (at least theoretically) manage multiple organizational dimensions simultaneously.
- **Cons**
  1. Matrix Model is unclear in decision-making and accountability.
  2. Individuals are unclear who they have to report.
  3. Accountability for results is diffuse.
- **Ways to manage cons**
  1. This model needs clarity around roles and decision-making (*e.g.*, RAPID tool)
  2. It has to define management responsibilities for each boss.
  3. There should be HR performance management system that will reinforce the work together.

Many organizations end up with a hybrid developed from one dominant model

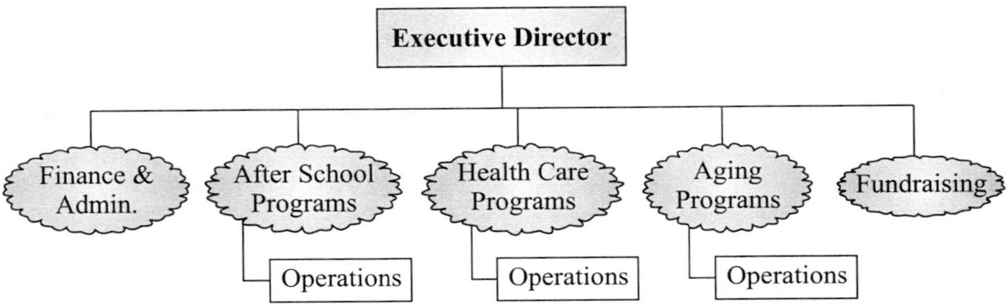

Fig. 3.20

A common model is a programmatic focus combined with a functional model for centralized support functions for skill, cost and control reasons. By using these models, ensure that "operating units" get the support they need to avoid creating duplicated support positions.

3.3 FIVE APPROACHES TO ORGANIZATIONAL DESIGN

These approaches also explain "**how to design an effective organization structure**".

Managers must make choices about how to group people together to perform their work. Five common approaches—functional, divisional, matrix, team and networking—help managers determine departmental grouping (grouping of positions into departments). The five structures are basic organizational structures, which are then adapted to an organization's needs. All five approaches combine varying elements of mechanistic and organic structures.

1. **Functional Structure:** The functional structure groups positions into work units based on similar activities, skills, expertise and resources which is shown in Fig. 3.21. Production, marketing, finance and human resources are common groupings, within a functional structure. As the simplest approach, a functional structure features well-defined channels of communication and authority responsibility relationships. Not only can this structure improve productivity

by minimizing duplication of personnel and equipment, but it also makes employees comfortable and simplifies training as well.

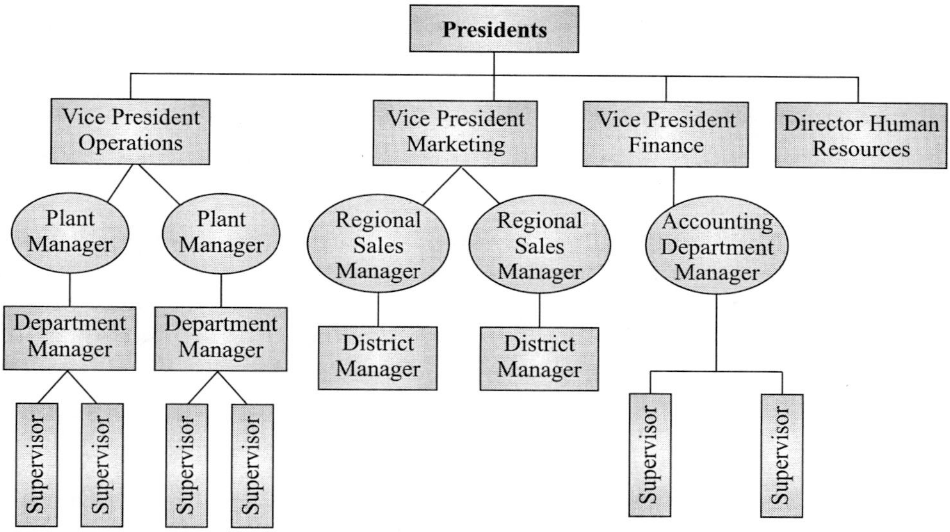

Fig. 3.21

But the functional structure has many downsides that may make it inappropriate for some organizations. Here are a few examples:

1. The functional structure can result in narrowed perspectives because of the separateness of different department work groups. Managers may have a hard time relating to marketing, for example, which is often in an entirely different grouping. As a result, anticipating or reacting to changing consumer needs may be difficult. In addition, reduced cooperation and communication may occur.
2. Decisions and communication are slow to take place because of the many layers of hierarchy. Authority is more centralized.
3. The functional structure gives managers experience in only one field—their own. Managers donot have the opportunity to see how all the firm's department work together and understand their interrelationships and interdependence. In the long run, this specialization results in executives with narrow backgrounds and little training handling top management duties.

2. **Divisional Structure:** Because managers in large companies may have difficulty keeping track of all their company's products and activities, specialized departments may develop. These departments are divided according to their organizational outputs. Examples include departments created to distinguish among production, customers service, and geographical categories. This groupings of departments is called divisional structure which is shown in Fig. 3.22.

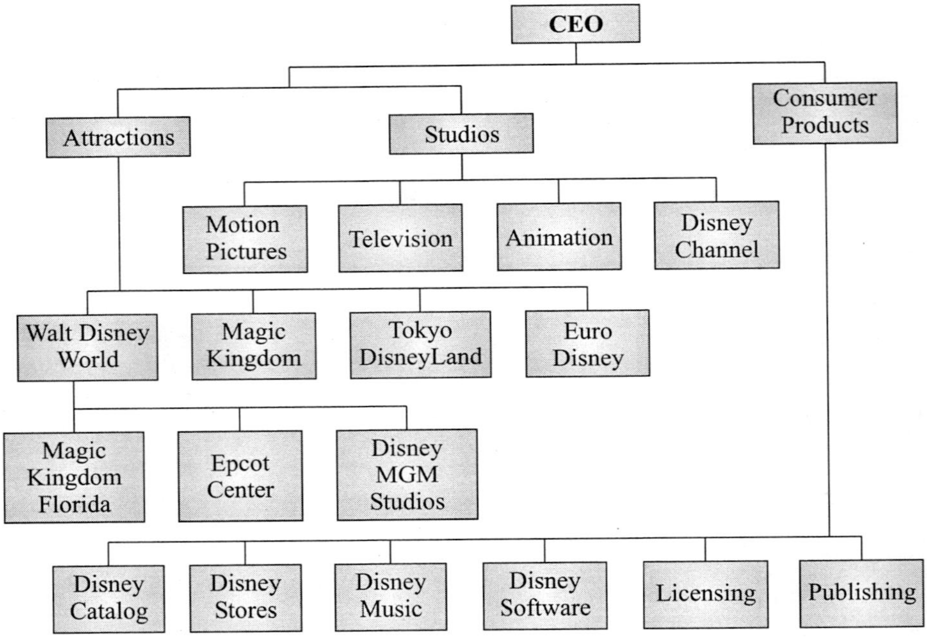

Fig. 3.22 The Divisional Structure-Disney in the early 1990s.

These departments allow managers to better focus their resources and results. Divisional structure also makes performance easier to monitor. As a result, this structure is flexible and responsive to change. However, divisional structure does have its drawbacks. Because managers are so specialized, they may waste time duplicating each other's activities and resources. In addition, competition among divisions may develop due to limited resources.

3. **Matrix Structure:** The matrix structure combines functional specialization with the focus of divisional structure which is shown in Fig. 3.23.

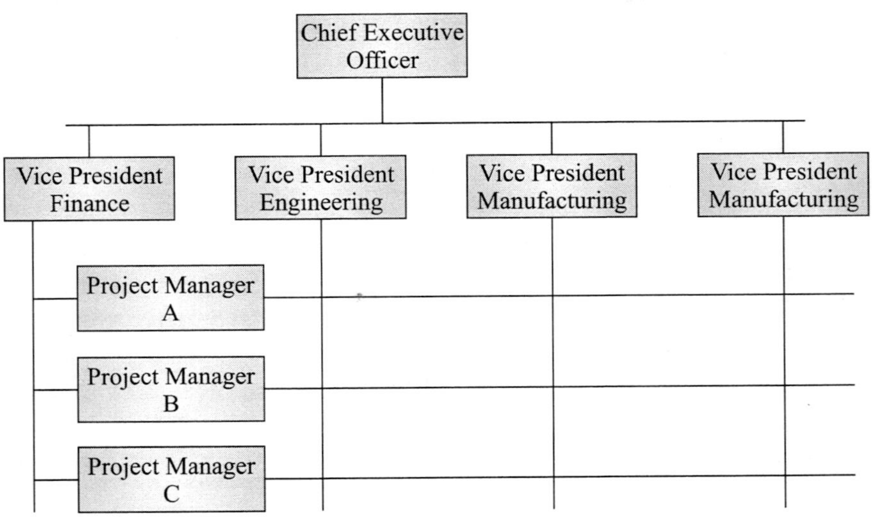

Fig. 3.23 The matrix structure

This structure uses permanent cross-functional teams to integrate functional expertise with a divisional focus. Employees in a matrix structure belong to atleast two formal groups at the same time—a functional group and a product, program or project team. They also report to two bosses—one within the functional group and the other within the team.

This structure not only increases employee motivation, but it also allows technical and general management training across functional areas as well. Potential **advantages** include:

1. Better cooperation and problem solving.
2. Increased flexibility.
3. Better customer service
4. Better performance accountability
5. Improved strategic management

Predictably, the matrix structure also has potential **disadvantages**. Here are a few of this structure's drawbacks:

1. The two-boss system is susceptible to power struggles, as functional supervisors and team leaders via with one another to exercise authority.
2. Members of the matrix may suffer task confusion when taking orders from more than one lose.
3. Teams may develop strong team loyalties that cause a loss of focus on larger organization goals.
4. Adding the team leaders, a crucial component, to a matrix structure can result in increased costs.

4. **Team Structure:** Team structure organizes separate functions into a group based on overall objective which is shown in Fig. 3.24. These cross-functional teams are composed of members from different departments who work together as needed to solve problems and explore opportunities. The intent is to break down functional barriers among departments and create a more effective relationship for solving ongoing problems.

   The team structure has many potential **advantages,** including the following:

   1. Interdepartmental barriers break down.
   2. Decision-making and response time speed up.
   3. Employees are motivated.
   4. Level of managers are eliminated.
   5. Administrative costs are lowered.

   The **disadvantages** include:

   1. Conflicting loyalties among team members.
   2. Team-management issues.
   3. Increased time spent in meetings.

Managers must be aware that how well team members work together often depends on the quality of interpersonal relations, group dynamics, and their team management abilities.

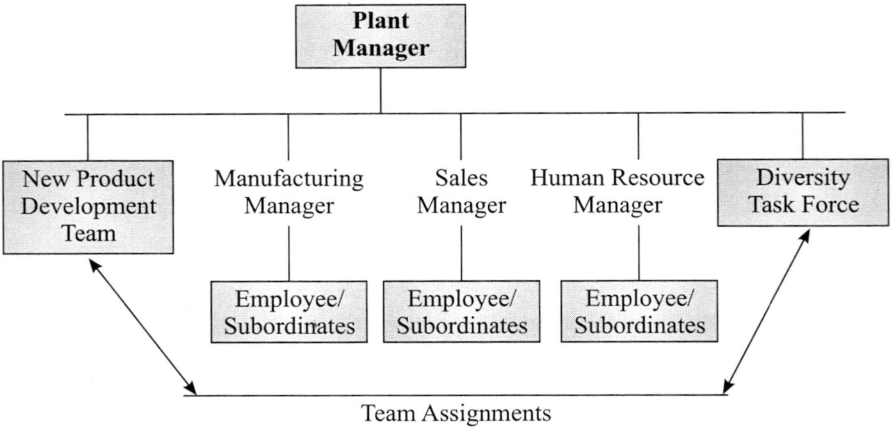

Fig. 3.24 The Team Structure

5. **Network Structure:** The network structure relies on other organizations to perform critical functions on a contractual basis which is shown in Fig. 3.25. In other words, managers can contract out specific work to specialists. This approach provides flexibility and reduces overhead because the size of staff and operations can be reduced. On the other hand, the network structure may result in unpredictability of supply and lack of control because managers are relying on contractual workers to perform important work.

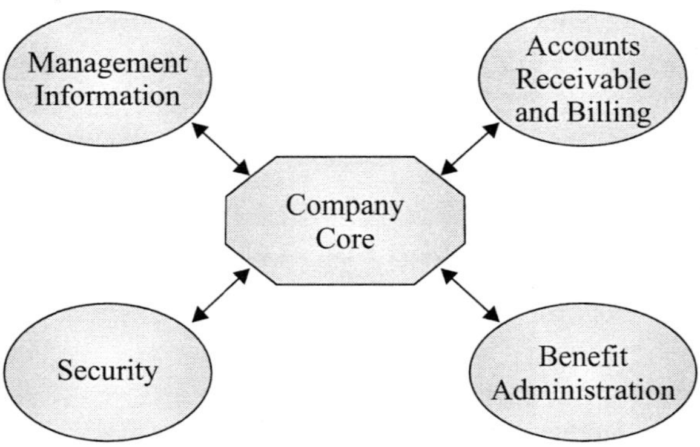

Fig. 3.25 The network structure

3.4 JOB DESIGN AND ENRICHMENT

3.4.1 What is Job Design?

Job design means to decide the contents of a job. It fixes the duties and responsibilities of the job, the methods of doing the job and the relationships between the job holder (manager) and his superiors, subordinates and colleagues.

Job design also gives information about the qualifications required for doing the job and the reward (financial and non-financial benefits) for doing the job. Job design is mostly done for manager's jobs. While designing the job, the needs of the organization and the needs of the individual manager must be balanced. Needs of the organization include high productivity, quality of work etc. Needs of individual managers include job satisfaction. That is, they want the job to be interesting and challenging. Jobs must not be made highly specialized because they lead to boredom.

Job design is very **important** function of staffing. If the jobs are designed properly then highly efficient managers will join the organization. They will be motivated to improve the productivity and profitability of the organization. However, if the jobs are designed badly, then it will result in absenteeism, high labour turnover, conflicts, and other labour problems.

3.4.2 Objectives of Job Design

There are clearly many alternative designs for any given job. For this reason, an understanding of what the job design is supposed to achieve is particularly important. As before, the five performance objectives give us a guide to what is relevant in job design decisions:

1. **Quality:** The ability of staff to produce high-quality products and services can be affected by job design. This includes avoiding errors in the short term, but also includes designing jobs which encourage staff to improve that job itself in such a way as to make errors less likely.
2. **Speed:** Sometimes speed of response is the dominant objective to be achieved in job design. For example, the way in which the jobs of emergency service personnel are organised (the range of tasks for which they are trained, the sequence of activities in their approved procedures, the autonomy which they have to decide on appropriate action, and so on) will go a long way to determine their ability to respond promptly to emergencies and perhaps same lives.
3. **Dependability:** Dependable supply of goods and services is usually influenced, in some way, by job design. For example, in the postal services, working arrangements, multiskilling, accurate use of sorting equipment through good staff machine interface design, and the 'design' of postal staff's clothing, can all aid dependable delivery of letters and particles.
4. **Flexibility:** Job design can affect the ability of the operation to change the nature of its activities. New product or service flexibility, mix flexibility, volume flexibility and delivery flexibility are all dependent to some extent on job design. For example, staff who have been trained in several tasks (multi-skilling) may find it easier to cope with a wide variety of models and new product or service introductions.
5. **Cost:** All the elements of job design described above will have an affect on the productivity, and therefore, the cost of the job. Productivity in this context means the ratio of output to labour input: for example, the number of customers served per hour or the number of products made per worker. In addition, job design will influence two other particularly important objectives.
6. **Health and safety:** Whatever else a job design achieves, it must not endanger the well-being of the person who does the job, other staff of the operation, the customers who might be present in the operation, or those who use any products made by the operation.

7. **Quality of Working Life:** The design of any job should take into account its effect on bob security, intrinsic interest, variety, opportunities for development, stress level and attitude of the person performing the job.

3.4.3 Factors Affecting Job Design

A well defined job will make the job interesting and satisfying for the employee. The result is increased performance and productivity. If a job fails to appear compelling or interesting and leads to employee dissatisfaction, it means the job has to be redesigned based upon the feedback from the employees.

Broadly speaking the various factors that affect a job design can be classified under three heads. They are

1. Organizational factors
2. Environmental factors
3. Behavioural factors.

1. **Organizational Factors:** Organizational factors that affect job design can be work nature or characteristics, work flow, organizational practices and ergonomics.
   (*a*) **Work Nature:** There are various elements of a job and job design is required to classify various tasks into a job or a coherent set of jobs. The various tasks may be planning, executing, monitoring, controlling etc. and all these are to be taken into consideration while designing a job.
   (*b*) **Ergonomics:** Ergonomics aims at designing jobs in such a way that the physical abilities and individual traits of employees are taken into consideration so as to ensure efficiency and productivity.
   (*c*) **Workflow:** Product and service type often determines the sequence of work flow. A balance is required between various product or service processes and a job design ensures this.
   (*d*) **Culture:** Organizational culture determines the way tasks are carried out at the work places. Practices are methods or standards laid out for carrying out a certain task. These practices often affect the job design especially when the practices are not aligned to the interests of the union.
2. **Environmental Factors:** Environmental factors affect the job design to a considerable extent. These factors include both the internal as well as external factors. They include factor like employee skills and abilities, their availability, and their socioeconomic and cultural prospects.
   (*a*) **Employee availability and abilities:** Employee skills, abilities and time of availability play a crucial role while designing the jobs. The above mentioned factors of employees who will actually perform the job are taken into consideration. Designing a job that is more demanding and above their skill set will lead to decreased productivity and employee satisfaction.
   (*b*) **Socio Economic and cultural expectations:** Jobs are nowadays becoming more employee centered rather than process centered. They are therefore designed keeping the employees into consideration. In addition the literacy

level among the employees is also on the rise. They now demand jobs that are to their linking and competency and which they can perform the best.

3. **Behavioural Factors:** Behavioural factors or human factors are those that pertain to the human need and that need to be satisfied for ensuring productivity at workplace. They include the elements like autonomy, diversity, feedback, etc. A brief explanation of some is given below:

   (*a*) **Autonomy:** Employees should work in an open environment rather than one that contains fear. It promotes creativity, interdependence and leads to increased efficiency.

   (*b*) **Feedback:** Feedback should be an integral part of work. Each employee should receive proper feedback about his work performance.

   (*c*) **Diversity:** Repetitive jobs often make work monotonous which leads to boredom. A job should carry sufficient diversity and variety so that it remains as interesting with every passing day. Job variety/diversity should be given due importance while designing a job.

   (*d*) **Use of skills and abilities:** Jobs should be employee rather than process centered. Though due emphasis needs to be given to the latter but jobs should be designed in a manner such that an employee is able to make full use of his abilities is able to make full use of his abilities and perform the job effectively.

3.4.4 Methods of Job Design

There are various methods in which job design can be carried out. These methods help to anlayse the job, to design the contents of and to decide how the job must be carried out. These methods are as follows in Fig. 3.26.

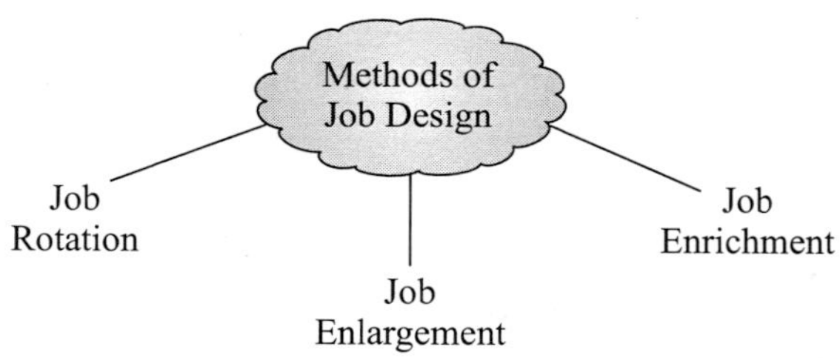

Fig. 3.26

1. **Job Rotation:** Job rotation involves shifting a person from one job to another, so that he is able to understand and learn what each job involves. The company tracks his performance on every job and decides whether he can perform the job in an ideal manner. Based on this he is finally given a particular posting.

   Job rotation is done to decide the final posting for the employee. *e.g.*, Mr. A is assigned to the marketing department whole he learns all the jobs to be performed for marketing at his level in the organization. After this, he is shifted to the sales department and to the finance department and so on. He is finally placed in the department in which he shows the best performance. Job rotation gives an idea about the jobs to be performed at every level. Once a person is able to understand this he is in a better understanding of the working of organization.

- **Advantages of Job Rotation:**
  (*a*) **Avoid monopoly:** Job rotation helps to avoid monopoly of job and enable the employee to learn new things and therefore enjoy his job.
  (*b*) **Provides an opportunity to broaden one's knowledge:** Due to job rotation the person is able to learn different job in the organization this broadens his knowledge.
  (*c*) **Avoiding fraudulent practice:** In an organization like bank jobs rotation is undertaken to prevent employees from doing any kind of fraud *i.e.,* if a person is handling a particular job for a very long time he will be able to find loopholes in the system and use them for his benefit and indulge (participate) in fraudulent practices job rotation avoids this.
- **Disadvantages of Job Rotation**
  (*a*) **Frequent interruption:** Job rotation results in frequent interruption of work. A person who is doing a particular job and get it comfortable suddenly finds himself shifted to another job or department. This interrupts the work in both the departments.
  (*b*) **Reduces uniformity in quality:** Quality of work done by a trained worker is different from that of a new worker. When a new worker "I" shifted or rotated in the department, he takes time to learn the new job, makes mistakes, in the process and affects the quality of the job.
  (*c*) **Misunderstanding with the union member:** Sometimes job rotation may lead to misunderstanding with member of the union. The union might think that employees are being harassed and more work is being taken from them. In reality this is not the case.

2. **Job Enlargement:** Job enlargement is another method of job design when any organization wishes to adopt proper job design it can opt for job enlargement. Job enlargement involves combining various activities at the same level in the organization and adding them to the existing job. It increases the scope of the job. It is also called the horizontal expansion of job activities. Job enlargement can be explained with the help of the following example. If Mr. A is working as an executive with a company and is currently performing three activities on his job after job enlargement or through job enlargement we add four more activities to the existing job so now Mr. A performs seven activities on the job.

   It must be noted that the new activities which have been added should belong to the same hierarchy level in the organization. By job enlargement we provide a greater variety of activities to the individual so that we are in a position to increase the interest of the job and make maximum use of employee's skill. Job enlargement is also essential when policies like VRS are implemented in the company.

   - **Advantages of job enlargement**
     (*a*) **Variety of skills:** Job enlargement helps the organization to improve and increase the skills of the employee due to organization as well as the individual benefit.
     (*b*) **Improve earning capacity:** Due to job enlargement the person learns many new activities. When such people apply for jobs to other companies they can bargain for more salary.

(*c*) **Wide range of activities:** Job enlargement provides wide range of activities for employees. Since a single employee handles multiple activities the company can try and reduce the number of employees. This reduces the salary bill for the company.

- **Disadvantages of Job Enlargement**

(*a*) **Increases work burden:** Job enlargement increases the work of the employee and not every company provides incentives and extra salary for extra work. Therefore the efforts of the individual may remain unrecognized.

(*b*) **Increasing frustraction of the employee:** In many cases employees end up being frustrated because increased activities donot result in increased salaries.

(*c*) **Problem with union members:** Many union members may misunderstand job enlargement as exploitation of worker and may take objection to it.

3. **Job Enrichment:** Job enrichment is a term given by Fredrick Herzberg. According to him, a few motivators are added to a job to make it more rewarding, challenging and interesting. According to Herzberg, the motivating factors to an existing job to make it more interesting. The motivating factors can be:

(*a*) Giving more freedom.

(*b*) Encouraging participation.

(*c*) Giving employees the freedom to select the method of working.

(*d*) Allowing employees to select the place at which they would like to work.

(*e*) Allowing workers to select the tools that they require on the job.

(*f*) Allowing workers to decide the layout of plant or office.

Job enrichment gives lot of freedom to the employee but at the same time increases the responsibility. Some workers are power and responsibility hungry. Job enrichment satisfies the needs of employees.

- **Advantages of Job enrichment**

(*a*) **Interesting and challenging job:** When a certain amount of power is given to the employees it makes the job more challenging for them, we can say that job enrichment is a method of employee empowerment.

(*b*) **Improves decision-making:** Through job enrichment we can improve the decision-making ability of the employee by asking him to decide on factory layout, method and style of working.

(*c*) **Identifies future managerial caliber:** When we provide decision-making opportunities to employees, we can identify which employee is better than other in decision-making and mark employees for further promotion.

(*d*) **Identifies higher order needs of employees:** This model identifies higher order needs of the employee. Abraham Maslow's theory of motivation speaks of these higher order needs, *e.g.*, ego and esteemed needs, self actualization etc. These needs can be achieved through job enrichment.

(*e*) **Reduces work load of superiors:** Job enrichment reduces the work load of senior staff. When decisions are taken by juniors the seniors work load is reduced.

- **Disadvantages of Job Enrichment**
  (*a*) Job enrichment is based on the assumptions that workers have complete knowledge to take decisions and they have the right attitude. In reality this might not be the case due to which there can be problems in working.
  (*b*) Job enrichment has negative implications *i.e.,* along with usual work decision-making work is also given to the employees and not many may be comfortable with this.
  (*c*) Superiors may feel that power is being taken away from them and given to the junior's. This might lead to ego problems.
  (*d*) This method will only work in certain situations. Some jobs already give a lot of freedom and responsibility, this method will not work for such jobs.
  (*e*) Some people are internally dissatisfied with the organization. For such people no amount of job enrichment can solve the problem.

3.5 JOB ENRICHMENT

Most of us want interesting, challenging, jobs where we feel that we can make a real difference to other people's live. As it is for us, it is for the people who work with or for us. So why are we so many jobs so boring and monotonous? And what can you do to make the jobs you offer more satisfying? By reducing recruitment costs, increasing retention of experienced staff and motivating them to perform at a high level, you can have a real impact on the bottom line.

One of the key factors in good job design is job enrichment, most notably promoted by psychological Fredrick Herzberg in his 1968 article "One More Time: How Do You motivate Employees"?. This is the practice of enhancing individual jobs to make the responsibilities more rewarding and inspiring for the people who do them.

Job enrichment involves providing an employee with more responsibility for a job and challenges the individual's skill at work. Enrichment involves increasing the decision-making authority and encouraging the employee with their tasks. Providing employees with meaningful tasks will increase their performance compared to just "piling" on a lot of work and increasing the same type of work-load.

Job enrichment directly relates to motivation factor and employee satisfaction. The concept of job enrichment was developed by Fredrick Herzberg in the 1950s. Previous studies have shown that job enrichment causes significant increases in employee job satisfaction, job involvement and increased internal motivation. Decreases in absenteism was also found to be a result of increased job enrichment.

Job enrichment is a type of job redesign intended to reverse the effects of tasks that are boredom, lack of flexibility, and employee dissatisfaction. The underlying principle is to expand the scope of the job with the greater variety of tasks, vertical in nature, that require self-sufficiency.

Since the goal is to give the individual exposure to tasks normally reserved for differently focused or higher positions, merely adding more of the same responsibilities related to an employee's current position is not considered job enrichment.

The basis for job enrichment practices is the work done by Fredrick Herzberg in the 1950's and 60's, which was further refined in 1975 by Hackman and Oldham using what they called the Job Characteristics Model. This model assumes that if five core job characteristics are present, three psychlogical states critical to motivation are produced, resulting in positive outcomes. Figure 3.27 illustrates this model.

Job Characteristics Model

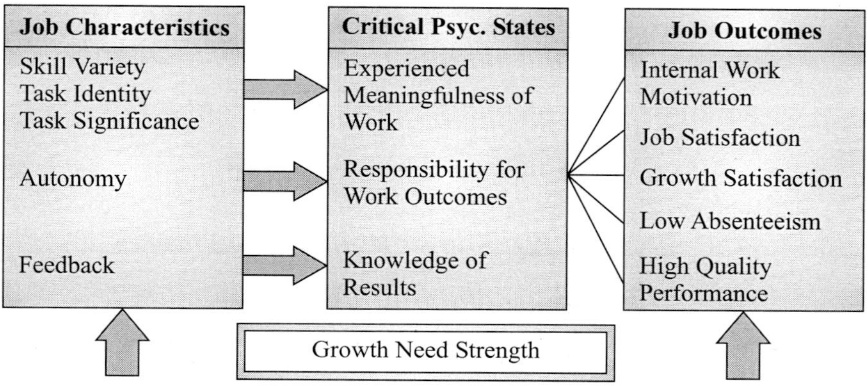

Fig. 3.27

Job enrichment can only be truly successful if planning includes support for all phases of the initiative. Ohio State University Extension began a job enrichment program in 1992 and surveyed the participants five years later. The results, broken down into three sub-buckets of data beyond the main grouping of advantages/disadvantages as shown in table, indicate the University had not fully considered the planning and administrative aspects of the program. While the benefits are seemingly obvious, program fail not because of a lack of benefits, but rather due to implementation problems. These problems can include a perception of too great a cost, lack of long-term commitment of resources, and potential job classification changes shown in Fig. 3.28.

ADVANTAGES	**Renewal**	reduction in boredom, making new contacts, thinking about new career options, change in viewpoint.
	Exploration	trying new skills, developing new relationships, testing management and administrative skills.
	Specialization	re-education, in-depth exploration, using special skills, meeting a need that was under-served.
DISADVANTAGES	**Balance**	time allocation, tough decisions about work load, taking time away from regular assignments, balancing work load.
	Clarity	lack of information needed to be successful, lack of clear plans and goals, guidance in using time.
	Orientation	lack of structured introduction to special assignment, not being taught new responsibilities, difficulty in learning new networks, lack of instruction, sink or swim approach.

Ohio State University Extension Job Enrichment Survey Results.

Fig. 3.28

3.5.1 Job Enrichment Options

The central focus of job enrichment is giving people more control over their work (lack of control is a key cause of stress, and therefore of unhappiness). Where possible, allow them to take on tasks that are typically done by supervisors. This means that they have more influence over planning, executing, and evaluating the jobs they do.

In enriched jobs, people complete activities with increased freedom, independence, and responsibility. They also receive plenty of feedback, so that they can assess and correct their own performance.

Here are some strategies which can be used to enrich jobs in the workplace

1. **Rotate Jobs:** People is given the opportunity to use a variety of skills and perform different kinds of work. The most common way to do this is through job rotation. The workers are moved through a variety of jobs that allow to see different parts of the organization, learn different skills and acquire different experiences. This can be very motivating, especially for the people in jobs that are very repetitive or that focus on only one or two skills.
2. **Combine Tasks:** Combine work activities to provide a more challenging and complex work assignment. This can significantly increase "task identity" because people see a job through from start to finish. This allows workers to use wide variety of skills, which can make the work seem more meaningful and important. For example, you can convert an assembly line process, in which each person does one task, into a process in which one person assembles a whole unit. This model is applied wherever people or groups that typically perform only one part of an overall process. Consider expanding their roles to give them responsibility for the entire process, or for a bigger part of the process.
3. **Identify Project-Focused Work Units:** It breaks the typical functional lines and form project-focused units. For example, rather than having all the marketing people in one department, with supervisors directing who works on which project, the departments could be split into specialized project units-specific storyboard creators, copywriters, and designers could all work together for one client or one campaign. Allowing employees to build client-relationship is an excellent way to increase autonomy, task identity and feedback.
4. **Create Autonomous Work Teams:** This is job enrichment at the group level. Set a goal for a team, and make team members free to determine work assignments, schedules, rest breaks, evaluation parameters, and the like. You may even give them influence over choosing their own team members. With this method, supervisory position are significantly cut back, and people will given leadership and management skills.
5. **Implement Participative Management:** Allow team members to participate in decision-making and get involved in strategic planning. This is an excellent way to communicate to members of a team that their input is important. It can work in any organization—from a very small company, with an owner/boss who is used to dictate everything to a large company with a huge hierarchy. When people realize that what they say is valued and makes a difference, they will be likely be motivated.

6. **Redistribute Power and Authority:** Redistribute control and grant more authority to workers for making job-related decisions. As supervisors delegate more authority and responsibility, team members autonomy, accountability, and task identity will increase.
7. **Increase Employee-Directed Feedback:** Make sure that people know how well, or poorly, they are performing their jobs. The more control is given for evaluating and monitoring the performance, the more enriched jobs will be. Rather than having quality control department go around and point out mistakes, consider giving each team responsibility for their own quality control. Workers will receive immediate feedback, and they will learn to solve problems, take initiative and make decisions.

3.5.2 Implementing a Job Enrichment Program

- **Step One:** Find out where people are dissatisfied with their current work assignment. There is little point to enrich the job and change the work environment if the wrong jobs are enriched and the wrong changes are made. Like any motivation initiative, determine what the people want before their work begin. Surveys are good means of doing this. There should not be any mistake made on the basis of any presumption (that you know what people want). Go to the source—and use that information to build enrichment options.
- **Step Two:** Consider which job enrichment options can be provided. There is no need to drastically redesign the entire work process. The way that the enriched job is designed, must strike the balance between operational need and job satisfaction. If significant changes are needed, consider establishing a "job enrichment task force"—perhaps use a cross-section of employees, and give the responsibility for deciding which enrichment options make the most sense.
- **Step Three:** The program is designed and communicated. If the significant changes are made, people should know about it, what is being done and why. Work with the managers to create an enriching work environment that includes lot of employee participation and recognition. Remember to monitor the efforts, and regularly evaluate the effectiveness of what is being provided.

3.6 JOB EVALUATION

Job evaluation is a systematic way of determining the value/worth of a job in relation to other jobs in an organization. It tries to make a systematic comparison between jobs to assess their relative worth for the purpose of establishing a rational pay structure. Job evaluation is the process of determining the relative worth of the job. It is process which is helpful even for framing compensation plans by the personnel manager.

Job evaluation is a practical technique, designed to enable trained and experienced staff to judge the size of one job relative to others. It doesnot directly determine pay levels, but will establish the basis for an internal ranking of jobs.

3.6.1 Features of Job Evaluation

The purpose of job evaluation is to produce a defensive ranking of jobs on which a rational and acceptable pay structure can be built. The important features of job evaluation may be summarised thus:

(*a*) It tries to assess jobs, not people.

(*b*) The standards of job evaluation are relative, not absolute.

(*c*) the basic information on which job evaluation are made is obtained from job analysis.

(*d*) Job evaluation are carried out by groups, not by individuals.

(*e*) Some degree of subjectivity is always present in job evaluation.

(*f*) Job evaluation does not fix pay scales, but merely provides a basis for evaluating a rational wage structure.

3.6.2 Some Principles of Job Evaluation

1. Clearly defined and identifiable jobs must exist. These jobs will be accurately described in an agreed job description.
2. All jobs in an organization will be evaluated using an agreed job evaluation scheme.
3. Job evaluators will need to gain a thorough understanding of the job.
4. It is concerned with jobs, not people. It is not the person that is being evaluated.
5. The job is assessed as if it were being carried out in a fully competent and acceptable manner.
6. Job evaluation is based on judgement and is not scientific. However if applied correctly it can enable objective judgements to be made.
7. It is possible to make a judgement about a job's contribution relative to other jobs in an organization.
8. The real test of the evaluation results is their acceptability to all participants.
9. Job evaluation can aid organizational problem solving as it highlights duplication of tasks and gaps between jobs and functions.

3.6.3 Process of Job Evaluation

The process of job evaluation involves the following steps:

(*a*) **Gaining acceptance:** Before undertaking job evaluation, top management must explain the aims and uses of the programme to the employees and unions. To elaborate the programme further, oral presentation could be made. Letters, booklets could be used to classify all relevant aspects of the job evaluation programme.

(*b*) **Creating job evaluation committee:** It is not possible for a single person to evaluate all the key jobs in an organization. Usually, a job evaluation committee consisting of experienced employees, union representatives and HR experts is created to set the ball rolling.

(*c*) **Finding the jobs to be evaluated:** Every job need not be evaluated. This may be too taxing and costly. Certain key jobs in each department may be identified. While picking up the jobs, care must be taken to ensure that they represent the type of work performed in the department.

(*d*) **Analysing and preparing job description:** This require the preparation of a job description and also an analysis of job needs for successful performance.

(*e*) **Selecting the method of evaluation:** The most important method of evaluating the jobs must be identified now, keeping the job factors as well as organizational demands in mind.

(*f*) **Classifying jobs:** The relative worth of various jobs in an organization may be found out after arranging jobs in order of importance using criteria such as skill requirements, experience needed, under which conditions job is performed, type of responsibilities to be shouldered, degree of supervision needed, the amount of stress caused by the job etc. Weights can be assigned to each such factor. When we finally add all the weights, the worth of a job is determined. The points may then be converted into monetary values.

(*g*) **Reviewing periodically:** In the light of changes in environmental conditions (technology, products, services, etc.) jobs need to be examined closely. For example, the traditional clerical functions have undergone a rapid change in sectors like banking, insurance and railways, after computerisation. New job descriptions need to be written and the skill needs of new jobs need to be duly incorporated in the evaluation process. Otherwise, employees may feel that all the relevant job factors—based on which their pay has been determined—have not been evaluated properly.

3.6.4 Methods of Job Evaluation

There are three basic methods of job evaluation-ranking, classification, and factor compensation. While many variations of these methods exist in practice, the three basic approach are described here:

(*a*) **Ranking Method:** The simplest method of job evaluation is the ranking method. According to this method, jobs are arranged from highest to lowest, in order of their value or merit to the organization. Jobs also can be arranged according to the relative difficulty in performing them. The jobs are examined as a whole rather than on the basis of important factors in the job; and the job at the top of the list has the highest value and obviously the job at the bottom of the list will have the lowest value.

Jobs are usually ranked in each department and then the department ranking are combined to develop an organizational ranking. The following table is a hypothetical illustration of ranking of jobs which is shown in Fig. 3.28.

Rank	*Monthly Salaries*
1. Accountant	Rs. 3,000
2. Accounts clerk	Rs. 1,800
3. Purchase Assistant	Rs. 1,700
4. Machine-Operator	Rs. 1,400
5. Typist	Rs. 900
6. Office Boy	Rs. 600

Fig. 3.29 Array of Jobs according to the Ranking Method

The variation in payment of salaries depends on the variation of the nature of the job performed by the employees. The ranking method is simple to understand and practice and it is best suited for a small organization. Its simplicity, however, works to its disadvantage in big organizations because ranking are difficult to develop in a large, complex organization. Moreover, this kind of ranking is highly subjective in nature and may offend many employees. Therefore, a more scientific and fruitful way of job evaluation is called for.

(*b*) **Classification Method:** According to this method, a predetermined number of job groups or job classes are established and jobs are assigned to these classifications. This method places groups of jobs into job classes or job grades. Separate classes may include office, clerical, managerial, personnel, etc. following is a brief description of such a classification in an office.

(*a*) *Class I-Executives:* Further classification under this category may be office manager, deputy office manager, office superintendent, departmental supervisor, etc.

(*b*) *Class II-Skilled workers:* Under this category may come the purchasing assistant, cashier, receipts clerk, etc.

(*c*) *Class III-Semi-skilled:* Under this category may come stenotypists, machine-operators, switchboard operators etc.

(*d*) *Class IV-Semi-skilled:* This category comprises daftaris, file clerks, office boys, etc.

The job classification method is less subjective when compared to the earlier ranking method. The system is very easy to understand and acceptable to almost all employees without hesitation. One strong point in favour of the method is that it takes into account all the factors that a job comprises. This system can be effectively used for a variety of jobs.

The weaknesses of the job classification method are:

(*i*) Even when the requirements of different jobs differ, they may be combined into a single category, depending on the status a job carries.

(*ii*) It is difficult to write all-inclusive descriptions of a grade.

(*iii*) The method oversimplifies sharp differences between different jobs and different grades.

(*iv*) When individual job descriptions and grade descriptions donot match well, the evaluators have the tendency to classify the job using their subjective judgements.

(*c*) **Factor Comparison Method:** A more systematic and scientific method of job evaluation is the factor comparison method. Through it is the most complex method of all, it is consistent and appreciable. Under this method, instead of ranking complete jobs, each job is ranked according to a series of factors. These factors include mental effort, physical effort, skill needed, supervisory responsibility, working conditions and other relevant factors (for instance, know-how, problem solving abilities, accountability, etc). Pay will be assigned in this method by comparing the weights of the factors required for each job, *i.e.,* the present wages paid for key jobs may be divided among the factors weighed by importance (the most important factor, for instance, mental effort, receives the highest weight). In other words, wages are assigned to the job in comparison to its ranking on each job factor.

The steps involved in factor comparison method may be briefly stated thus:

1. Select key jobs (say 15 to 20), representing wage/salary levels across the organization. The selected jobs must represent as many departments as possible.
2. Find the factors in terms of which the jobs are evaluated (such as skill, mental effort, responsibility, physical effort, working conditions, etc.)
3. Rank the selected jobs under each factor (by each and every member of the job evaluation committee) independently.
4. Assign money value to each factor and determine the wage rates for each key job.
5. The wage rate for a job is apportioned along the identified factors.
6. All other jobs are compared with the list of key jobs and wage rates are determined.

- **Merits of Factor Comparison Method**
  1. This method is analytical and objective.
  2. This method is reliable and valid as each job is compared with all other jobs in terms of key factors.
  3. Money values are assigned in a fair may based on an agreed rank order fixed by the job evaluation committee.
  4. This Method is flexible as there is no upper limitation on the rating of a factor.
- **Demerits of Factor Comparison Method**
  1. This method is difficult to understand, explain and operate.
  2. Its use of the same criteria to assess all jobs is questionable as jobs differ across and within organizations.
  3. This method is time consuming and costly.

(*d*) **Point Method:** This method is widely used currently. Here, jobs are expressed in terms of key factors. Points are assigned to each factor after prioritizing each factor in the order of importance. The points are summed up to determine the wage rate for the job. Jobs with similar point totals are placed in similar pay grades. The procedure involved may be explained thus:

(*i*) Select key jobs. Identify the factors common to all the identified jobs such as skill, effort, responsibility, etc.

(*ii*) Divide each major factor into a number of sub factors. Each sub factor is defined and expressed clearly in the order of importance, preferably along a scale.

- **Merits and Demerits of Point Method:** The Point method is a superior and widely used method of evaluating jobs. It forces raters to look into all keys factors and sub-factors of a job. Point values are assigned to all factors in a systematic way, eliminating bias at every stage. It is reliable because raters using similar criteria would get or less similar answers. "The methodology underlying the approach contributes to a minimum of rating error". It accounts for differences in wage rates for various jobs on the strength of job factors. Jobs may change over time, but the rating scales established under the point method remain unaffected.

On the negative side, the point method is complex. Preparing a manual for various jobs, fixing values for key and sub-factors, establishing wage rates for different grades, etc., is a time-consuming process. According to Decenzo and Robbins, "the key criteria must be carefully and clearly identified, degrees of factors have to be agreed upon in terms that mean the same to all rates, the weight of each criterion has to be established and point values must be assigned to degrees". This may be too taxing, especially while evaluating managerial jobs where the nature of work (varied, complex, novel) is such that it cannot be expressed in quantifiable numbers.

3.6.5 Future of Job Evaluation

As organization constantly evolve and new organizations emerge there will be challenges to existing principles of job evaluation. Whether existing job evaluation techniques and accompanying schemes remain relevant in a faster moving and constantly changing world, where new jobs and roles are invented on a regular basis, remains to be seen. The formal points systems, used by so many organizations is often already seen to be inflexible. Sticking rigidly to an existing scheme may impose barriers to change. Constantly, updating and writing new jobs together with the time that has to be spent administering the job evaluation schemes may become too cumbersome and time consuming for the benefits that are derived.

3.7 MERIT RATING

According to dictionary, merit rating is defined as, "employee rating achieved through a periodic employee evaluation system, often used as the basis for pay increases and/or promotion".

Evaluation of individual merits of the employees is known as merit rating. Systematic analysis and classification of jobs on the basis of their characteristics is defined as job evaluation whereas the evaluation of the merits of the workers and on the basis of that, the classification of the workers is known as **merit rating.** By keeping the performance records of every worker and by assessing their performance in terms of some norms or standards, merit rating is done by organisations.

Merit rating is associated with performance appraisal of an employee. This is a systematic approach for evaluating the performance of an employee on the job, which he performs. This is also called as performance appraisal, personnel rating and employee evaluation. Merit-rating is a formal, objective personality, contributions and potentials of employees in a working organization.

3.7.1 Job Evaluation Vs Merit-Rating

Job evaluation and merit-rating are compared in the following ways:

Job Evaluation	*Merit-Rating*
1. It evaluates a job or work.	1. It evaluates a worker.
2. It is for the purpose of fixing a base-wage for a job.	2. It is for the purpose of deciding reward for exceptional merit of worker.
3. It is independent of operator or worker. It is impersonal in nature.	3. It is independent of job. It is impersonal in nature.
4. Useful for decision regarding wage and salary administration, skill match, etc.	4. Useful for decision regarding training, placement, promotion, counselling etc.
5. It considers requirement of job.	5. It considers ability and performance of individual.

3.7.2 Objects of Merit-Rating

The objects of merit-rating are:

1. for each individual worker, assessment of the standard of performance,
2. providing a basis for rewarding for high merit without detailed work study being applied,
3. providing a basis for determining the remuneration of the indirect workers, the performances of whom cannot be easily determined,
4. choosing suitable worker for a job; job evaluation mentions the job requirement of the worker, while merit-rating mentions what qualities the worker should possesses. The selection will be perfect if these two fit in each other,
5. providing basis for increment, promotion etc.,
6. finding out the abilities and defects of each worker,
7. improving labour relations and reducing labour turnover.

3.7.3 Limitations of Merit-Rating

The limitations of merit-rating are:

1. Labour unrest may arise due to wrong rating.
2. There may not be any relevancy of the factors considered for merit rating.
3. Previous records of the employees may influence the rating and thereby justice may be lost.
4. Employees may oppose the rating or may not recognize the same.
5. Workers may not be satisfied with the incentives allowed on the basis of merit rating.

For merit rating purpose, the following factors may be usually in the employee: Attendance, discipline, co-operation, knowledge, skill, experience, aptitude, sense of responsibility and judgement, quantity and quality of output etc.

3.8 THE CONCEPT OF 360 DEGREE FEEDBACK

3.8.1 What is 360 Degree

360-Degree Appraisal is basically a Multi-Rater Appraisal and Feedback System. Almost every Fortune 500 Company is using this in some form or the other. In this system:

1. The candidate is assessed periodically (once in a year and some times even half yearly) by a number of assessors including his boss, immediate subordinates, colleagues, internal customers and external customers.
2. The assessment is made on a questionnaire specially designed to measure behaviors considered as critical for performance.
3. Others do the appraisal anonymously and the assessment is collected by an external agent (consultant) or specially designated internal agent.
4. The assessment is consolidated; feedback profiles are prepared and given to the participant after a workshop or directly by his boss or the HRD department in a performance review discussion session.

T.V Rao and team as Top Management Styles and organization Effectiveness program started 360-degree feedback in mid-eighties at the Indian Institute of Management, Ahmadabad; it was not branded till 1998 as 360-degree appraisal. Once it was branded in the US as 360-degree appraisal, it caught the attention of many corporations and today a large percentage of Fortune 500 companies use it in some form or the other. Experiences with it as a development tool have been good, though experiences as an appraisal tool are mixed.

3.8.2 Advantages of 360-Degree Feedback

(*a*) Provides objective, well-rounded views, expressed by a number of colleagues.

(*b*) Increases involvement of people at all levels of the University, as well as the University community.

(*c*) Direct reports and peers, rather than managers/supervisors are better to judge some skills, such as leadership.

(*d*) Increases understanding of the behaviors required to improve both individual and organizational effectiveness.

(*e*) Individuals can better manage their own performance and careers.

(*f*) Evaluators are able to respond confidentially and anonymously.

(*g*) Provides reliable and user-friendly feedback reports.

Feedback can provide organizational value, allowing us to identify needs and set goals for organizational development and training within the University.

CASE STUDY

Work Redesign at BHEL, Hardwar

Nitish R. De

Bharat Heavy Electricals Limited (BHEL) is one of the largest public enterprises in India, with six major manufacturing plants and several divisions. The enterprise employs over 60,000 employees in the manufacturing, marketing, R&D, servicing, and corporate headquarters units. It also services overseas markets. The corporation has been making spectacular progress since 1972.

The Hardwar unit, located in the north, is mainly concerned with the manufacture of heavy electrical equipment such as steam and hydroturbines, generators, and other related equipment. It employs over 10,000 employees and has a self-contained township. Although over 10 years old, the unit was not coming up as well as was expected and productivity was not, by any means, satisfactory. Technical collaboration existed with a large Soviet firm located in Leningrad, though gradually the design and manufacturing activities have been taken over by the Indian counterparts as they gathered more and more experience and confidence.

The Executive Director of the unit, having attended a 'quality of work-life' seminar, was keen that some action should be initiated in the plant. The National Labour Institute team, consisting of Nitish R. De, Subhash Gakkhar, V. Nilakant, and Rukmini Rao, started dialogues with the managers, the trade union leaders (four identifiable trade unions in operation), and the supervisory staff. The dialogues, held sometimes in camera and sometimes in the open, occasionally in separate groups and at times in joint sessions, gradually moved the groups towards a resolution that a pilot project would be started at a favourable work site.

Off the Mark: Survey and Study

It may also be mentioned that a comprehensive survey was undertaken in the unit to diagnose the training needs of the middle management in the plant and another survey to explore the linkages between the perception of work-life and the lifestyle in the community. The first study indicated that, on 10 key dimensions, significant levels of differences existed in terms of knowledge and information gaps.

This study revealed that in some of the key areas the middle management personnel were lacking in knowledge and there were information gaps. This situation was certainly not conducive to effective performance of their supervisory functions. The other study to establish interlinkages between the quality of family and community life and the quality of life in the workplace indicated a clear dichotomy between the life of the workers at the workplace and their life with their families.

A section of Block V, newly created, with 25 workmen working on fabrication of the upper part of the condenser unit, was considered to be a favourable site: compact in character, with a reasonable layout, and a positive attitude among the manager and the shop-floor trade union leaders. There were also other compelling reasons to select the group. The condenser was an expensive piece of equipment. Second, for the setting up of a power station, it was necessary that a condenser unit should be placed at the site before the steam turbine was installed and as such it had to be manufactured and despatched at least two months ahead of the completed steam turbine. Third, the productivity in the shop was not of a high order.

(The workers had a series of dialogues with internal and external consultants and they agreed to undertake the work redesign experiment. First, a joint study of the work flow was undertaken, which is presented in Fig. 1.

The involvement of some of the key workers in the study also provided them with excitement, if not insight, at being involved in an action–research project. The workforce consisted of the categories shown in Fig. 2.

The study of the social system of work imposed by the organization indicated that each worker was concerned with his or her own trade and that none of them identified himself or herself with the product itself. Second, there was invariably forced idle time because when a particular worker was working at a spot, another worker who was required to do his job in close proximity had to wait till the first worker had finished his or her job. Third, there was uneven demand on the services of the material supplies group, crane operator and the riggers.) Productivity, when the studies were being undertaken in April and May 1975, was certainly low.

Part of the low productivity was on account of the high rate of absenteeism during the summer months.

First Lap: The Action Plan

When all the data, generated by the workers from their own experiences, were analysed, they decided on two steps:

1. To set up a task force with representatives of each category of workers and the supervisor. The manager, under whom the shop was placed, would also participate in the meeting if the group so wanted. An industrial engineer was also associated with the group as a resource person. The task group had a membership of eight. It was also decided that while two members would continue because of their leadership abilities, the others would rotate (except for the supervisor).
2. To devise a new work system which would take care of the workers' motivation as well as overcome the persisting culture of low productivity.

After deliberations, the task force, with the concurrence of the employees concerned, evolved the work system detailed in Fig. 3. According to this design, the direct production groups would each consist of one welder, three fitters, and one fettler. Each group would take charge of the totality of its task and would gradually take up one another's skills by undergoing on-the-job training assisted by the supervisor, the industrial engineer, and the fellow workers themselves. This would also be the case with the crane operator and the riggers. It was decided that the gas-cutters and helpers on the one hand and material supplies group on the other would be integrated into the new work system at a later stage. As the system started working in May–June 1975, several positive and negative phenomena showed up. On the positive side, a welder now started working as a fitter and if he or she did not know the art of reading elementary drawings, which was a necessity for a fitter's job, arrangements were made by the management to provide for on-the-job training. The same procedure was followed in the case of fitters, fettlers, gas-cutters, and others.

With more experience and confidence, the workers working along with the task force brought about a second redesign of their work organization in the month of

September 1975. It was decided that the workforce would be distributed in two shifts in the following manner.

Table 3.1 Division of Workforce

Shift 1		*Shift 2*	
Fitters	5	Fitters	4
Welders	5	Welders	6
Gas-cutter	1	Gas-cutter	1
Fettler	1	Fettler	1

In addition, there would be crane operators in both the shifts. Each shift group became an integrated group and it was decided that one shift would fabricate the right side of the upper part of the condenser unit and the other shift would do the same with the left side of the condenser unit. As and when several such parts were to be assembled, the two groups would together weld them in order to make a complete upper part of the condenser unit.

The Winning Edge: Implementation

While the new system started in Block V in late May 1975, in Block II the bay which was concerned with the manufacture of the lower part of the condenser unit started the same process in June. The senior supervisor of the shop took the initiative and 39 workmen with similar trade experiences were involved. They also designed a similar small work-group module. Here also the task force consisted of eight members with the provision of monthly rotation.

Data indicate that there have been steady increases in productivity although there have been occasional setbacks. Progressively, (the old culture of one man–one function was replaced by the acquisition of multiple skills leading to the development of a group system of working with internal monitoring of group norms, internal control of work flow and work allocation, identification with the product and its quality, and what, from the management point of view, was considered significant, the gradual drop in personal idle time on account of loitering, spending time outside the workplace without any justifiable reasons, etc. The commitment of the workers and, along with them, that of the supervisor and the manager was distinctly visible. The old culture that higher status, such as that of a welder or fitter, would stand in the way of taking up a low-status job even for a part of the time could be overcome.)

It is worth mentioning that the target of seven condensers for the year 1975–6 was completed by the workers of Block V by January 1976, with the result that the management had to give them other items of work from February 1976. The demand for thermal power units for the year 1976–7 having gone down, the need for condenser fabrication also was much less. Therefore, the workers in Block II and Block V had to take up other products, in the manufacture of which they did not have much experience. Both the groups responded to the challenge with optimism and alacrity. Table 3.2 indicates the product mix for these two groups, which were traditionally fabricating condenser units.

Table 3.2 indicates that, while in Block II, two major items were being produced during the year 1976–7, and as such the two groups working in two shifts had a steady flow of manufactured products, Block V—also working in two groups distributed in two shifts—had taken diverse product lines varying from month to month. This switchover somewhat adversely affected productivity and caused initial disruption in group working. However, the tension between the management and the workers that would normally arise in such a situation was absent. Both groups worked together on this challenge in November 1976 without conflict.

Table 3.2 Product Mix

Month	*Block*	*Items*
February 1976	II	Lower part of condenser fabrication
	V	Four water boxes—rear and front
March 1976	II	Lower part of condenser fabrication
	V	Tooling items—dies, punches, fixtures, etc.
April 1976	II	Lower part of condenser fabrication
	V	Hydrogenerator stator frames
May 1976	II	Hydrogenerator stator frames
	V	Condenser stand for tool room
June 1976	II	Hydrogenerator stator frames
	V	Lower ring for HT—stainless steel welding
July 1976	II	Hydrogenerator stator frames
	V	Rotor spiders for HG
August 1976	II	Lower part of condenser fabrication
	V	Hydrogenerator stator frames
September 1976	II	Condenser fabrication and hydrogenerator stator frames
	V	Spherical valve body (stainless steel deposits), LC—rear and front part
October 1976	II	Condenser fabrication and HG stator frames
	V	Water boxes—rear and front, LPC upper part
November 1976	II	Condenser fabrication and hydrogenerator frames
March 1977	V	Upper part of condenser fabrication (3 in number), 2 LPC—upper part front and rear, and any urgent work

Note: Block V group also carried out, as and when necessary, fitting and assembly of the heater section for the upper part of the condenser unit.

It appears that despite the new demands on the workers for involvement in new product lines, their urge to maintain multiple skills still continued. The minutes of meetings of the task force indicate a high degree of orientation towards work-interest issues such as delay in the repair of cranes, repair of welding sets, improving the system of on-the-job training, a joint search mechanism in identifying the forces contributing to lower productivity, and a search for ways to reduce job tension as against the usual union–management type of meeting in which interest-related issues assume a divisive bargaining picture. In fact, examining the records of the agenda items and the minutes of meetings of the task forces of Block II and Block V from

April 1975 to October 1976, one can discern the distinct qualitative shift towards problem-solving orientation with a view to looking at a problem as a collective one instead of making a scapegoat of the other group.

Goal in Sight: Spreading the Word

The work reorganization experiences led to a new supervisory role in the form of liaison with the input and output departments, service units, and involvement with central planning. This became possible because the work groups took substantial control over the production process, including routine inspection and maintenance activities in addition to implementing the norms of discipline. All these experiences encouraged the management and the task forces to set up three project teams to deliberate on and submit reports on the three following themes:

1. Multi-skill training, its role in employee satisfaction and higher productivity, and encouraging workers to acquire skills in different trades.
2. Changing role of supervisors, particularly in respect of coordination, planning, and training.
3. Diffusion of the scheme of work redesign to white-collar areas such as personnel, finance, and medical departments.

The project teams consisting of managers and the workers from the three blocks where the work redesign experiments were on (Blocks II, IV, and V) submitted their reports to the management.

The diffusion process had been a matter of active consideration by the workers and the management, but the initiative in starting new projects in shops was left to the employees of the concerned shops. In September 1975, Block IV came forward and, through a series of meetings, improve-ments started taking place. The management also decided that time spent reworking on account of certain design changes would not affect the bonus earnings of the workers.

The Final Hurdle and Beyond

The task force became somewhat confident of the success of the experiment and then, in the month of September 1976, it decided that there would be a almost complete switchover of jobs between fitters and electricians. This premature enthusiasm brought down the work efficiency to 37.5 per cent and it obviously affected the bonus earnings. The task force, on reflection, decided that the misplaced enthusiasm should be curbed and, in October 1976, it decided to concentrate on training programmes which would gradually lead to skill diffusion among electricians and fitters.

From September 1976, the work redesign culture was being introduced in Block I (turbo-rotor col section and detailed assembly section), Block IV (panel fabrication), Block III (rotor assembly production and blading section), and in the personnel department, where a matrix type of organization had been introduced. In the personnel department, there were three divisions—one dealing with the plant operation and maintenance, another with the design organization, and the third with the policy matters located in the office of the executive director. The personnel department attached to the plant, consisting of 22 people, organized itself into three operating groups: one dealing with all the problems of supervisory, clerical, and ministerial staff; another group with the skilled category of workers; and the third with the semi-skilled and unskilled groups of workers. A functional division of work was replaced by a total service package, but in order to ensure that the three groups functioning simultaneously did not come to contradictory conclusions and policy recommendations, a three-man coordinating team was set up. According to the reports of the office assistants, who belonged to the clerical category, this new system of work was much more satisfying than the fragmented type of work they were traditionally accustomed to.

REFERENCE

De, Nitish R. (1984), *Alternative designs of human systems,* New Delhi: Sage.

EXERCISES

Questions for Discussion

Read and reflect on the case in the light of motivation management and answer the following questions:

1. What principles of work redesigning were used in BHEL to deal with the problem of low productivity?
2. Which aspects of role efficacy increased as a result of the changes introduced in the organization?
3. How was consensus achieved on the changes made in the organization?
4. Did frustration decrease in the organization as a result of work redesigning? Discuss with some examples.
5. Describe the leadership style of the plant. How did it contribute to the success of the experiment?

Short Questions Answers

1. What is organizational analysis?
2. What is organizational structure?
3. Discuss planning approach to organizational analysis?
4. Explain models of organizational analysis?
5. Explain different types of organizational planning structures?
6. Discuss design of organizational structure?
7. Describe key elements of organizational structure?
8. Discuss various components of organization design?
9. Discuss grouping models with its advantages and disadvantages?
10. Explain how to design an effective organization structure?
11. Explain five approaches to organizational design?
12. What is job design?
13. Discuss the objectives of job design?
14. Explain the factors affecting job design?
15. Discuss the methods of job design?
16. What is job enrichment?

17. Discuss four steps for implementing job enrichment program?
18. What is job evaluation?
19. Give the features of job evaluation?
20. Discuss some principles of job evaluation?
21. Explain the process of job evaluation?
22. Discuss the methods of job evaluation?
23. What will be the future of job evaluation?
24. What is merit rating?
25. Differentiate between job evaluation and merit rating?
26. Discuss the objects of merit rating?
27. Discuss the limitations of merit rating?

Long Questions Answers

1. What is organizational analysis? Explain models of organizational analysis?
2. Discuss the planning approach to organizational analysis?
3. What is organizational structure? Explain different types of organizational planning structures?
4. Discuss components of organization design? Explain how to design an effective organization structure?
5. What is job design? Discuss the methods and factors affecting job design?
6. What is job enrichment? Discuss steps for implementing job enrichment programme?
7. What is job evaluation? Discuss the process and methods of job evaluation?
8. What is merit rating and discuss its limitations?

■■■

CHAPTER 4

Motivation and Productivity

4.1 INTRODUCTION

The term 'motivation' has been derived form the word 'motive'. Motive is anything that initiates or sustains (maintain) activity. It is an inner state that energises, activates/ movers and that directs/channels behavior toward goals. Motive is the psychological force with in an individual that sets him in motion for the achievement of certain goals/satisfaction of certain needs.

According to Likert, "Motivation is the core of management". It is an important function which every manager performs for enthusing people to work for the accompalishment of organizational objectives.

4.2 CONCEPT AND NATURE OF MOTIVATION

Motivation is the work a manager performs to inspire and encourage people to take required action. According to Scott, "Motivation is a process of stimulating people to action to accompalish desired goals".

Motivation is a personal and internal feeling. This feeling arises from needs and wants. As we all know that, human needs are unlimited. Fulfilment of one set of needs give rise to other needs. Therefore, motivation is a continuous process. The success of an organization ultimately depends on how effectively managers are able to motivate their subordinates.

According to Robert Dubin, "Motivation is complex of forces starting and keeping a person at work in an organization. Motivation is something that moves the person to action, and continues him in the course of action already initiated".

The role of motivation is to develop and intensify the desire in every member of the organization to work effectively and effeciently in his position.

In the words of Dalton E. McFarland, "Motivation refers to the way in which urges, drives, desires, aspirations, strivings or needs direct, control or explain the behaviour of human beings".

Motivation is also the result of interaction of the individual and the situation. Certainly, individuals differ in their basic motivational drive. But the same student who finds it difficult to read a textbook for more than 25 minutes may finish a Harry Potter book in one afternoon. For this student, the change in motivation is driven by the situation. We should always keep in mind that the level of motivation varies both between individuals and within individuals at different times.

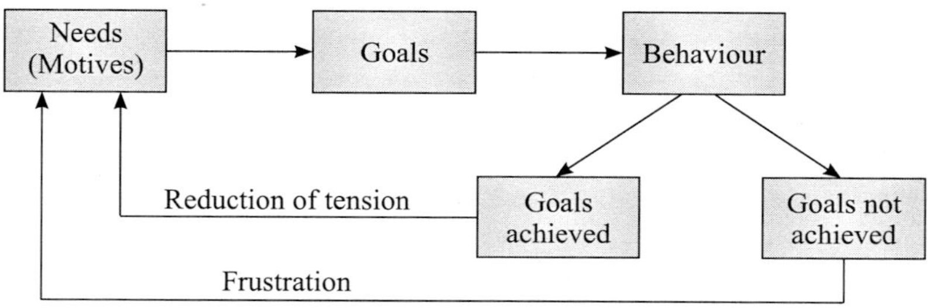

Fig. 4.1 A simple model of motivation

As in Fig. 4.1, the unsatisfied needs of a person are the starting point in the motivation process. The unsatisfied need results in tension within the individual and motivates him to search for the ways to relieve this tension. He develops certain goals for himself, if he is successful in achieving his goals, certain other needs will emerge which will lead to setting a new goal. But if the goal is not achieved, the individual will engage himself in either constructive or distructive behaviour. This process never stops. It keeps on working within an individual.

4.3 NATURE OF MOTIVATION

1. **Internal Feeling:** Motivation is an internal feeling of an individual. It cannot be observed directly, we can observe an individual's actions and then interpret his behaviour in terms of underlying motives. This leaves a wide margin of error. Our interpretation maynot reveal the individual's true motivation.
2. **Continuous Process:** Motivation is a continuous process that produces goal directed behaviour. The individual tries to find alternatives to satisfy his needs.
3. **Complex Process:** Motivation is a complex process. Individuals may differ in their motivation even though they are performing the same type of job. For example—if two men are engaged in cutting stones for constructing a temple, one may be motivated by the amount of wages he gets and the other by the satisfaction he gets by performing his job.
4. **Dynamic Process:** Needs and motives of an individual changes from time to time, even though he may continue to behave in the same way. For example—a temporary worker may produce more in the beginning to become permanent. When made permanent, he may continue to produce more, this time to get promotion.

5. **Different from Satisfaction:** Motivation is different from satisfaction. Motivation implies a drive (conclusion) towards an outcome while satisfaction involves outcomes already experienced. Satisfaction is the contentment experienced when a want has been satisfied.

4.4 OBJECTIVES AND IMPORTANCE OF MOTIVATION

Rensis Likert has called motivation as the "core of management". Motivation is an important function which every manager performs for actuating (influencing) the people to work for the accomplishment of objectives of the organization. Issuance of well—conceived instructions and orders doesnot mean that they will be followed. A manager has to make appropriate use of various techniques of motivation to enthuse the employees to follow them. Effective motivation succeeds not only in having an order accepted but also in gaining a determination to see that it is executed efficiently and effectively.

Motivation is an effective instrument in the hands of manager for inspiring the workforce and creating a confidence in it. By motivating the workforce, management creates, "will to work", which is necessary for the achievement of organizational goals.

Motivation involves getting the members of the group to pull weight effectively, to give their loyalty to the group, to carry out properly the purpose of the organization.

The following results may be expected if the employees are properly motivated:

1. The workforce will be better satisfied if management provides them with opportunities to fulfil their psychological and physiological needs. The workers will cooperate voluntarily with the management and will contribute their maximum towards the goals of the enterprise.
2. Workers will tend to be as efficient as possible by improving upon the skills and knowledge so that they are able to contribute to the progress of the organization. This will also result in increased productivity.
3. The rates of labour turnover and absenteeism among the workers will be low.
4. There will be good human relations in the organization as friction among the workers themselves and between the workers and management will decrease.
5. The number of complaints and grievances will come down. Accident rate will also be low.
6. There will be increase in the quality and quantity of products. Wastage and scrap will be less. Better quality of products will also increase the public image of the business.

4.5 THORIES OF MOTIVATION

Approaches understanding motivation differ because many individual theorists have developed their own views and theories of motivation. They approach motivation from different starting points, with different ideas in mind and from different backgrounds. Some important theories are described below:

4.5.1 Maslow's Need Hierarchy Model

A.H. Maslow developed a conceptual framework for understanding human motivation which has been widely aclaimed. He defined a person's effectiveness as a function of matching man's opportunity with the appropriate position of hierarchy of needs. Process of motivation begins with an assumption that behavior, atleast in part, is directed towards the achievement of satisfaction of needs. Maslow proposed that human needs can be arranged in a particular order from the lower to the higher as shown in the Fig. 4.2. The need hierarchy is as follows:

1. Basic physiological needs.
2. Safety and security needs.

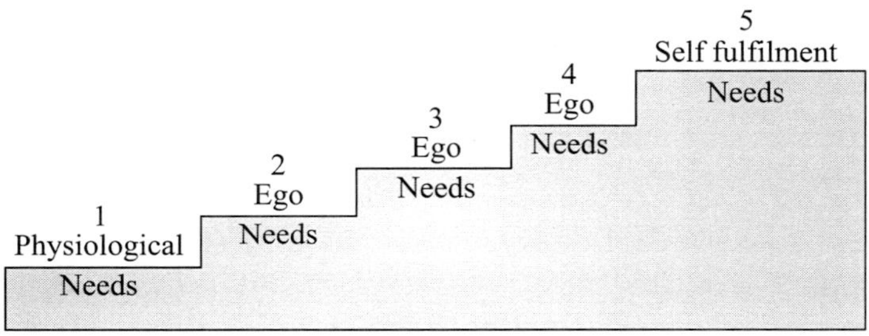

Fig. 4.2 Maslow's need hierarchy

3. Social needs.
4. Esteem and Status needs.
5. Self-fulfilment needs.

1. **Basic Physiological Needs:** The needs that are taken as the starting point for motivation theory are the so-called physiological needs. These needs relate to the survival and maintenance of human life. They include such things as food, clothing, shelter, air, water and other necessities of life.
2. **Safety and Security Needs:** After satisfying the physiological needs, people want the assurance of maintaining a given economic level. They want job security, personal bodily security, security of source of income, provision for old age, insurance against risks etc.
3. **Social Needs:** Man is social being. He is therefore, interested in conversation, sociability, exchange of feelings and grievances, companionship, recognition, belongingness etc.
4. **Esteem and Status Needs:** These needs embrace such things as self-confidence, independence, achievement, competence, knowledge and success. They are also known as egoistic needs. They are concerned with prestige, and status of the individual.
5. **Self-Fulfilment Needs:** The final step under the need priority model is the need for self-fulfilment or the need to fulfil what a person considers to be his mission in life. It involves realising one's potentialities for continued self—development and for being creative in the broadest sense of the word. After his other needs are fulfilled, a man has the desire for personal achievement. He wants to do something which is challenging and since this challenge gives him enough dash and initiative to work, it is beneficial to

him in particular and to the society in general. The sense of achievement gives him psychological satisfaction.

Maslow proposed that the needs have a definite sequence of domination. Second need doesnot dominate until the first need is reasonably satisfied and the third need doesnot dominate until the first two needs have been reasonably satisfied and so on.

The other side of the need hierarchy is that man is wanting animal, he continues to want something or the other. He is never fully satisfied. If one need is satisfied, the other need arises. As said above (according to Maslow), needs arise in a certain order of preference and not randomly. Thus, if one's lower level needs (physiological and security needs) are unsatified, he can be motivated only by satisfying his lower level needs and not satisfying his higher level needs. Another point to note is that once a need or a certain order of needs are satisfied, it ceases to be a motivating factor. Man lives for bread alone as long as it is not available. In the absence of air one cannot live, it is plenty of air which ceases to be motivating.

Appraisal of Maslow's Need Hierarchy—Maslow's need hierarchy has application to the lower level workers in a country like India where the basic needs of the workers are not satisfied. It points out that people are motivated by unfulfilled needs. It also postulates that as lower-level needs are fulfilled, upper-level needs replace them. Thus, Maslow's theory seem to be a very simple description of the complex process of motivation of human beings.

The limitations of the Maslow's theory are as under:

1. Every individual may have a different need hierarchy which maynot follow the sequence suggested by Maslow. For instance, an individual may have social or egoistic needs even though his safety need is not satisfied as yet.
2. It is wrong to presume that only one need is satisfied at one time. Man's behaviour at any time is mostly guided by multiplicity of motives. However, while others may be of secondary importance. Moreover, at different levels of needs, the motivation will be different.
3. Money can act as a motivator only for physiological and social needs, not for satisfying higher level needs. Employees are enthusiastically motivated by what they are seeking, more than by what they already have. They may react cautiously in order to keep what they already have, but they move forward with enthusiasm when they are seeking something else. In other words, man works for bread alone as it is not available.
4. There are always some people for whom the need for self-esteem seems to be more prominent than that of love. There are also creative people in whom the drive for creativeness seems to be more important. In certain people, level of operation may be permanently lower. For example, a person who has experienced chronic unemployment may continue to be satisfied for the rest of his life if only he can get enough food and clothes. Another cause of reversal of need hierarchy is that when a need has been satisfied for a long time, it may be undervalued.
5. It is doubtful that once a need is satisfied it loses its motivating force. It is also doubtful that satisfaction of one need automatically activates the next need in the hierarchy. Some persons will not aspire after their lower-order needs have been satisfied. Human behaviour is the outcome of several needs

acting simultaneously. The same need maynot lead to the same response in all individuals.

Needs are not the only determinant of human behaviour. People may engage in behaviour that are in no way concerned with the satisfaction of their needs. In practice, behaviour is influenced by needs, expectations, perception etc. It is also influenced by the cultural background of people.

4.5.2 Herzberg's Motivation—Hygiene Model

A significant development in motivation was distinction between motivational and maintenance factors in job situation. A research was conducted by Herzberg and his associates based on the interview of 200 engineers and accountants who worked for eleven different firms in Pittsburgh area.

These men were asked to recall specific incidents in their experience which made them feel either particularly good or particularly bad about jobs. The findings of the research was that good feelings in the group under test were keyed to the specific tasks that the men performed rather than to background factors such as money, security/ working condition and when they felt bad, it was because of some disturbance in these background factors which had caused them to believe that they were being treated unfairly. This led to draw a distinction between what are called as 'motivators' and 'hygiene factors'. To this group of engineers and accountants, the real motivators were opportunities to become more expert and to handle more demanding assignments. Hygiene factors served to prevent loss of money and efficiency. Thus, hygience factors provide no motivation to the employees, but the absense of these factors serves as dissatisfiers.

Some job conditions operate primarily to dissatisfy employees when they are absent, but their present doesnot motivate employees in a strong way. Many of these factors are traditionally perceived by management as motivators, but the factors are really more powerful as dissatisfiers. They are called maintenance factors in job because they are necessary to maintain a reasonable level of satisfaction among the employees.

Their absence proves to be a strong dissatisfier. They are also known as dissatisfiers or hygiene factors because they support employees mental health. Another set of job conditions operates primarily to build strong motivation and high job satisfaction among the employees. These conditions are motivational factors.

Table 4.1 Maintenance and Motivational Factors

S.No.	*Maintenance/Hygiene Factors*	*Motivational Factors*
1.	Company policy and administration	Achievement
2.	Technical supervision	Recognition
3.	Inter personal relations with superior	Advancement
4.	Inerpersonal relations with peers	Work itself
5.	Interpersonal relations with subordinates	Possibility of growth
6.	Salary	Responsibility
7.	Job security	
8.	Personal life	
9.	Working conditions	
10.	Status	

Hygiene factors include such things like wages, fringe benefits, physical conditions and over all company policy and administration. The presence of these factors at a satisfactory level prevents job dissatisfaction, but they donot provide motivation to the employees. So, they are not considered as motivational factors.

Motivational factors, on the other hand, are essential for increasing the productivity of the employees. They are also known as satisfiers and include such factors as recognition, feeling of accompalishment and achievement, opportunity of advancement and potential for personal growth, responsibility and sense of job and individual importance, new experience and challenging work etc.

Criticism of Herzberg's theory—Herzberg theory has been criticised on the following grounds:

1. Herzberg drew conclusions from a limited study including engineers and accountants. Engineers, accountants and other professionals may like responsibility and challenging jobs. But the general body of workers are motivated by pay and other financial benefits.
2. In the Herzberg's study, the interviewees were asked to report exceptionally good/exceptionally bad moments. This methodology is defective because there is a common bias among human beings to take more credit for good things and put the blame on others for bad things.
3. Herzberg gave too much emphasis on job enrichment (challenging and interesting work). But job enrichment is not the only answer. Off—the job satisfaction of the workers is also very important. Herzberg did not attach much importance to pay, status or interpersonal relationships which are generally held as important contents of satisfaction.
4. The distinction between maintenance and motivating factors is not definite. What is a maintenance factor (*e.g.*, pay) for a worker in the developed country like United States may be a motivator for a worker in a developing country like India. Thus, Herzberg ignored the dominating influence of situational variables.

Comparison of Herzberg and Maslow Model:

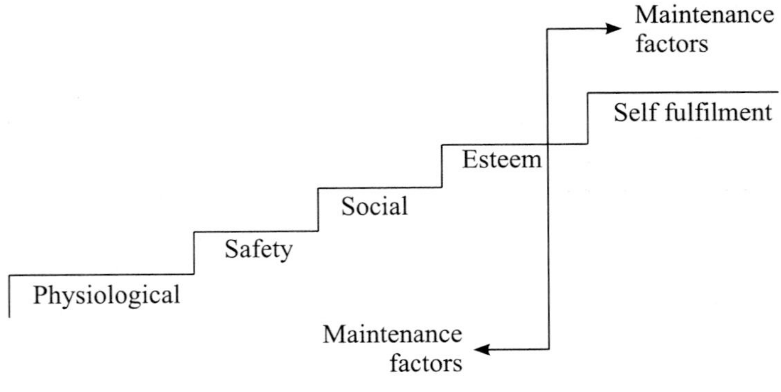

Fig. 4.3 Relationship between Maslow's and Herzberg's models

There appears to be a great similarity between Herzberg's and Maslow's model. A close examination of Herzberg's model indicates that when he actually says is that some employees may have achieved a level of social and economic progress in the society and for them higher level needs of Maslow (esteem and self-actualisation) are the primary motivators. However, they still must satisfy the lower level needs for the maintenance of their current state.

Thus, we can say that money might still be a motivator for operative employees and for some managerial employees. Herzberg's model adds to the Maslow's need hierarchy model because it draws a distinction between the two groups of factors, that is motivational and maintenance, and points out that the motivational factors are often derived from the job itself. Most of the maintenance factors come under comparatively lower order needs. In economically advanced countries, such needs of the employees are fulfilled and hence cease to be motivators.

As shown in Fig. 4.4, Maslow's physiological, security and social need come under Herzberg's maintenance factors where as self-fulfilment under motivating factors. It may further be noted that a part of esteem need comes under maintenance factors and another part under motivatonal factors. The esteem needs are divided because there are some distinct differences between status perse and recognition. Status tends to be a function of the position one occupies. This position may be gained through family ties or social pressures and so this maynot be a reflection of personal achievement or earned recognition. Recognition is gained through competence and achievement. It is earned and granted by others. That is why, status is classified with physiological, safety and social needs as a hygiene factor, while recognition is classified with esteem as a motivator.

S. No.	*Maslow*	*Herzberg*
1.	Needs are arranged in sequential order.	There is no hierarchy.
2.	It is descriptive theory.	It is prespective theory.
3.	Unsatisfied need energize behaviour, this behaviour causes performance.	Gratified needs cause performance.
4.	Any need can be a motivator if it is relatively unsatisifed.	Only higher order needs server as motivators.
5.	Takes a general view of the motivational problems of all workers.	Takes a microview and deals with work oriented motivational problems of higher level professional employees.
6.	Need hierarchy theory is based on intuition and not on any emperical study.	Motivation-hygiene theory is based on study of empirical data collected from professional accountants and engineers.

4.5.3 McGregor's 'Theory X' and 'Theory Y'

Douglas McGregor proposed two distinct sets of assumptions about what motivates people—one basically negative, labelled Theory X and other basically positive, labelled Theory Y.

Theory X: Theory X indicates the traditional approach to managerial motivation and control. It represents old stereotyped (unchanging and monotonous) and authoritarian management style of motivation. The underlying assumptions of the theory are as follows:

1. The average human being is basically lazy and has an inherent dislike to work. He will avoid work, if he can.
2. Most people lack ambition. They are not interested in achievement. They like to be directed.
3. Most people have little capacity for creativity in solving organizational problems.
4. Most people are indifferent to the organizational goals.
5. Most people must be closely controlled and often threatened to achieve organizational goals.
6. Motivation of average human beings occur at the physiological (food, clothing, shelter etc.) and safety levels.

These negative assumptions of human behaviour underlie the traditional mechanisation of people and process. The world is supposed to be full of unskilled workers, peons and messengers and to manage them is largely a matter of vigilance and strict supervision. Management merely thinks of catering to their physical and safety needs with some fringe benefits, keeping the implied threat of punishment handy in case of need.

Carrot and Stick approach: The carrot and stick approach of motivation comes from the old story that the best way to make a donkey move is to put a carrot in front of him or jab him with the stick from behind. The carrot is the reward for moving and the stick is the punishment for not moving—the carrot and stick approach of motivation takes the same view. In motivating people for behaviour that is desirable, some carrots, rewards are used such as money, promotion and other financial and non-financial factors; some sticks, punishments, are used to push the people for desired behaviour or to refrain from undesired behaviour.

Thus, carrot and stick approach to motivation is followed. Theory X suggests that threats of punishment and strict control are the ways to manage people. It was practised during the days the scientific management approach gained prominence and human beings were treated like machines.

The above assumptions have been challenged by the human rationalists because employees are treated merely as a commodity or passive factor of production.

McGregor questioned the assumptions of Theory X which followed carrot and stick approach to motivate the people and suggested autocratic style of leadership. He felt that management by direction and control is a questionable method for motivating such people whose physiological and safety needs have been satisfied and whose social, esteem and self-actualisation needs are important. In his own words, "The 'carrot and stick' theory of motivation which goes along with Theory X works reasonably under certain circumstances". The means for satisfying man's physiological and (within limits) safety needs can be provided or with held by management. Employment itself is such a means, and so are wages, working conditions and benefits. By these means, the individual can be controlled so long as he is struggling for subsistence. Men tends to live for bread alone when there is little bread. But the 'carrot and stick' theory doesnot work at all, once the man has reached an adequate subsistence level and after that he is motivated primarily by the higher level needs.

Theory Y: After challenging the validity of Theory X, McGregor developed an alternative theory of human behaviour which is known as Theory Y. This theory assume that people are not unreliable and lazy by nature. If they are properly motivated, they could really be creative.

The main task of the management is to unleash the potential in the employees. An employee who is properly motivated can achieve his goals by directing his own efforts and thus, he can help in accompalishing the organizational goals.

The assumptions of McGregor's Theory Y are as follows:

1. Work is as natural as play, if the conditions are favourable. The average person doesnot inherently dislike work.
2. External control and threat of punishment are not the only means for bringing about efforts towards organizational objectives. The average human being will exercise self-direction and self-control in the service of objectives to which he is committed.
3. Committment to objectives is a function of rewards associated with their achievement. The most significant of such reward, for example—the satisfaction of ego and self-actualisation needs can be direct products of efforts towards organization objectives.
4. The average human being learns under proper conditions, not only to accept but also to seek responsibility. Avoidance of responsibility, lack of ambition and emphasis on security are generally the consequences of exercise, not inherent in human characteristics.
5. The capacity to exercise a relatively high degree of imagination, ingenuity and creativity in the solution of organizational problems is widely, not narrowly, distributed in the population.
6. The intellectual potentialities of the average human being are the only partially utilised under the conditions of modern industrial life.

Theory Y assumes that goals of the organization and those of the individuals are not necessarily incongruent. The basic problem in most of the organization is that of securing commitment of the workers to the organizational goals. Workers commitment is directly related to the satisfaction of their needs. Thus, this theory places emphasis on satisfaction of the needs of the workers. It doesnot rely heavily on the use of authority as an instrument of command and control. It assumes that workers exercise self-direction and self-control in the direction of the goals to which they feel themselves committed. Because of these reasons, 'Theory Y' is realistic and frequently used at different levels in most of the organizations.

In support of the assumptions emboided in 'Theory Y', McGregor cited a few practices where in the subordinates are given a freedom to direct their activities, to assume responsibility and importantly, to satisfy their egoistic needs. These practices include decentralisation, and delegation, job enlargement, participation and consultative management, management by objective.

S. No.	Theory X	Theory Y
1.	Based on the assumption that people are basically lazy and so shirk work.	Based on the assumption that people like work as children like play.
2.	People dont take initiative. They like to be directed.	People like to take initiative. They seek self-direction and self-control.
3.	People avoid responsibility whenever possible.	People assume responsibility gladly if conditions are favourable.
4.	For getting things done, people must be supervised strictly.	People donot require close and strict supervision for properly performing their jobs.
5.	Autocratic style of leadership is likely to be more effective.	Democratic or participative style of leadership is likely to be more effective.
6.	Applicable to illiterate, unskilled and lower-level workers.	Applicable to educated and skilled employees who occupy higher position in the organization.
7.	People lack self-motivation.	People are self-motivated. Believe in
8.	Belive in mental sickness and so negative motivation of employees.	mental health and so positive and intrinsic motivation of employees.

Appraisal of McGregor's Contribution: Douglas McGregor's Theory X and Theory Y represent extremes to draw the fencing within which the organizational man is seen to behave. No man would belong completely to either Theory X or Theory Y. He possesses the traits of both in varying degrees under different situations. Thus, these theories are important tools in understanding the behaviour of human beings and in designing the incentive schemes to motivate the employees. Neither of the two theories is fully applicable to all situations and to all types of human beings.

However, Theory X is more applicable to unskilled and uneducated employees where as theory Y is more applicable to skilled and well educated employees who are mature enough and understand the resposibility.

Therefore, management should use an amalgamation of both the theories to motivate the different kind of employees at different level in the organisation.

How to Motivate Employees?

We have gone through various motivation theories in detail in this chapter. If you are a manager concerned with motivating your employees, how do you apply these theories?

The following suggestions may help you in solving the puzzle to some extent:

1. **Recognise Individual Differences:** Employess are not homogeneous. They have different needs. Employees also differ in terms of attitudes, personalities and other variables. So, recognise these differences and handle the motivational issues carefully.
2. **Match People to Job:** People with high growth needs perform better on challenging jobs. Achievers will do best when the job provides opportunities to participatively set goals and when there is autonomy and feedback. At the same time, not everybody is motivated in jobs with increased autonomy, variety and responsibility. When the right job is given to the right person, the organisation benefits in various ways.

3. **Use Goals:** The employee knows what he is doing by providing specific goals. Also, we should let people know what we expect from them. Make them understand that they can achieve the goals in a smooth way. If you expect resistance to goals, invite people to participate in the goal setting process.
4. **Individual Rewards:** Use rewards selectively, keeping the individual requirement in mind. Some employees have different needs; what act as a motivator for one may not work for another. So, rewards such as pay, promotion, autonomy, challenging jobs, participative management must be used keeping the mental make-up of the employee in view.
5. **Link Rewards to Performance:** Make rewards contingent on performance. To reward factors other than performance (favouritism, regionalism, yes—sir culture, etc.) will only act to strengthen those other factors. Employee should be rewarded immediately after achieving the goals. At the same time, managers should look for ways to increase the visibility of rewards. Publicise the award of performance bonus, lumpsum payments for showing excellence, discussing reward structure with people openly these will go a long way in increasing the awareness of people regarding the reward-performance linkage.
6. **Check the System for Equity:** The inputs for each job in the form of experience, abilities, effort, special skills must be weighed carefully before arriving at the compensation package for employees. Employees must see equity (justice) between the rewards or recognition obtained from the organisation and the efforts put in by them.
7. **Don't Ignore Money:** Money is a major reason why most people work. Money is not only a means of satisfying the economic needs, but also a measure of one's power, prestige, independence, happiness and so on. Money can buy many things. It can satisfy biological needs (food, shelter, sex, recreation, etc.) as well as security, social and esteem needs.

4.6 CONCEPT OF LEADERSHIP

"Leader" is one who guides and directs other people. He interprets the objectives of the group and guides it towards their accomplishment. Because he has others subordinate to him and subject to his command, the leader must give their efforts a direction and purpose. Managers at all levels must perform the function of leadership to lead the subordinates towards a definite direction.

In fact, leadership is an essential element of managing for influencing people to achieve mutually compatible goals. That is, leadership is the process of influencing the behaviour of others to work willingly and enthusiastically for achieving predetermined goals. It is a psychological process of influencing followers or subordinates and providing guidance to them. It is always related to a situation which means leader may be effective in one situation while ineffective in others. To be effective, a leader should change his leadership style depending upon the requirement of the situation.

4.6.1 Meaning of Leadership

It is the quality of the behaviour of a person whereby he is able to persuade others to seek group goals enthusiastically. It is the influence relationship in which one person (the leader) influences the behaviour of others (the led or followers) in a given situation to work together on related tasks to attain what the former desires. Leadership is that part of a manager's job by which he influences the behaviour of his subordinates towards the desired goals.

"Leadership is the process by which an executive imaginatively directs, guides and influences the work of others in choosing and attaining specified goals by mediating between the individuals and the organization in such a manner that both will attain maximum satisfaction". — **Theo Haimann**

Leadership is the process of influencing the behaviour and work of others in group efforts towards the realization of specified goals in a given situation. In other words, leadership is a function of three factors:

1. The leader
2. The led (follower)
3. The situation.

Managerial Leadership is the process of influencing a group of subordinates to attain organizational objectives. The interaction between the leader and his followers is based on interpersonal relations.

4.6.2 Features of Leadership

Certain features of leadership are as follows:

1. Leadership is a continuous process of behaviour, it is not one-shot activity.
2. It may be seen in terms of relationship between a leader and his followers (individuals and or groups) which arises out of their functioning for common goals.
3. By exercising his leadership, the leader tries to influence the behaviour of individuals or group of individuals around him to achieve common goals.
4. The followers work willingly and enthusiastically to achieve those goals. Thus, there is no coercive force which induces the followers to work.
5. Leadership gives an experience of help to the followers to attain common goals. It happens when the leader feels the importance of individuals, give them recognition and conveys them about the importance of activities performed by them.
6. Leadership is exercised in a particular situation, at a given point of time, and under specific set of circumstances. It implies that leadership styles may be different under different situations.

4.7 VARIOUS APPROACHES TO LEADERSHIP STYLE

Now, let us discuss few approaches of leadership as shown in Fig. 4.4.

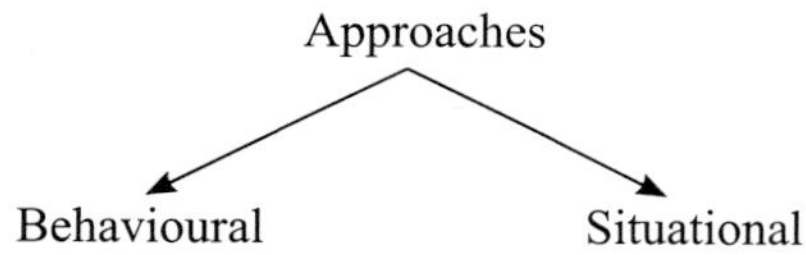

Fig. 4.4

4.7.1 Behavioural Approaches to Leadership Style

The total pattern of explicit and implicit leaders actions as seen by employees is called leadership style. The behavioural pattern which a leader exhibits is known as his style of leadership.

Let us first take an overview on style of leadership. Different leadership style exist among leaders in different times and in different situations. The leadership style in particular is determined by the leader's personality, experience and value system, nature of followers and nature of environment. There are three important leadership styles which are based on the use of authority.

These are:

(*i*) Autocratic Leadership

(*ii*) Participative leadership

(*iii*) Free rein leadership

(*i*) **Autocratic Leadership:** It is also known as "authoritarian or directive" style of leadership. The autocratic leader gives orders which must be obeyed by the subordinates. He centralizes decision-making power in himself. He takes decision for the group without consulting the group members and simply tells the group what the members have to do. The autocratic leader gives personal praise or criticism to each member on his own initiative and remain aloof from the group for most of the time.

There are further three categories of autocratic leaders as follows:

(*a*) ***Strict Autocrat*:** He relies on negative influences and give orders which the subordinate must accept. He follows negative motivational style to get the work done. Negative motivation include imposing penalty, criticizing subordinates and so on.

(*b*) ***Benevolent Autocrat*:** Benevolent means kind. An autocratic leader may also follow positive style of using his power to disburse (to pay money) rewards to the subordinates. A benevolent autocrat is effective in getting higher productivity in many situations and developing effective human relationships.

(*c*) ***Manipulating Autocrat*:** Such a leader makes the subordinates feel that they are participative in decision making, but he takes all the decisions himself.

Autocratic leader have proved to be successful in many cases. In particular, subordinates who depend upon the boss and do not want to take any initiative get satisfaction from this style.Autocratic style has also proved successful in cases when there is a need of quick decisions. Autocratic style is not liked by the people who are enlightened and want to participate in decision making. Autocratic style of leadership may endanger the organisational efficiency. Another drawback is that it does not help develop future leaders in the organisation. Autocratic leader is shown in Fig. 4.5.

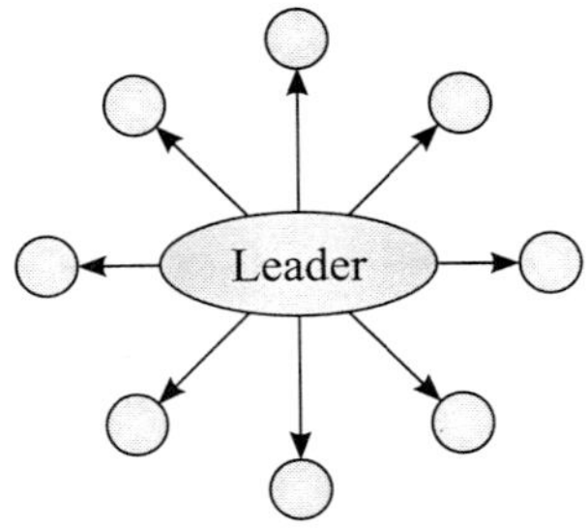

Fig. 4.5 Autocratic leader

(*ii*) **Laissez Faire/Free Rein Leader:** A free rein leader doesnot lead, but leaves the group entirely to itself. He is represented by the chairman of the board who leaves all responsibility for most of the work to his subordinates. The free rein manager avoids using power. He depends largely upon the group to establish its own goals and work out its own problems. Group members work themselves and provide their own motivation. The manager serves as a contact man with the outsiders to bring for his group the information and resources it needs for the accomplishment of its goals.

This type of leadership is evident in research laboratories (for example) where the scientists are fairly free to conduct their research and make their decision. Similarly, in the college, the principal doesnot interfere in the faculty teaching methods, but only assigns the courses to be taught. From then onwards, the faculty members are free to decide about the method of teaching, books to be recommended and various teaching aids to be used.

The possible advantage of free-rein style of leadership are as under:

(*a*) It creates an environment of freedom, individuality as well as team spirit.

(*b*) With a free and informal work environment, it is highly conducive to create work.

(*c*) It is very suitable where people are highly motivated and achievement oriented.

Laissez Faire leader is shown in Fig. 4.6.

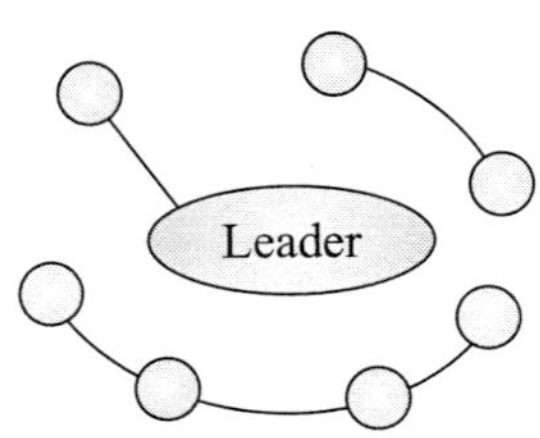

Fig. 4.6 Free-rein leader

(*iii*) **Participative or Democratic Leaders:** A democratic leader is one who gives order after consulting the group. He sees to it that policy are worked out in group discussions and with the acceptance of the group. He never asks people to do things without making the long-term plans on which they are working and makes it clear that praise or blame will be shared by all the group members.

His decisions are not unilateral because they arise from consultation and participation by the followers. Unlike an autocratic manager who controls through the official authority, a participative manager exercises control mostly by using forces within the group. It increases the acceptance of management's ideas and reduces resistance to change. It increases their morale. It also leads to reduction in the number of grievances of the workers.

It may lead to the following advantages:

(*a*) Active participation in decision-making by the employees ensures greater productivity and satisfaction.

(*b*) Workers develop a greater sense of self esteem, due to importance given to their ideas and their contribution.

(*c*) Workers become more committed to changes that may be brought about by policy changes, since they themselves participated in bringing about these changes.

(*d*) The leadership induces confidence, cooperation and loyalty among the employees.

(*e*) The morale of the employees is increased.

Figure 4.7 shows participative leader.

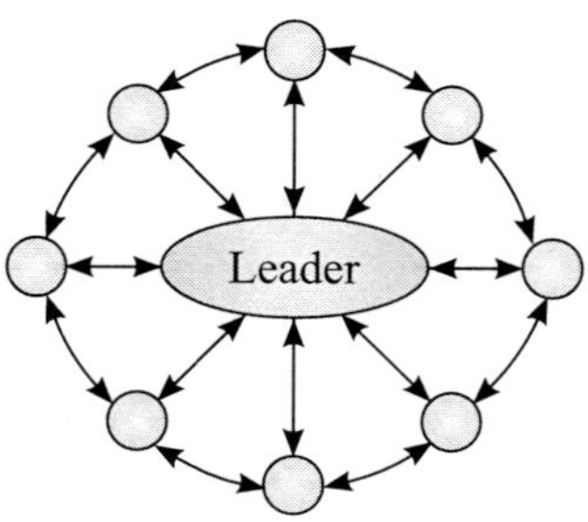

Fig. 4.7 Participative leader

Now, comes behavioural approaches to leadership style.

The behavioural pattern which the leadership exhibits is called leadership style. It represents a consistent combination of philosophy, skills, traits and attitudes that are exhibited in a person's behaviour.

Now, let us discuss variety of styles that differ on the basis of motivation, power or orientation towards task and people.

1. Positive Leader and Negative Leader

Leaders approach people to motivate them in many ways. If the approach emphasizes reward-economic or otherwise the leader uses positive leadership. Better employee education, greater demand for independence and other factors have made satisfactory employee motivation more dependent on positive leadership.

If the emphasis is placed on the threat, fear, harshness and penalties, the leader is applying negative leadership. This approach can get acceptable short term performance in many situations, but it has high human costs.

Negative leaders act domineering and superior with people. To get the work done, they hold over their personnel such penalties as loss of job, reprimand in the presence of others and a few days off without a pay. They display authority in the false belief that it frightens everyone into productivity. They are bosses more than a leader.

2. Autocratic, Consultive and Participative Leaders

Consultive leaders approach one or more employees and ask them for inputs prior to make a decision. These leaders may then choose to use or ignore the information and advice received, however. If the inputs are seen as used, employees are likely to feel as though they had a positive impact, if the inputs are consistently rejected, employees are likely to feel that their time has been wasted. Autocratic and participative leaders have been explained in 4.5.1.

3. Employee Orientation and Task Orientation

Two different leadership styles used with the employees are consideration and structure, are also known as employee orientation and task orientation.

Considerate leaders are concerned about the human needs of their employees. They try to build teamwork, provide psychological support and help employees with their personal problems.

Structured, task oriented leaders, on the other hand, believe that they get results by keeping people constantly busy, ignoring personal issues and emotions and crying them to produce.

The most successful managers are those who combine relatively high consideration and structure, giving somewhat more emphasis to consideration.

In several types of environment such as truck manufacturing, railroad construction and insurance offices, the strongly considerate leader was shown to have achieved somewhat higher job satisfaction and productivity. Subsequent studies confirm this general tendency and report desirable side effects such as lower turnover, and reduced stress within the group. Conversely, turnover, stress and other problems seemed likely to occur if a manager was unable to demonstrate considerations.

4.8 BLAKE AND MOUTON'S MANAGERIAL GRID

Robert R. Blake and Jane S. Mouton developed the managerial grid as a tool for identifying a manager's own style. The grid is based on the leadership style dimensions of concern for people and concern for production, which essentially mirrors the dimensions of consideration and structure. The grid clarifies, on two 9-point scales, how the two dimensions are related. It also establishes a uniform language and framework for communication about appropriate leadership style. It is shown in Fig. 4.8.

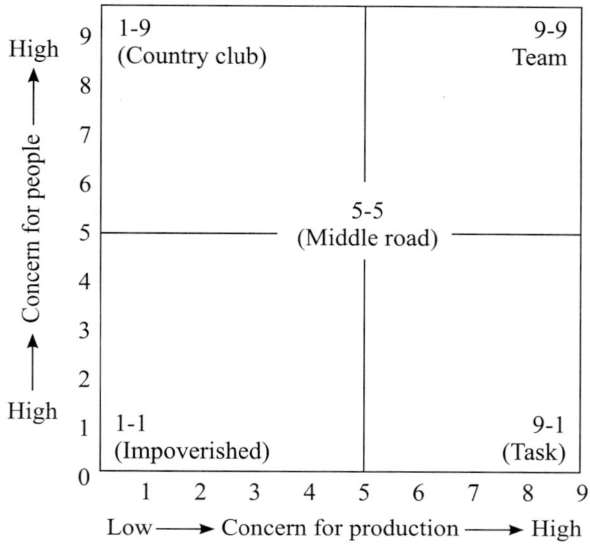

Fig. 4.8 Managerial grid

1. **The 1, 9 Managerial Grid (Country Club):** The 1, 9 managerial grid is also known as country club which is defined as an association of people. The 1, 9 leaders are high in concern for people but so low in concern for production that output is typically low. They are country club leaders. Since 1, 9 managers believe that attitudes and feelings are of atmost importance, their emphasis is on promoting good feelings. The 1, 9 manager may be highly motivated to work hard, but that motivation is based on a need for approval and acceptance rather than on a commitment to the production goals of the organisation. That is, minimum concern for production coupled with maximum concern for people.
2. **The 9, 1 Managerial Style (Task):** The 9, 1 leaders are overly concerned with production to the exclusion of their employees need. The 9, 1 leaders tend to be authoritarian bosses. Under this style, the leader has maximum concern for production (level 9) combined with the minimum concern for people (level 1). He attempts to maximize production by using authority and power to achieve control over subordinates. These leaders decide on work assignments, provide subroutinates with detailed instructions and supervise closely to make sure that their directives are properly carried out. Subordinates are expected to carry out orders with an unquestioning obedience. They are taken as merely a means for doing the task assigned to them. Little attention is given to their development or to communicate with them beyond the issuing of orders or instructions.
3. **The 1, 1 Managerial Style (Impoverished):** The 1, 1 leader does not place adequate emphasis on either dimension and would predictably fail. Such a manager exerts minimum influence on the contacts with the group members. He express little concern for the production or people. In a supervisory position, he is most likely to be found executing messenger carrier functions communicating orders from the layer above to the layer below.

He is an expert in passing on blame to others for failure in such a way that he set himself free from the responsibilities, yet he rarely initiates criticism spontaneously. His criticism is strictly in self-defence. He wants minimum involvement in the organisations purpose and with its people.

4. **The 5, 5 Managerial Style (Mid of the road):** A more desirable balance of the two dimensions is from 5, 5 to 9, 9 – with the latter assumed by Blake and Mouton to be the most effective style. This grid can help the individuals to identify not only their primary leadership style but also their backup style. The backup style is the one, managers tend to use when their normal style does not get results. In general, managers tend to be more autocratic and concerned with production when their primary style is unsuccessful. The 5, 5 style seeks to maintain a balance between the people dimension and the production dimension. A basic assumption of this style is that people work willingly and do as they are told to do if they have been explained the reasons for doing that work. However, just enough is communicated so that people have only a general sense of what is going on. If too much is told, it is feared that they might resist.

   Enough concern is shown for people so that adequate production may be achieved. This is seen in the 5, 5 approach to management development, communication and performance reviews. Meetings are held to listen to the suggestions of the subordinates and to create a sense of belonging.

5. **The 9, 9 Managerial Style (Team):** The 9, 9 style meets the "people" need to be involved and committed to work. A major difference between 9, 9 style and other managerial style is in goal setting and its use as a basic management approach to a large variety of problems. The 9, 9 orientation aims at integrating the people and production dimensions of work under conditions of high concern for growth. The key is the involvement and participation of those responsible for it in planning and execution of work. This brings about team spirit that leads to high performance.

Each of the 5 styles given by Blake and Mouton points the relative concern for production and people and implies that the most desirable leadership is 9, 9 that is maximum concern for both production and people.

It may be noted that the five positions emphasised in the managerial grid are rarely found in their pure form in actual life. That means manager may have a style of 8, 2 or 4, 6 or same other. Neverthless, managerial grid is widely used as a means of managerial training and identifying various combination of leadership styles.

4.8.1 Situational Theory of Leadership

Situational leadership approach was applied, for the first time, in 1920 in the armed force of Germany with the objective to get good generals under different situations.

The prime attention in situational theory of leadership (also known as contingency theory) is given to the situation in which leadership is exercised.

Contingency Approaches to Leadership Style

1. **Fiedler's Contingency Theory:** Fred Fiedler developed a contingency theory which described successful leadership as a direct result of matching

manager's style with situational variables. Fiedler grouped situational variables into three areas of concern:

(*i*) Leader-member relations, that is, the degree to which the followers accept their leader.

(*ii*) Task structure, or the degree to which the task is spelled out or routinised.

(*iii*) Position Power of the leader, that is, the degree of formal authority attributed to the leader's position.

These three variables determine whether a given situation is favourable or unfavourable to the leader. The favourableness of situation may be defined as "the degree to which the situation enables the leader to exert the influence over his group".

	Leader follower relatinos	**Task structure**	**Position power**	**Situational favourableness**
1	Good	High	Strong	High
2	Good	High	Weak	
3	Good	Low	Strong	
4	Good	Low	Weak	
5	Poor	High	Strong	
6	Poor	High	Weak	
7	Poor	Low	Strong	
8	Poor	Low	Weak	Low

Task-oriented style	Relationship-oriented considerate style	Task-oriened style
Favourable leadership situation	Situation intermediate in favourableness for leader	Unfavourable leadership situation

Fig. 4.9 Fiedler's situational combinations

Fiedler perceived eight possible combinations of the 3-situational variables as shown in Fig. 4.9. As a leadership situation varies from high to low on these variables, it falls into one of the 8 combinations/situations. The most favourable situation is one in which the leader follower relations are very good, the leader enjoys great position power and the task structure is well-defined. On the other hand, the most unfavourable situation signifies that the leader is disliked, has little position power and the task is unstructured.

Fiedler attempted to determine the most effective leadership style for each of the above 8 situations. He came to the conclusion that:

1. Task-oriented leaders tend to be most effective in situations that are either very favourable or very unfavourable to them.
2. Relation-oriented leaders tend to be most effective in situations that are intermediate in favourableness.

1. Implications of Fiedler's Model

This model is and will probably remain a rich source of leadership and its effectiveness.

From the view point of organisational implications, Fiedler's model offers two implications:

1. No leadership style is appropriate for all situations. Therefore, there is nothing like the best style. The managers can adopt the leadership style according to the requirements of the situations.
2. There is a need for matching leaders and job situations to achieve organisational effectiveness. Fiedler says that, "if we wish to increase organisational and group effectiveness, we must learn not only to train leaders more effectively but also how to build an organisational environment in which the leader can perform well".

2. Hersey Blanchard's Situational Model

Hersey Blanchard's situational leadership model takes a different perspective of situational variables as compared to Fiedler's model. Hersey and Blanchard feel that the leader has to match his leadership style according to the need of maturity of subordinate which moves in stage and has a cycle.

Therefore, this theory is also known as life-cycle theory of leadership. There are two basic considerations in this model, leadership styles and maturity of subordinates.

(*a*) **Leadership Style:** Leadership styles may be classified into four categories based on the combination of two considerations, relationship behaviour and task behaviour.

Relationship behaviour is determined by socio-emotional support provided by the leader. Task behaviour is seen in terms of the amount of guidance and direction provided by the leader. Combination of these two-dimension results in four leadership is shown in Fig. 4.10.

High	High relationship and low task	High relationship and high task
Low	Low relationship and low task	Low relationship and high task
	Low	High

Task behaviour

Fig. 4.10 Leadership behaviour (styles)

(*b*) **Subroutines Maturity:** Maturity in this model has been used in the context of ability and willingness of the people for directing their own behaviour. Ability refers to the knowledge and skills of an individual to do the job and is called job maturity.

Willingness refers to the psychological maturity and has much to do with confidence and commitment of the individual. These variables of maturity should be considered only in relation to a specific job to be performed. That is to say, an individual or a group is not mature or immature in any total sense.

All persons tend to be more or less mature in relation to a specific task, function, or objective that a leader is attempting to accomplish through their efforts. In addition to, assessing the level of maturity of individuals in the group, he has to assess the maturity level of the group as a whole, particularly if the group interacts frequently together in the same work area.

When both components of maturity, ability and willingness are combined, we can get four combinations:

(*a*) Low ability and low willingness – low maturity.

(*b*) Low ability and high willingness – low to moderate maturity

(*c*) High ability and low willingness – moderate to high maturity

(*d*) High ability and high willingness – high maturity

Combining leadership Styles and Maturity

If we combine leadership styles and maturity, we may arrive at the relationship between the two as shown in Fig. 4.11.

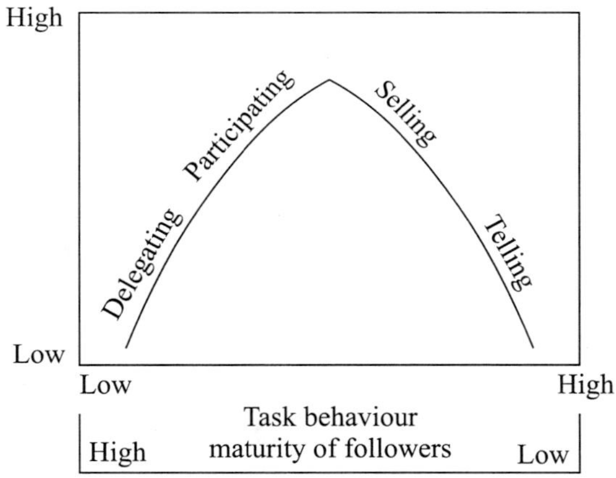

Fig. 4.11 Hersy-Blanchard's model of situational leadership styles

Thus, there are four leadership styles, each being appropriate to a specific level of maturity. The four leadership styles are:

(*a*) **Telling:** Where the subordinates have low maturity, that is neither they have ability nor they are willing to do, they require telling leadership style. It emphasis directive behaviour and involves high task behaviour and low relationship behaviour.

(*b*) **Selling:** For subordinates of moderate maturity who have high willingness but lack ability, selling leadership style is appropriate. The subordinates require both supportive and directing behavior which is marked by high task and high relationship behaviour.

(*c*) **Participating:** Subordinates with moderate to high maturity who have ability to do but lack willingness require high external motivating force. In such situation, participating leadership style with low task behaviour and high relationship behaviour is more appropriate.

(*d*) **Delegating:** Subordinates with high maturity that is, they have both ability and willingness to work, hardly require any leadership support. The most appropriate leadership style in such a situation is delegating which involves low task behaviour and low relationship behaviour.

In other words, situational theory emphasis that, "there is no one best style of leadership universally applicable to all situations" and that the leader has to change his style of leadership from situation to situation. If the leader adopts the same style under all situations, he maynot be successful.

For example, Winston Churchill was the most effective and successful Prime Minister of Britain during the period of 2nd World War, but he was flop afterwords when the situation changed.

Though this theory states leadership ability of an individual in a given situation and measurs his leadership potentialities, it is silent on the point whether this individual will fit in another situation.

The situational theory suffers from the following weaknesses:

(*i*) It gives much emphasis on situation aspect and overlooks the qualities needed in a successful leader.

(*ii*) Leadership is a subjective consideration in which qualities of head and heart of a leader play their part. But this theory overlooks this aspect.

4.9 COORDINATION

The basic function of coordination in an enterprise is the same as that of an orchestra conductor who directs the activities of the orchestra party in such a manner that it produces harmony in music. Likewise the coordinator of an enterprise also directs the activities of the group in such a manner that it brings harmonious and unified actions to achieve common purpose. Like the orchestra conductor, a manager also performs the function of securing and maintaining unity of direction throughout the organization.

Coordination has been viewed by different management experts in different ways. Henri Fayol considers coordination as *a function of Manager*. Louis A. Allen also regards coordination as one of the separate managerial functions. James D. Mooney considers coordination as *the first principle of the organization*. Ralph C. Davis looks upon coordination primarily as a *vital phase of controlling*.

George R. Terry and Theo Haimann consider coordination as a permeating function of management passing through the managerial functions of planning, organizing, staffing, directing and controlling. Thus, Terry and Theo Haimann are not in favour of regarding coordination as a separate function of management. According to them, coordination is a part of the managerial functions of planning, organizing, staffing, directing and controlling. It traverses the entire process of managing. In short, it is the *essence of management*.

4.9.1 Definition of Coordination

Coordination is the orderly synchronization of efforts of the subordinates to provide the proper amount, timing and quality of execution so that their united efforts lead to the stated objective. According to Terry, *"Coordination deals with the task of blending efforts in order to ensure successful attainment of objective. It is accomplished by means of planning, organizing, actuating and controlling."* Thus, Terry considers coordination as a permeating function of management passing through the managerial functions of the planning, organizing, directing and control.

In the words of Mooney and Railey, "*Coordination is the orderly arrangement of group efforts to provide unity of action in the pursuit of a common purpose.*" It is the effort to ensure a smooth interplay of the functions and forces of the different component parts of the organization to the end that its purpose will be realised with a minimum of friction and a maximum of collaborative effectiveness. In other words, coordination means the orderly synchronization of efforts of the people working in the organization or the achievement of organizational objectives.

4.9.2 Nature of Coordination

Coordination leads to blending the activities of different individuals and group of individuals for the achievement of certain objectives. The features of coordination are as follows:

(*a*) It is an essential managerial activity. It is needed at all levels of management.

(*b*) It involves an orderly arrangement of group efforts.

(*c*) It is a continuous process carried on by the managers.

(*d*) Its purpose to secure unity of action towards common objectives.

(*e*) It is the essence of managing. Each of the managerial function is an exercise in managing.

4.9.3 Coordination and Cooperation

Co-ordination is an orderly arrangement of efforts to provide unity of action in the fulfillment of common objective whereas co-operation denotes collective efforts of persons working in an enterprise voluntarily for the achievement of a particular purpose. It is the willingness of individuals to help each other. Co-ordination is an effort to integrate effectively energies of different groups whereas cooperation is sort to achieve general objectives of business.

Though these two are synonymous but they are different as below:

Differences between Co-ordination and Co-operation

Basis	*Coordination*	*Cooperation*
Meaning	It is an orderly arrangement of group efforts in pursuit of common goals.	It means mutual help willingly.
Scope	It means mutual help willingly.	It is termed as a part of coordination.
Process	The function of co-ordination is performed by top management.	The functions of co-operation are prepared by persons at any level.
Requirements	Co-ordination is required by Employees and departments at work irrespective of their work.	Co-operation is emotional in nature because it depends on the willingness of people working together.

Relationship	It establishes formal and informal relationships.	It establishes informal relationship.
Freedom	It is planned and entrusted by the central authority and it is essential.	It depends upon the sweet will of the individuals and therefore it is not necessary.
Support	It seeks wholehearted support from various people working at various levels.	Co-operation without co-ordination is fruitless and therefore it may lead to unbalanced developments.

Therefore, existence of co-operation may prove to be effective condition or requisite for co-ordination. But it does not mean that co-ordination originates automatically from the voluntary efforts of the group of members. It has to be achieved through conscious and deliberate efforts of managers, therefore to conclude we can say that co-operation without co-ordination has no fruit and co-ordination without co-operation has no root.

4.9.4 Relationship Between Coordination and Cooperation

S. No.	*Basis*	*Coordination*	*Cooperation*
1.	*Definition*	Coordination is a deliberate effort by a manager.	Cooperation is voluntary attitude of organizational members.
2.	*Purpose*	It is an orderly arrangement of group efforts to provide unity of action in the pursuit of common objectives.	It denotes collective efforts of the group to accomplish a particular objective.
3.	*Essential*	It is essential where a group of people work together for a common purpose.	It is voluntary in nature. It arises out of the desire of the people to work together.
4.	*Relations*	It is achieved through both formal and informal relations.	Cooperation arises out of informal relations.
5.	*Result*	Coordination seeks whole-hearted support of employees and departments.	Cooperation without coordination is fruitless.

4.9.5 Objectives of Coordination

The main purposes which are sought to be realized through coordination are as follows:

(*a*) **Reconciliation of Goals:** This can be done by the coordination only. The conflict of goals arises because everybody perceives the organizational goals differently and tries to achieve them in his own way. This may lead to confusion and chaos in the organization. Therefore, coordination is necessary to bring unity of action in the organization.

(*b*) **Total Accomplishment:** Through coordination it is possible to bring about total accomplishment which will be far in excess of the sum of the individual parts. It has been observed that the total accomplishment of ten employees of a department whose efforts are properly coordinated will be far greater than the mathematical sum of their individual accomplishment. This happens because through coordination, duplication of efforts is prevented and the time and energy thus saved are better utilized in more creative tasks.

(*c*) **Economy and Efficiency:** Through coordination, it is possible to bring about economy and efficiency in the organization. Coordination will avoid duplication of efforts due to which there will be economy in labour, time and equipment. When the activities are properly integrated, there will be lest delays which will bring efficiency in the business organization.

(*d*) **Good Personal Relations:** Coordination is achieved through systematic efforts. Good coordination gives job satisfaction to the employees which keep their morale high. Moreover, there are good human relations because the authority-responsibility relationships are clear. The conflict between line and staff personnel can also be avoided through proper coordination of their efforts.

(*e*) **Retention of Managerial and other Personnel:** It has been pointed out that sound coordination has a significant effect on the development and retention of good personnel in business. If the total organization is so designed and patterned that each executive and employees derives job satisfaction, there will be tendency on the part of the executive and employees and employees to stay with the organization. The absence of this will create suffocating conditions in which it would be difficult for any person to stay for long in the organization.

4.9.6 Coordination as The Essence of Management

Coordination is the essence of managership. Coordination is not something which can be ordered by a manager. Instead, it is something which the manager attempts to achieve while performing his functions of planning, organizing, staffing, directing and controlling. Thus, every managerial function is an exercise in coordination. How coordination is achieved by performing the managerial functions is explained below:

1. **Coordination through Planning:** The planning stage is the ideal time to bring about coordination by properly integrating the various plans through mutual discussion, exchange of ideas etc. For instance, if the advertising manager is to plan his advertising campaign, it would be helpful if he discusses the plan with the Production Manager, the Sales Manager, the Finance Manager, and so on. In this way, as he performs his managerial function of planning, he is attempting to achieve coordination by seeking cooperation of all the managers who are directly or indirectly concerned with it.

2. **Coordination through Organizing:** Mooney considers coordination as the very essence of organizing. In fact, when a manager group assigns the various activities to subordinates, and when he creates departments, coordination is one thing which is uppermost in his mind. By placing related activities in the same administrative unit, coordination will be facilitated. Coordination is an essential part of organizing. Unless the manager bears in mind how various groups are to function together, the organization structure will not be a successful one. In order to obtain coordination while organizing, the manager must look at it both vertically and horizontally.

Poor coordination is caused quite often by a lack of understanding of who is to perform what or due to the fact that the manager did not clearly delegate authority and responsibility. This may lead to the duplication of efforts instead of their synchronization.

3. **Coordination through Staffing:** The staffing function of management involves manpower planning, employment, training, wage determination, performance appraisal, etc. All these sub functions are performed in such a manner that there are right person on different jobs. This will help in achieving coordination in assigning tasks to various individuals.
4. **Coordination through Directing:** When a manager directs, he is also performing the function of coordination. The very essence of giving orders, instructions, coaching and teaching subordinate means to coordinate their activities in such a manner that the overall enterprise objective will be achieved in the most efficient way. Thus, manager's expertness in effectively directing his subordinates will bring about coordination.

   The manager should use free and open discussion and group decision-making techniques with subordinates in order to achieve the necessary coordination. It is vital that various subordinates have an opportunity for free and open discussion and for an interchange of ideas so that the details of a particular directive are properly understood. In short, as the manager performs his directing function, he will invariably coordinate.
5. **Coordination through Controlling:** While performing the function of controlling, the manager comes to know whether or not current activities are in keeping with the desired activities. Such frequent evaluation of operations helps to synchronize the efforts of the subordinates. If, while controlling, manager finds that the performance is not as planned or as directed, he should immediately take remedial action so that whatever deviations have occurred can be remedied. By doing this, he brings about coordination. The very nature of the controlling function brings about coordination and leads the organization to the desired goals.

Good communication will be of immeasurable help in the coordination of various activities. There must be continuous and free flow of communication which will give the subordinates the necessary information needed for coordination. Personal contact is probably the most effective means of communication to achieve coordination. However, there are many other additional devices such as written communication, reports, procedures, bulletins and numerous modern mechanical devices which will ensure a speedy dissemination of the necessary information to the subordinates who must have this information to achieve proper coordination.

4.9.7 Types of Coordination

Coordination can be classified into two broad categories, one on the basis of its shape in the organization and the other on the basis of its scope and coverage. On the former basis, it can be classified into vertical and horizontal coordination and on the latter basis into internal and external coordination. Each of these has been discussed below:

1. **Vertical and Horizontal Coordination:** The term 'vertical coordination' is used when coordination is to be achieved between various links or different

levels of the organization. Vertical coordination is needed to ensure that all the levels in the organization act in harmony and in accordance with organization policies and programmes. It is the functions of the top executives to bring about this coordination. Vertical coordination is secured through delegation of authority and with the help of directing and controlling. There is no doubt that the delegated authority will carry great weight but vertical coordination cannot be achieved by the mere weight of authority itself. This should rather come about as a by-product of the superior efficient and expert performance of the managerial functions.

The term horizontal coordination is used when coordination is to be achieved between departments on the same level in the managerial hierarchy. Thus, when coordination is brought about between production department, sales department, personnel department, etc., it is said to be horizontal communication. Coordination between interdependent managers of different departments is facilitated in a small enterprise due to proximity of working arrangements, close contacts, and short lines of communication. However, when a large company is involved, the problems become more complicated, and it is the fine art of the superior manager to achieve coordination in the horizontal sense.

2. **Internal and External Coordination:** Coordination may be internal and external to the organization. Coordination is internal when it is achieved between different departments, sections and units of an enterprise. It is both vertical and horizontal.

A business enterprise is a system in itself. It acts and reacts with the environment continuously. The various factors with whom it has interaction include government, customers, suppliers and competition. An enterprise has to keep proper coordination with these. Such type of coordination is known as external coordination and it is essential for the survival of the enterprise. External coordination also involves interaction with other business, economic and research institutions to have the benefits of latest information and technological advances.

Difficulties in Coordination

Henri Fayol stated that in a well coordinated enterprise the following facts are to be observed:

1. Each department should work in harmony with the rest.
2. Each department, division, and sub-division should be precisely informed as to the share must take in the common task.
3. The working schedule of the various departments and sub-divisions should be constantly attuned to circumstances.

In practice, these requirements are not always fulfilled. Ineffectiveness of coordination is apparent because of the following difficulties:

(*a*) Lack of cooperation and understanding between and among individuals, groups and departments.

(*b*) Lack of good impersonal relations.

(*c*) Failure in accomplishing objectives according to time and work schedule.

(*d*) Lack of direction and consequently aimless individual efforts.

(*e*) Functioning of departments in the organization as water-tight components

(*f*) Lack of initiative and loyalty towards the organization.

4.9.8 Techniques of Effective Coordination

The following steps should be taken for achieving effective coordination:

(*i*) **Clearly Defined Goals:** The goals of the enterprise should be laid down clearly. Every individual in the enterprise should understand the overall objectives and the contribution by his job to these objectives. Unity of purpose is a must for achieving proper coordination.

(*ii*) **Clear Lines of Authority and Responsibility:** There is a line of authority in every enterprise which indicates as to who is accountable to whom. The line of authority and responsibility should be clearly defined to achieve coordination. Clear-cut authority relationships help in reducing conflicts among different positions (particularly line and staff) which is essential for sound coordination.

(*iii*) **Precise and Comprehensive Programmes and Policies:** Laying down well-defined programmes and policies is another measure for achieving effective coordination. This brings uniformity of action because everybody understands the programmes and policies in the same sense.

(*iv*) **Cooperation:** Coordination must be accompanied by cooperation. The individuals in the organization must be willing to help each other voluntarily. Cooperation can be brought about by keeping harmonious relations among the people in the organization by encouraging informal contacts to supplement formal communication and using committees for exchange of ideas and views at the top level.

(*v*) **Effective Communication:** Effective communication is the key to proper coordination. The channels of the communication used in the enterprise should be reliable so that they are able to create proper understanding in the minds of receivers. Personal contacts should be encouraged as it is the most effective means of communication for achieving coordination.

(*vi*) **Effective Leadership and Supervision:** Management can achieve better coordination through effective leadership and supervision. Effective leadership ensures coordination both at the planning and the implementing stage. Effective supervision is necessary to guide the activities of individuals in the proper direction.

4.9.9 Principles of Coordination

Mary Parker Follett laid down the following principles of coordination:

(*i*) **Principles of Direct Contact:** The activities of different individuals can be coordinated effectively through direct personal contacts. This helps in exchanging the opinions and ideas in a better way and clarifying the misunderstandings more easily.

(*ii*) **Principles of Early Start:** Coordination may be achieved easily during the early stages of planning and policy-making. It becomes difficult to secure coordination at the execution stage.

(*iii*) **Principles of Reciprocal Relationship:** The principles states that all the factors in a situation like men, material, and environment are reciprocally related. For instance, when *A* works with *B*, each finds himself influenced by the other and both are influenced by the other persons and factors in the total situation.

(*iv*) **Principle of Continuity:** Coordination should be a continuous process starting with planning and running through the other managerial processes. It is something which must go on all the time. It should be viewed as a never-ending process and every manager should strive for it constantly.

4.10 JAPANESE MANAGEMENT TECHNIQUES

In the post-World War II era a set of Japanese cultural patterns and managerial practices came to be known collectively as the Japanese management style or Japanese management techniques. Many of these techniques were credited with helping vault the Japanese economy to its status as the world's second largest, behind only the United States, and with making Japanese businesses, particularly in the manufacturing sector, more competitive than their international counterparts. In the wake of Japan's prolonged and arduous struggle with recession throughout much of the 1990s, however, many observers both inside and outside Japanese called into question the effectiveness of some traditional Japanese management practices. As a result, at the dawn of the 21st century Japanese management techniques are more than ever in a state of flux, as scholars and business leaders alike reconsider which practices work and which don't.

4.10.1 Historical Context

Although Japanese management techniques and economic strategies came to be recognized in Western countries only during the postwar period, their origins are considerably older. Most directly, their origins can be traced to at least the latter part of the 19th century, when a Western influenced modernization program began under the new monarchy created in the 1868 Meiji restoration. In part as a response to the bitter European colonization experiences of its Asian neighbors, the new Japanese government began to open the Japanese economy and society to controlled outside influences in order to stave off any Western conquerors.

Some recognizably modern practices arose during the Meiji period. Even then, when the Japanese economy was still shedding the trappings of feudalism after centuries of closure to foreigners and slow technological development, heavy emphasis was placed on developing domestic imitations and innovations ones tern goods, rather than relying on imports. The practice was summarized well under a slogan of the era, "Japanese spirit, Western technology."

This ambition to preserve the character of the Japanese culture and the autonomy of the economy can be seen in 20th-century practices at both the macro- and microeconomic levels. In the national economy it is evidenced by long-standing restrictions (direct and indirect) on imports into Japan and the concomitant trade surplus Japan has maintained for years. At the company level, the same motive helps explain the prevalence of the Japanese *keiretsu,* the large and complex families of interdependent companies centered around their own banks (*e.g.*, Sumitomo, Hitachi, Mitsubishi). In theory, at least, these firms can avoid "importing" their raw materials, components, or even capital from "foreign" (*i.e.*, unaffiliated) companies by sourcing these goods from within their extensive organizations.

4.10.2 Profile of Traditional Japanese Management Practices

Rooted in these and other historical traditions, some of the other key practices commonly associated with Japanese management techniques include:

- in-house training of managers
- consensual and decentralized decision-making
- extensive use of quality control methods
- carefully codified work standards
- emphasis on creating harmonious relations among workers
- lifetime employment and seniority-based compensation

It is important to note that these are generalizations according to a conventional formula. There have always been variations, and, as noted above, some aspects of these practices have been increasingly reconsidered in recent years.

Management Education

The education of managers in Japan traditionally takes place on a relatively informal basis within firms. The percentage of Japanese chief executives who have attended university is high, similar to that in the United States and Western Europe. However, very few Japanese executives have attended graduate schools compared to their U.S. and European counterparts. In fact, only one Japanese university offers a degree analogous to an MBA, a key credential for managers in the United States.

Formal education for managers is also not well developed at the undergraduate level. Undergraduate education is not viewed by firms as a means of attaining business skills, and firms base their hiring decisions less on a recruit's knowledge than on general attributes such as character and ambition. Firms do not hire recruits to fill specific occupations. Rather, recruits are expected to be malleable, identifying with the general interests of the firm rather than with their specific role within it. The mentor system is widely used in the early training of management recruits and involves middle-level and senior managers serving as teachers and role models.

The emphasis on in-house education is related to the traditional lifetime employment system, in which management recruits are hired each April following university graduation; they typically would stay with the firm until retirement. The lifetime-employment system makes it probable that a firm will benefit from its investment in training, and also enables the firm to develop long-range plans for training recruits.

Management training is based on regular rotation through a broad range of a firm's operations. Management recruits also frequently begin their careers as ordinary workers on a production line. The pattern of regular rotation enables managers to develop a detailed understanding of a number of varied operations, and thus over time to attain a rich general knowledge of the firm.

Linked with the lifetime-employment system is the emphasis on seniority in compensation and promotionften over what Americans would take to be "qualifications" for the job. This results in a higher average age and less variation in age among top executives in Japan. Compared to the United States and Europe, for instance, relatively few company presidents are under age 50. This practice is believed to equip Japanese executives with an intricate knowledge about their particular business.

Japanese managers typically take a more long-term interest in their firms than do their American counterparts, partly a result of the lifetime employment and seniority systems. In the United States, managers are typically compensated on the basis of their divisions' performance. This bonus system is not used for Japanese managers, as it is considered detrimental to a long-term perspective and an interest in the firm as a whole.

Capital and Priorities

The long-term view of Japanese managers is also based on sources of finance. While American firms rely heavily on capital from the stock markets, Japanese firms tend to rely more heavily on borrowing from banks and generally have much higher debt-to-equity ratios. Consequently, Japanese managers are under less pressure to maximize short-term earnings to please shareholders. By contrast, in the United States there is intense market pressure for companies to meet quarterly earnings expectationsven exceed themr else face a sell-off of their shares. In general, Japanese firms are more likely to focus on productivity, growth, and market share, whereas U.S. firms are more inclined to concentrate first on profitability.

Corporate Governance

While directors from outside the company are common in the United States, they are rare in Japan. The decision-making process in Japanese firms is highly decentralized. In publicly held U.S. corporations, power is concentrated in a board of directors, with each director having one vote. In Japan, both middle and senior management serve as directors. Japanese directors typically retain production-line responsibilities. For example, in the early 1970s, 14 of Hitachi's 20 directors were engineers. This represents another facet of the strong production orientation of Japanese management.

The Ringi System

The traditional decision-making process in Japanese firms is referred to as the *ringi* system. The system involves circulating proposals to all managers in the firm who are affected by an impending decision. Proposals are generally initiated by middle managers, though they may also come from top executives. In the latter case, an executive will generally give his idea to his subordinates and let them introduce it. Managers from different departments hold meetings and try to reach an informal consensus on the matter. Only after this consensus is reached will the formal document, or *ringi-sho,* be circulated for approval by the responsible managers.

The ringi system requires long lead times, and thus is problematic in a crisis. In recent years the focus on speeding up decision making has made this approach unpopular at many firms. Nonetheless, one of its underlying principles remains prevalent. That is, when a decision proves beneficial, the middle-level managers who initially advocated it receive credit; when a decision proves unsuccessful, responsibility is taken by top-level executives. This practice is intended to promote aggressiveness in younger managers.

Enterprise Unions

One distinctive characteristic of *labor-management relations* in Japan is the enterprise union, which is organized around a single plant. Consequently, any given company may have several enterprise unions representing various portions of its workforce. Enterprise unions generally belong to a larger federation, but the balance of power is at the local level. Japanese unions are distinct not only because of their highly decentralized nature, but also because they represent both white-collar and blue-collar workers, with union membership open to managers up to the section chief level. The fact that many upper-level managers have moved up through union ranks and may have even served as union officials highlights the generally less antagonistic relationship between labor and management in Japan. Combined with a relatively narrow income gap between managers and workers and the willingness of manager recruits to work on production lines as part of their training, the open membership policies of Japanese unions contributes to the fairly harmonious interaction between unions and management.

Union membership is generally associated with lifetime employment guarantees. Membership varies widely by firm size, and relatively few workers in firms with fewer than 100 employees receive lifetime employment guarantees. Nonetheless, in large firms the lifetime employment guarantee creates an environment in which workers are less likely to feel threatened by technological change. As a consequence, changes in the production process are likely to be undertaken by management and workers on a cooperative basis. More generally, since semiannual bonuses and annual wage negotiations are based on a firm's competitive strength, workers have a large stake in their firm's long-term success.

Quality Circles

The extensive use of *quality circles* is another distinguishing characteristic of Japanese management. The development of quality circles in Japan in the early 1960s was inspired by the lectures of American statisticians W. Edwards Deming and J.M. Juran, in which they discussed the development of wartime industrial standards in the United States. Noting that American management had typically given line managers and engineers about 85 percent of responsibility for quality control and only 15 percent to workers, Deming and Juran argued that these proportions should be reversed. Production processes should be designed with quality control in mind, they contended, and everyone in the firm, from entry level workers to top management, should be familiar with statistical control techniques and undergo continuing education on quality control. In general, Deming and Juran argued that quality control should focus on prevention, with the ultimate goal being to improve the production process until no defective parts or products are produced. Quality circles were one method of reaching these goals.

In Japan, quality circles consist of groups of about 10 workers who meet weekly, often on their own time. The groups typically include foremen, who usually serve as circle leaders. Quality circles focus on concrete aspects of the operations in which they are directly involved, using tables and graphs to communicate the statistical details of their quality issues. In one common format, problems are categorized by materials, manpower, and machines.

Quality circles provide a means for workers to participate in company affairs and for management to benefit from worker suggestions. Indeed, employee suggestions play an important role in Japanese companies. Two associations, the Japanese Association of Suggestion Systems and the Japan Human Relations Association, were developed to encourage this process. Japanese employee suggestions reportedly create billions of dollars' worth of benefits for companies.

Scientific Management

Japanese management techniques have been strongly influenced by the tenets of scientific management. Like quality circles, scientific management originated in the United States, only to be more systematically adopted in Japan. The pioneering figure of scientific management is Frederick Jackson Taylor (1856-1915). Taylor is best known for his time and motion studies of workers as part of an effort to optimize and standardize work efforts, but he also argued for a system of bonuses to reward workers based on productivity. These ideas were implemented by Japanese firms as early as 1908, and a translation of his *Principles of Scientific Management* sold 2 million copies in Japan.

In the post-World War II years, carefully codified work standards and the use of semiannual bonuses for workers became common practices in Japan. Consistent with the Japanese emphasis on teamwork, bonuses are generally allotted to a work group rather than an individual worker. Scientific management emphasizes the role of management in the production process. This is reflected in the more hands-on approach in Japanese management training, as well as the relatively high share of managers directly involved in the production process.

Pervasiveness of Engineers

As with managers, Japanese industrial engineers are more directly involved with production processes than their counterparts in the United States. In his book *The Japanese Industrial System,* Charles J. McMillan explained that most Japanese companies make few distinctions between engineers and blue-collar workers, although engineers do tend to earn more. They work closely alongside production workers. In addition, Japan produces up to three times as many engineers a year as the United States. Japan's emphasis on production oriented engineering is consistent with its dominant competitive strategy in the postwar yearsndeed, since the Meiji eraf focusing on improving existing products or processes rather than developing completely new ones.

New Directions

While many of the patterns just described continue unabated at some Japanese companies, a variety of forces have caused them to change, often toward Western practices. Since the 1980s, for example, the predominance of seniority-based raises has been gradually giving way to a Western style regime of merit-based pay. Indeed, as of 1995, three-quarters of Japanese companies surveyed allocated at least some of their reward pay based on skills or achievements as opposed to tenure. And more than a few Japanese companies have attacked seniority more directly, explicitly revising policies to diminish or even eliminate it as a criterion in the compensation structure. This trend may be evidence of a cultural shift from valuing length of service to valuing quality of service.

Also mirroring Western trends, labor union membership in Japan has dropped considerably since the 1970s, falling from 35.4 percent of the workforce in 1970 to just 22.4 percent by 1998, according to figures compiled by the Japanese Ministry of Labor. Union participation remains the highest in large companies (those with 1,000 or more workers), where in 1998 membership was still nearly 57 percent. This share was down from 68 percent in 1987, the first year statistics by company size were kept.

Other traditional Japanese practices appear more enduring, notably lifetime employment. Although Japan's economic troubles have meant that some employees have lost their jobs, a continuing commitment to the principle of lifetime employment seems to remain at many companies and in the society as a whole. Still, younger workers (*e.g.*, those under age 30) are decidedly less loyal to companies than in decades past, and there is growing evidence of a rise in professional identification over corporate identification among workers (*i.e.*, "I'm a tax accountant" instead of "I'm a Toyota worker").

Nonetheless, even at the depths of the Japanese recession during the late 1990s Asian financial crisis, companies went to great lengths to avoid outright layoffs. One of the most common practices instead was to reassign workers, either within the corporate family or to other companies, such as vendors the company does business with. These transfers (known as *shukko*) could be temporary, in which case the worker is still officially employed by the company that has loaned him or her out, or permanent, where the company essentially finds a new job for the employee at another company. Employees who were never considered part of the lifetime staff, such as part-time help, usually didn't enjoy such privileges.

Although most agree that Japanese management has been moving in new directions, academics who study Japanese management practices are divided on how profound the shifts in the Japanese business paradigm really are. Indeed, the gamut of opinions has ranged from declaring the death of the Japanese management system to asserting its overarching continuity and strength. A number of observers see a continued convergence with Western practices, but many believe that, as in the past, the adoption of Western principles and practices will never be wholesale, but will blend with prevailing norms and beliefs in Japanese business and the broader culture.

4.11 ORGANIZATIONAL CONTROL TECHNIQUES

Control techniques provide managers with the type and amount of information they need to measure and monitor performance. The information from various controls must be tailored to a specific management level, department, unit, or operation.

To ensure complete and consistent information, organizations often use standardized documents such as financial, status, and project reports. Each area within an organization, however, uses its own specific control techniques, described in the following sections.

Financial Controls

After the organization has strategies in place to reach its goals, funds are set aside for the necessary resources and labor. As money is spent, statements are updated to reflect how much was spent, how it was spent, and what it obtained. Managers use these financial statements, such as an income statement or balance sheet, to monitor the progress of programs and plans. **Financial statements** provide management with information to monitor financial resources and activities. The**income statement** shows the results of the organization's operations over a period of time, such as revenues, expenses, and profit or loss. The *balance sheet*shows what the organization is worth (assets) at a single point in time, and the extent to which those assets were financed through debt (liabilities) or owner's investment (equity).

Financial audits, or formal investigations, are regularly conducted to ensure that financial management practices follow generally accepted procedures, policies, laws, and ethical guidelines. Audits may be conducted internally or externally. **Financial ratio analysis** examines the relationship between specific figures on the financial statements and helps explain the significance of those figures:

- **Liquidity ratios** measure an organization's ability to generate cash.
- **Profitability ratios** measure an organization's ability to generate profits.
- **Debt ratios** measure an organization's ability to pay its debts.
- **Activity ratios** measure an organization's efficiency in operations and use of assets.

In addition, *financial responsibility centers* require managers to account for a unit's progress toward financial goals within the scope of their influences. A manager's goals and responsibilities may focus on unit profits, costs, revenues, or investments.

Budget Controls

A budget depicts how much an organization expects to spend (expenses) and earn (revenues) over a time period. Amounts are categorized according to the type of business activity or account, such as telephone costs or sales of catalogs. Budgets not only help managers plan their finances, but also help them keep track of their overall spending.

A budget, in reality, is both a planning tool and a control mechanism. Budget development processes vary among organizations according to who does the budgeting and how the financial resources are allocated. Some budget development methods are as follows:

- **Top-down budgeting:** Managers prepare the budget and send it to subordinates.
- **Bottom-up budgeting:** Figures come from the lower levels and are adjusted and coordinated as they move up the hierarchy.
- **Zero-based budgeting:** Managers develop each new budget by justifying the projected allocation against its contribution to departmental or organizational goals.
- **Flexible budgeting:** Any budget exercise can incorporate flexible budgets, which set "meet or beat" standards that can be compared to expenditures.

Marketing Controls

Marketing controls help monitor progress toward goals for customer satisfaction with products and services, prices, and delivery. The following are examples of controls used to evaluate an organization's marketing functions:

- **Market research** gathers data to assess customer needs—information critical to an organization's success. Ongoing market research reflects how well an organization is meeting customers' expectations and helps anticipate customer needs. It also helps identify competitors.
- **Test marketing** is small-scale product marketing to assess customer acceptance. Using surveys and focus groups, test marketing goes beyond identifying general requirements and looks at what (or who) actually influences buying decisions.
- **Marketing statistics** measure performance by compiling data and analyzing results. In most cases, competency with a computer spreadsheet program is all a manager needs. Managers look at *marketing ratios,* which measure profitability, activity, and market shares, as well as *sales quotas,* which measure progress toward sales goals and assist with inventory controls.

Unfortunately, scheduling a regular evaluation of an organization's marketing program is easier to recommend than to execute. Usually, only a crisis, such as increased competition or a sales drop, forces a company to take a closer look at its marketing program. However, more regular evaluations help minimize the number of marketing problems.

Human Resource Controls

Human resource controls help managers regulate the quality of newly hired personnel, as well as monitor current employees' developments and daily performances.

On a daily basis, managers can go a long way in helping to control workers' behaviors in organizations. They can help direct workers' performances toward goals by making sure that goals are clearly set and understood. Managers can also institute policies and procedures to help guide workers' actions. Finally, they can consider past experiences when developing future strategies, objectives, policies, and procedures.

Common control types include performance appraisals, disciplinary programs, observations, and training and development assessments. Because the quality of a firm's personnel, to a large degree, determines the firm's overall effectiveness, controlling this area is very crucial.

Computers and Information Controls

Almost all organizations have confidential and sensitive information that they don't want to become general knowledge. Controlling access to computer databases is the key to this area.

Increasingly, computers are being used to collect and store information for control purposes. Many organizations privately monitor each employee's computer usage to measure employee performance, among other things. Some people question the appropriateness of computer monitoring. Managers must carefully weigh the benefits against the costs—both human and financial—before investing in and implementing computerized control techniques.

Although computers and information systems provide enormous benefits, such as improved productivity and information management, organizations should remember the following limitations of the use of information technology:

- **Performance limitations:** Although management information systems have the potential to increase overall performance, replacing long-time organizational employees with information systems technology may result in the loss of expert knowledge that these individuals hold. Additionally, computerized information systems are expensive and difficult to develop. After the system has been purchased, coordinating it—possibly with existing equipment—may be more difficult than expected. Consequently, a company may cut corners or install the system carelessly to the detriment of the system's performance and utility. And like other sophisticated electronic equipment, information systems do not work all the time, resulting in costly downtime.
- **Behavioral limitations:** Information technology allows managers to access more information than ever before. But too much information can overwhelm employees, cause stress, and even slow decision making. Thus, managing the quality and amount of information available to avoid information overload is important.
- **Health risks:** Potentially serious health-related issues associated with the use of computers and other information technology have been raised in recent years. An example is carpal tunnel syndrome, a painful disorder in the hands and wrists caused by repetitive movements (such as those made on a keyboard).

Regardless of the control processes used, an effective system determines whether employees and various parts of an organization are on target in achieving organizational objectives.

Direct Supervision and Observation

'Direct Supervision and Observation' is the oldest technique of controlling. The supervisor himself observes the employees and their work. This brings him in direct contact with the workers. So, many problems are solved during supervision. The supervisor gets first hand information, and he has better understanding with the workers. This technique is most suitable for a small-sized business.

Personal Observations

This technique acts as a great tool for controlling and bringing the system back to having a very sound health. In many cases, the eyes act as the best tool for noticing the overbalancing in the people, their working, and the environment. A lot of time is saved by this technique. The personal observation avoids much of the paper work. One of the best methods used here is the 'Wandering Around' method. In this method, wandering around is taken as a regular exercise and the work monotony is used. People are not allowed to get stuck into their cabins for long period of time as with this, a gap is created between the relationships and the environmental attachment.

Break Even Analysis

Break Even Analysis or Break Even Point is the point of no profit, no loss. For *e.g.* when an organization sells 50000 cars it will break even. It means that, any sale below this point will cause losses and any sale above this point will earn profits. The Break-even analysis acts as a control device. It helps to find out the company's performance. So the company can take collective action to improve its performance in the future. Break-even analysis is a simple control tool.

Return on Investment (ROI)

Investment consists of fixed assets and working capital used in business. Profit on the investment is a reward for risk taking. If the ROI is high then the financial performance of a business is good and vice-versa.

ROI is a tool to improve financial performance. It helps the business to compare its present performance with that of previous years' performance. It helps to conduct inter-firm comparisons. It also shows the areas where corrective actions are needed.

Management by Objectives (MBO)

MBO facilitates planning and control. It must fulfill following requirements:

1. Objectives for individuals are jointly fixed by the superior and the subordinate.
2. Periodic evaluation and regular feedback to evaluate individual performance.
3. Achievement of objectives brings rewards to the individuals.

It acts as a great tool for the control purpose. By having a clear picture of the various objectives in mind, one can easily plan the various controls. Management can be observed as the people working for the measurable goals, deciding the output at the end of an activity.

Systems and the Procedures

Manuals form very necessary documents for bringing some rigidity in the operation. It helps in creating a judicial system for an organization. Management audits can be conducted sometimes in this technique to reduce the lacunae that may be present in an organization.

Management Audit

Management Audit is an evaluation of the management as a whole. It critically examines the full management process, *i.e.* planning, organizing, directing, and controlling. It finds out the efficiency of the management. To check the efficiency of the management, the company's plans, objectives, policies, procedures, personnel relations and systems of control are examined very carefully. Management auditing is conducted by a team of experts. They collect data from past records, members of management, clients and employees. The data is analysed and conclusions are drawn about managerial performance and efficiency.

Management Information System (MIS)

In order to control the organization properly the management needs accurate information. They need information about the internal working of the organization and also about the external environment. Information is collected continuously to identify problems and find out solutions. **MIS** collects data, processes it and provides it to the managers. MIS may be manual or computerized. With MIS, managers can delegate authority to subordinates without losing control.

PERT and CPM Techniques

Programme Evaluation and Review Technique (**PERT**) and Critical Path Method (**CPM**) techniques were developed in USA in the late 50's. Any programmer consists of various activities and sub-activities. Successful completion of any activity depends upon doing the work in a given sequence and in a given time.

CPM / PERT can be used to minimize the total time or the total cost required to perform the total operations.

Importance is given to identify the critical activities. Critical activities are those which have to be completed on time otherwise the full project will be delayed.

So, in these techniques, the job is divided into various activities/sub-activities. From these activities, the critical activities are identified. More importance is given to completion of these critical activities. So, by controlling the time of the critical activities, the total time and cost of the job are minimized.

Self-Control

Self-control means self-directed control. A person is given freedom to set his own targets, evaluate his own performance and take corrective measures as and when required. Self-control is especially required for top level managers because they do not like external control.

The subordinates must be encouraged to use self-control because it is not good for the superior to control each and everything. However, self-control does not mean no control by the superiors. The superiors must control the important activities of the subordinates.

Social Controls

The managers possessing the objective view of the outsiders are helped a lot by this technique. The user's clubs of the customers, the supplier's club, the dealer's club, the Employees clubs etc. can be organized easily with this method.

Statistical Database Control

In this method, the generation of the Statistical data coupled with the planning system is generated, mainly for the management control system. Special reports like the market prospects, the customers survey findings etc. are used as a feedback for the different activities involved and these activities can be based on the criteria of per person or per machine or per activity. The data can involve the relations or the trends, which can be easily extrapolated. For controlling the deviations and to know about them, the audits like the statutory audits, the performance audits, the cost audits etc., can be carried out.

The Activities Network Control

It forms a very critical and a necessary part of the management tool. Here the performance of the various activities is compared and then depending on this, one comes to know about the various deviations and then accordingly the activities can be controlled. Various techniques like the Gantt Charts, the Program Evaluation and the Review Techniques etc are used for performing the various activities.

4.12 WHAT ARE REWARDS?

Rewards can take two forms.

Intrinsic rewards are derived from within the individual. Extrinsic rewards pertain to rewards that are given by another person.

4.12.1 Extrinsic Rewards

Extrinsic Rewards are a host of external things that managers can provide that may serve as incentives for employees to increase their productivity. These include:

1. Money-in the form of pay, bonuses, stock options etc.
2. Benefits-also in many different forms, including health assurance, vacations, sick leave, retirement accounts etc. Increasingly benefits are offered under some form of cafeteria plans, allowing employees flexibility in what can be selected and in the management of their own benefit package.
3. Flexible schedules
4. Job responsibilities and duties.

5. Promotions.
6. Change in status-conveyed either by changes in job title or in new and different job responsibilities.
7. Supervision of others.
8. Praise and feedback.
9. A good boss.
10. A strong leader.
11. Other inspirational people.
12. A nurturing organizational culture.

As this list demonstrates, extrinsic rewards are all tangible types of rewards. Intrinsic rewards stand in marked contrast to these.

4.12.2 Intrinsic Rewards

Intrinsic rewards are internal to the individual and are in many ways less tangible. In fact they are highly subjective, in that they represent how the individual perceives and feels about work and its value. Five types of intrinsic rewards that have been summarized by Manion include:

1. Healthy relationship-in which employees are able to develop a sense of connection with others in the workplace.
2. Meaningful work-where employees feel that they make difference in people's lives. This is typically a motivator for people to enter and stay employed in the health care industry.
3. Competence-where employees are encouraged to develop skills that enables them to perform at or above standards, preferably the latter.
4. Choice-where employees are encouraged to participate in the organization in various ways such as expressing their views and opinions sharing in decision making, and finding other ways to facilitate participatory approaches to problem solving, goal setting and the like.
5. Progress-where managers find the ways to hold employees accountable, facilitate their ability to make headway towards completing their assigned tasks, and celebrate when progress is made toward completing important milestones within the project.

Intrinsic rewards couple with the extrinsic ones, lead to high personal satisfaction and serve as motivators for most employees.

4.13 WHO MOTIVATES EMPLOYEES?

While rewards may serve as incentives and those who bestow rewards may seek to use them as motivators, the real motivation to act comes from within the individual. Maagers do exert a significant amount of influence over their employees, but they donot have the power to force the person to act. They can work to provide various types of incentives in an effort to influence an employee in any number of ways such as by changing job descriptions, rearranging work schedules, improving work conditions, reconfiguring team and a host of other activities. While these may have an impact on an employees level of motivation and willingness to act when all is said and done, it is the employees decision to take action or not.

4.14 IS EVERYBODY MOTIVATED?

As managers, we often assume that employees are motivated and will respond to inducements from managers. While this is perhaps a logical and rational approach from the manager's perspective, it is critical to understand that this is not always the case. While the majority of the employees do, in fact, want to do a good job and are motivated by any number of factors, other may not share the same drive or high level of motivation. Those folks may be merely by putting in time and may be more motivated by other things such as family, school, hobbies or other interests. Keeping this in mind is useful in helping managers understand employee behaviors that seem to be counterproductive.

CASE STUDY

Global Adaptation of Japanese Management Practices: What You May Be Able To Adopt

—Professor Mayumi Otsubo

Professor Otsubo lead a seminar at UNUM Corp., Portland, Maine on May 7, 1993. He began his seminar with 10 items that distinguish Japanese Management Practices, and then went into some detail about each of them. He speaks almost entirely from a manufacturing perspective, but it is not hard to see how many of his points apply just as well to service. His address: Mayumi Otsubo, Professor of Management, University of Shizuoka, 395 Yata, Shizuoka City, Shizuoka, Japan 422.

1. Market First

- Products are made for the U.S. market, not the Japanese market. Usually a product is made for the factory, that is, it is easiest for the factory to make. While this happens in Japan, it is a sure way to loose the competitive edge.
- From the beginning, the Japanese are export minded.
- They learn from their markets and their customer, and how the market is changing.
- As Romans do in Rome—localize products. An example is Ford built a single car for world export, the so called world car. It didn't work. McDonald's and Coke, however, taste slightly different in Japan than in the U.S. They succeeded;

with localization comes a loss of your identity. It is also necessary to learn how to manage people in different cultures. He suggests that the Japanese are struggling with how to manage white collar, information workers.

2. Long Term Commitment

- Relationships are critical to good business. Trust is more important than contracts, receipts and law courts. Trust takes time and patience, but so does legal wrangling.
- Learning—information and know-how accumulation.
- Stability and Flexibility especially with customers, suppliers, employees and investors. In down times Japanese companiesavoid layoffs and contract terminations. They use this time to send employees out to talk to customers, train or re-train them, clean the physical plant, and so on. The government supplies funds for training during slow economic times.
- The Japanese spend much more time up front with a potential customer or supplier before making a commitment.

3. Training and Education

- On the Job Training.
- Communication Plaza Concept—while this is fading in Japan, employees meet with the executive informally over lunch or dinner to listen to each other.

4. Get-in-Touch, Learning From Facts

- Mix with employees and customers. Japanese engineers, *e.g.*, go to the factories, and don't wear distinguishing jackets or hard-hats. An insight in Japanese culture...Japanese neighborhoods are less stratified in terms of economic class. CEOs may live next door to engineers and factory workers.
- Genbutso Genba (facts, figures and check)—Develop theories and check with the facts, learn from facts, see it and touch it.—Learn from competitors—Americans are creative, but would be smart to do more copying. Japanese are world class at copying and improving upon an idea, but would be smart to develop their creativity.

5. Effort Evaluation

- Process versus Results—this seems to be the key difference between Americans and Japanese. Americans are more results oriented, Japans focus on process improvements. Once they learn how to do something, they work on small improvements—they evaluate effort not results. Because Americans are process averse, they depend on manuals to tell what the results should be.
- Blue collar workers like the rewards in a process environment, *i.e.*, for their effort not the result, especially in a service environment.
- The down side of the process focus is to squelch creativity. Overall, the two cultures should learn from each other, and become more like each other. Neither completely process nor completely results focused.

6. Customer First and Shareholder Last

- The priority order of customers and suppliers is different for U.S. and Japanese businesses:

Japan	*America*
1. Customer 2. Employee 3. Supplier 4. Community 5. Country 6. Shareholder	1. Shareholder 2. Customer 3. Employee*

* but the employee is told the customer is number 1

- American entrepreneurs follow more closely the Japanese ranking.
- The Japanese firm is organized for the employee. It is a more human orientation. Here we build an organization, and find employees to fill it. [I heard from another source, the right hand person to the CEO is the CFO in American firms. In Japanese firms the right hand person is the VP of Human Resources—ed.]

7. Team Work

- The Japanese and Americans see two different meanings behind these words. In Japan team work means to help others, here it means functional maximization, that is to improve results.
- This leads to a difference in the roles on the team. In Japan the team leader is always asking team members to help more, here the team leader is responsible for results.

8. Flexibility

- General management transfers to any department, it is fuzzy and easy to adopt. Americans tend to specialize, which leads to rigidity and difficulty in changing.
- Japanese take the long range view, and ask How long can you wait? This can also lead to a lack of decisiveness.

9. Quality First, Cost Last

- This is the proper ranking of quality in the organization:

  1. Quality 2. Quality
  3. Quality 4. Cost
- Profit is the result of the pursuit of quality—as quality improves, costs go down.
- Quality includes products, services, machines, layout, policy, planning and organization.

10. Learn from the Best

- Always look at someone better. The problem is when you get to the top, you have to become more creative—this is Japan's challenge.

- Short comings of Japans practices are:
  - Lack of Decisiveness: they are not transparent. Decisions sometimes take too long and they loose their timing.
  - Individual Ability Ignored: miss creative opportunities.
  - Miss Strategic Opportunities: too much delay and time.

Following this presentation, Professor Otsubo took questions from the audience:

More on Process and Results: Must have a good mix of process and results. In an organization, some departments must focus on process and others on results. Example, accounting must be process focused, while finance is results focused. Quality can save costs because you don't need as many supervisors, and less people to fix problems.

America and Japan are Closing the Quality Gap: One insight into Japanese thinking…If the Yen is trading at 110 to the Dollar, Americans will focus on being profitable at that rate. The Japanese will look ahead, can we still be profitable at 80?

Long Term Changes in Japanese Management: Japan will become more results focused, and more creative. They will not change the focus on a human organization. After all, the organization must provide for human welfare and happiness.

Demographics: Japanese population over the age 65 is growing twice as fast as the group under 15. The shrinking population will drive more women into management after the year 2006. Japanese industry will focus on more automation and high value added. Maybe more foreign labor, but more likely to do more work in China and Russia.

Training: Like the U.S., many Japans workers are not well educated—their focus is on practical training such as reading, machine operation, and problem solving methods. Japan will focus on teaching their workers how to change—cross training and so on. A key difference in education between the U.S. and Japan is that here we get our education, and then go to a corporation to work. In Japan, one gets a job, and then gets their education. He thinks Japan ought to have more business schools given the new global focus.

Managing Knowledge Workers and Professionals: This is something he has been struggling with, but some of his thoughts are to eliminate controls—provide targets and a framework and ask them to put forward their best efforts. Management by results not process. [Sounds like the American paradigm is actually a plus here—Ed.]

Success in Japan. Focus on niche marketing. Take the time to build long term relationships.

Hoshin Kanri (process management): Improvement of management quality means corporate policies must be:

1. Written and Formalized
2. Integrated
3. Implemented
4. Flexible

5. Participation
6. Coordinated
7. Communication
8. Report and Review Periodically
9. Feedback
10. Suggestions and Recommendations.

CASE STUDY ON MOTIVATION

By Khurana Sharma on August 21, 2010

Motivation is a complex subject, although it seems very interesting to all of us, but it has many facets for different people.

I have come across many such incidents during my tenure as HR professional at various companies. Some of them are very complicated and some are just the result of some or the other conflicts. If all the reader here have some knowledge on the topic of motivation, this case study would surely be of some value as well as interesting to read.

Motivation has many theories written by various authors, one such theory is Reward and Punishment theory also known as Theory X, another name of which is Carrot and stick. Here in this case our main focus will be on this Theory X.

As a part of the company's Management Development Program, a group of managers from various functional areas have devoted several class sessions to a study of motivation theory and the relevence of such knowledge to the manager's responsbility for directing and controling the operations of his organizational unit.

One of the participant is Rohit sharma, who has been the supevisor of production department from last one year. During his past in the industrial unit he had no chance to attend any supevisory or development program and his duties have been concentrated on routine technical jobs.

The present plant mangager has seen Rohit grow from an apprentice to a supervisor and is yet to reconcile with the change of responsbilities and designation of Rohit Sharma.

Rohit prepared the schedule on a particular day for all machines and on a specific machine, which was on a top priority from another established company. When Rohit came for his round he was surprised to find that the opretor had loaded some other job rather than the job of the company which is on priority.

Looking at this Rohit got annoyed and following conversation took place:

Rohit to operator: Please do the job for the company that is on priority.

Operator: Plant manager has given me another job after that I will take up your work.

Rohit: I dont want to hear anything, plant manager has told me to stop everything else and take up the job of that company on priorty.

Operator: I have been taking up all the job immediately whenever you have asked. But what has been done regarding my increment which has been pending for over six months now.

Rohit: I'm not responsible for these problems related to increments, the Top Managment manages the decision on increment. I am helpless in this regard and suggest you to go and sort this problem with Plant Manager.

Operator: As a supervisor it is your duty to solve my problem as I report to you for all my work.

Rohit: Please do not tell me my duties and if you will not do this work on priority now, I would report this incidence as misbehaviour to the plant manager. Operator replied in a threat to go on strike and complained about the less salary and increment he gets.

Now during a session of Motivation at Management Development Program, Rohit made the following comments:

"Motivation theory makes sense in general, but there is no opportunity for us to apply these concepts in job situation. After all our shop level employees are unionised and have job security. Motivation theories helps me get the work done from my kids, but in a working enviorment we are working with adults and it seems to me this reward and punishment thing smacks of manipulation that just would not go over with people".

So, now it has become a complicated situation as Rohit does not seems to be interested in the Motivational theories at all.

Does the incident that happened between Operator and Rohit Sharma left such a mark in his mind that he started feeling threatened with Unions and shop floor people??

So, once we have gone through the case I would like the reader to provide the solution to the case study, I would surely be waiting for response and in times to come will share my views with all of you on the case...

EXERCISES

Short Questions Answers

1. What is motivation?
2. Discuss basic model of motivation?
3. Explain the nature of leadership?
4. What are the objectives and importance of motivation?
5. Explain some of the important theories of motivation?
6. Define the basic concept of Maslow Need Hierarchy Model?
7. Give some limitations of Maslow Need Hierarchy Model?
8. Explain the Herzberg's Motivation-Hygiene Model?
9. Compare Herzberg and Maslow model?
10. Explain McGregor's Theory X and Theory Y?
11. Give the difference between Theory X and Theory Y?
12. What is leader?
13. What do you mean by leadership?
14. Explain the characteristics of leadership?
15. Give the difference between leadership and managership?
16. Explain the importance of leadership?

17. Discuss some of the approaches to leadership style?
18. Explain various leadership styles?
19. Discuss variety of styles that differ on the basis of motivation, power or orientation towards task and people?
20. Explain Blake and Mouton's Managerial Grid?
21. Explain fred fiedler's contingency theory?
22. Explain Hersey Blanchard's Situational Model?
23. What is coordination?
24. Explain the nature of coordination?
25. Give the difference between coordination and cooperation?
26. Explain the objectives of coordination?
27. Explain various types of coordination?
28. Discuss some of the difficulties in coordination?
29. Discuss principles of coordination?
30. Discuss the techniques of effective coordination?

Long Questions Answers

1. Write short notes on:
   (*a*) Mc Gregor's Theory of Motivation
   (*b*) Maslow Need Hierarchy Model.
2. Define motivation. Discuss different theories of motivation?
3. (*a*) Why motivation of workforce is important? Explain Maslow's theory of motivation?
   (*b*) What is the objective of motivation? Discuss Mc Gregor's theory of motivation?
4. Explain nature of leadership? "There is no one best style of leadership universally applicable to all situations", comment?
5. Define leadership. What are the different approaches to leadership styles?
6. Why is leadership important? Explain different leadership styles?
7. (*a*) "Leaders are not born but made," comment?
   (*b*) Discuss the managerial grid as developed by Blake and Mouton?
8. What is coordination? Discuss the objectives and principle of coordination?
9. On what basis the coordination and cooperation are related?
10. "Coordination is the essence of managership". Explain how coordination is achieved?
11. Discuss types of coordination? Explain the steps which should be taken to achieve effective coordination?
12. Explain the techniques of organizational control?
13. Give the differences between coordination and cooperation?
14. What are rewards?
15. Explain the two forms of rewards?
16. Who motivates Employees? Is everybody motivated?

CHAPTER

5

HRD and Organization

5.1 HUMAN RESOURCE PLANNING

5.1.1 Introduction

Traditionally, 5Ms that is Men, Materials, Machines, Methods and Money are regarded as the most important factors of production. But men (manpower or human resources) are the only dynamic resources with the organisation. Other factors are useless without their effective use by the human resources. Thus, it is logical that there should be proper manpower or human resource planning in the organisation to use the other resources effectively.

Human Resource (HR) are the people who are generally contributing in the society and human resource management (HRM) is making the optimum use of the people who are contributing in the society. Human Resource or manpower may be defined as, "the total knowledge, skills, creative abilities, talents and aptitudes and beliefs of the individuals involved".

The human resource include not only human beings, but also their knowledge, skills, attitudes, aptitudes etc. Manpower planning is essential to put the plans into action of the organisation for the achievement of its goals or objectives.

5.1.2 Meaning of Human Resource Planning

Human Resource may be defined as the quantitative and qualitative measurement of workforce required in an organisation. Where as planning may be defined as determining the workforce requirements to achieve the objectives of the organisation.

In other words, human resource planning may be defined as a process by which the management ensures the right number of people and the right kind of people at the right place at the right time doing the right things for which they are best suited for the achievement of organisational objectives.

It is the process of developing and determining objectives, policies and procedures of procurement of manpower.

In the words of Dale S. Beach, "Manpower planning is a process of determining and ensuring that the organisation will have an adequate number of qualified personnel."

5.1.3 Nature of Human Resource Planning

The first stage in the staffing process is the estimation of human resource requirements which is known as manpower or human resource planning. It's purpose is to make right kind of person available so that there is no shortage or surplus of manpower in any department. It may have a time perspective from one year to five years depending upon the expansion programme of the organisation.

Human resource planning is concerned with getting the right number of personnel having required education and training. Both present as well as future requirements of manpower should be determined in order to lay down manpower plans for an enterprise.

The requirements of jobs can be known from job descriptions and job specifications. Job descriptions lay down the duties and responsibilities of the job, risk and working conditions associated with it. Where as, job specificatications show the requirements of a job in terms of qualities required to perform it.

Human Resource Planning is a "two-phased" process, in this management can plan future manpower requirement as well as develop manpower plans. It helps in proper recruitment and selection so that right type of people are available to man various positions in the organisations. To develop the required skills in the employees, it also facilitates the designing of training programmes for them. Thus, manpower planning plays an important role in the effective management of personnel.

5.1.4 Needs (Significance) of Human Resource Planning

The benefits of manpower planning are as follows:

1. *Right Kind of Personnel:* Manpower (human resource) planning help in procuring personnel with the qualification, skills, knowledge, work experience and aptitude for work.
2. *Smooth Operation:* It helps in personnel selection and development as it ensures that adequate persons are selected well in advance so that they may be developed for anticipated openings. This would ensure smooth operations of the organisation.
3. *Full Utilisation of Plant:* Shortage or surplus of manpower in various departments will be revealed by manpower planning. Corrective steps can be taken in time.
4. It reduces labour costs by avoiding surplus of manpower. Overstaffing can be known quickly and steps taken accordingly.
5. It facilitates growth and diversification of business. Suitable employees are made available to handle new jobs. Adequate arrangements can be made to ensure the availability of technical personnel.
6. It leads to a greater awareness of the importance of sound manpower management throughout the organisation.
7. It serves as tool to evaluate the effect of alternative manpower actions and policies.

5.1.5 Process of Human Resource Planning

The process of human resource planning involves the following steps:

1. Ascertaining (Determining) Current Inventory: Analysis of current manpower supply may be undertaken by department, by function or by occupation. Appropriate adjustments are need to be made in the light of any changes in weekly hours of work, holidays, leave entitlements etc. Assessment of demand for the operative personnel presents less problems of uncertainity and manpower supply can be adjusted accordingly. But projection of manpower requirements for supervisory and managerial levels presents a complex problem. The required talents are not available at a short notice. This explains the need to determine the present manpower inventory in the enterprise. This will also help in drawing employment and training plans to meet the needs of certain skills of the future.

2. Estimating Human Resource Demand or Requirements: Forecasting or estimation of manpower requirements may be done by finding the total workload and then converting it into the manpower required. Workload analysis proceeds with the projection of sales in the coming year. The budget of productive hours is compiled by the use of standard hours for direct labour.

The standard hours per unit of a product are then multiplied by the planned units of production to give the total planned hours or the period. This is divided by the number of actual working hours for an individual worker to show the number of workers required. This gives a crude estimate of manpower requirements. Allowances may have to be made for absenteeism, labour turnover and idle time to know the actual workforce required. This is known as "workforce analysis".

The whole procedure of manpower forecasting has been illustrated in the following example:

(*a*) Planned output for a year = 50,000 units

(*b*) Standard manhours per unit = 4 hours

(*c*) Planned hours for the year ($a \times b$) = 2,00,000 hours

(*d*) Productive hours per man per year = 2,000 hours

(*e*) Number of direct workers required (c/d) = 100 workers

(*f*) Allowance for absenteeism, labour turnover and idle time = 25 workers.

(*g*) Number of workers required ($e + f$) = 125 workers.

3. Projecting Supply of Human Resources: It is necessary to estimate the number and kind of employees that are expected to be with the organisation after one year, two years, three years and so on. Such projections should be based on careful assessment of organisations current human resources and consideration of employee movement through retirement, resignation etc. in future.

4. Comparing Forecast Demand with Projected Supply: The human resource planner must compare the forecast demand with the projected internal supply of human resource before coming to any conclusion. This exercise should be carried out department wise as well as skill wise in order to know the deficiencies or surplus of various types of personnel in future. The possibilities of transfer of personnel from one department to another and promotion of personnel to higher jobs must also be considered to determine net human resource requirements in the organisation as a whole.

5. Planning Employment Programme to Meet Human Resource Needs: Long term employment programme must be chalked out to deal with expected deficits of manpower. It will include steps like recruitment, selection, placement, performance appraisal, transfer and promotion. Recruitment, selection and placement will be discussed later in this chapter.

6. Training and Development Programme: The purpose of assessing the manpower inventory is to find out the training and development needs of the organisation. The operative employees should be given appropriate training so that they may learn the required skills. Appropriate development programmes should also be designed for the executives so that the required talents should be developed.

7. Evaluation of Human Resource Planning Effectiveness: Human resource planning is closely linked to the goals of human resource management and those of the organisation. Shortage or surplus of human resources in any department or at any level should be minimized. Human resource planning programme will be effective if it can anticipate the deficiencies of various skills well in time so that steps are taken for the recruitment of required human resources. Similarly, if some type of personnel are surplus, they should be adjusted after giving them appropriate training.

If it is not possible, they may be phased out through voluntary retirement scheme.

5.1.6 Introduction on Job Analysis

Job analysis is a systematic collection and compilation of data about each job in the organisation to redesign each job in such a manner so as to distinguish it from other jobs.

Job Analysis
(Systematic Collection and Compilation of Job data)

Job Description
(A statement of duties and responsibilities)

1. Job title/name
2. Job location
3. Job summary
4. Duties to be performed
5. Machines, tools and materials
6. Relationship with other jobs
7. Working environment

Job specification
(A statement of qualifications required to perform a specific task)

1. Educational experience
2. Training and experience
3. Physique and health
4. Personality
5. Mental ability
6. Maturity
7. Creativity
8. Aptitude etc.

5.2 RECRUITMENT AND SELECTION

5.2.1 Introduction

After manpower planning has been done, the personnel Human Resource Manager can proceed with identification of sources of recruitment and finding suitable candidates for employment. Both internal, and external sources of manpower are used depending upon the types of personnel needed.

The selection process starts with the receipt of applications for various jobs from the interested candidates. Totally unsuitable candidates are rejected at the screening stage. The human resource (HR) department take various kind of tests to determine if the candidates would be able to do their job effeciently. Those passing this stage are called for employment interview. Candidates found suitable for employment are required to go through medical examination and references checking. The employment process is completed when appointment letters are issued to the candidates found suitable for employment.

5.2.2 Definition of Recruitment

"Recruitment is the process of identifying the sources for prospective candidates to stimulate them to apply for the jobs in the organisation".

It includes seeking and attracting a pool of people from which qualified candidates for job vacancy can be choosen. The process of identification of different sources of personnel is known as recruitment. It is a positive process as it attracts suitable applicants who apply for available jobs.

The process of recruitment:

1. Identifies the different sources of labour supply,
2. Assesses their validity,
3. Chooses the most suitable source or sources, and
4. Invites applicants from the prospective candidates for the vacant jobs.

Recruitment is concerned with listing the candidates for consideration of selection to various jobs. It enables the management to select suitable employees for different jobs. Recruitment is done before selection or employment of workers. Recruitment is a positive process of searching the prospective employees and attracting then to apply for the vacancies.

5.2.3 Sources of Recruitment

The sources of recruitment of employees may be either internal or external. Internal sources of recruitment means the personnel already working in the organisation. External sources denote recruitment from sources outside the organisation.

Sources of Recruitment

Internal Source
- Transfers
- Promotions
- Re-employment of ex-employers

External Source
- Factory gate
- Advertising
- Employment Exchange
- Employment Agencies
- Recommendation
- Labour contractors

Fig. 5.1

1. **Internal Sources:** The internal sources of recruitment include:

   (*a*) ***Transfer:*** Transfer means shifting a worker from one job to another having similar status and responsibilities.

   According to Dale Yoder," A transfer involves the shifting of an employee from one job to another without special reference to changing responsibilities or compensation". That is Transfer means shifting of an employee from one job to another, one unit to another or shift to another and may involve a new geographical location.

   (*b*) ***Promotion:*** A promotion is the transfer of an employee to a job that pays more money and that enjoys some preferred status. According to Pigors and Myres," Promotion is the advancement of an employee to a better job in terms of greater responsibilities, greater skills, higher status and higher pay". The existing employees may be given training for promotion to get higher position in the enterprise. If a vacancy is filled from within the enterprise, it acts as a stimulating force for the employees to develop themselves.

   (*c*) ***Re-employment of Ex-employers:*** It refers to employing the employees who have left the organisation because of some reason. It may be on temporary or permanent basis.

Merits of Internal Sources

1. The internal candidates having intimate knowledge of the organisation structure can handle the new jobs easily.
2. Promotion will increase the morale of the employees.
3. Promotion will ensure sincere effort on the part of employees because they know that they may rise to senior positions by promotions.
4. The cost of training the insiders for the senior post will be less.
5. The cost of internal recruitment is very low.
6. Reliance on internal sources will enable the enterprise to make best use of the capabilities of its employees.

Demerits/Limitations of Internal Sources

1. Better qualified outsiders may not be reached by the management. Moreover, the person promoted to a senior job mayn't possess the required qualities.
2. The mobility of manpower is restricted. Once a man joins an organisation, he will stick to that organisation. Employee turnover to a certain extent is good for the health of the organisation.
3. In promotion, there may be clash between factors like seniority and ability. Such decision may lead to favouritism.
4. Further growth of the business is hindered by the limited talent of the insiders. The entry of fresh talent into the organisation will be restricted.

2. **External Sources:** Many a times, suitable candidates are not available from the internal sources. So, in order to meet the demands of the company, the external sources have to be used.

The personnel department may use the following external sources of recruitment:

(*a*) ***Direct Recruitment:*** The unemployed persons may contact the employment section of various enterprises to determine if they can be casually employed. Many firms follow the practice of recruitment at factory gate to fill casual vacancies, such workers are known as Budli workers. This method is very useful for recruiting unskilled workers. It does not involve cost of advertising the vacancies. Whenever the regular worker absent themselves in large number or whenever there is rush for work, this source of recruitment may be used.

(*b*) ***Casual Callers:*** In case of reputed organisations, many qualified people apply for employment on their own initiative , such applications are known as unsolicited application or casual callers. They serve as a good source of manpower. A proper file may be kept of such applications and the candidates may be called whenever the need arises. For example- an applicant visits various colleges and give his resume copy so that if any vacancy arrives, he can be called. A file is always maintained of all these applicants.

(*c*) ***Advertising:*** Advertising a job has become the fashion of the day with the large companies particularly when the vacancy is for a higher post or when there are a large number of vacancies. This helps in approaching the candidates spread over different parts of the country. Information about the job description and job specification can be given in the advertisement so that only the candidates who think themselves to be suitable may only apply. This method also increases the choice of the management. For example-advertisement given in the newspapers.

(*d*) ***Employment Exchanges:*** There is a network of employment exchanges in the country which is run by the government has also made it compulsory for the employers to fill some vacancies through the employment exchanges. Thus employment exchanges play an important role.

(*e*) ***Placement Agencies:*** Professional bodies have come into existence which provides their services to the enterprises for recruitment and selection of employees. These agencies specialise in the supply of particular categories of workers.

(*f*) ***Management Consultants/Head Hunters:*** Management consultancy firms help the organisations to recruit technical, professional and managerial personnel. They specialise in middle and top level executive placements. They maintain data bank of persons with different qualifications and skills and even advertise the jobs on behalf of their clients to recruit right number of personnel.

(*g*) ***Campus Recruitment:*** Colleges and institutes of management and technology have become a popular sources of recruitment for technical, professional and managerial jobs. Many big organisation maintain a close contact with the universities, vocational schools and management institutes to recruit qualified personnel for various jobs. Recruitment from educational institutions is the well established practice of business and other organisations. This is referred to as "campus recruitment".

(*h*) ***Recommendations by the Existing Employees:*** Some employers recruit employees on the basis of recommendations of the existing workers. The policy creates goodwill among the present employees and helps in finding reliable candidates. For example– Many multinational companies recruit the employees on the referral basis that is referral week is being conducted may be once or twice a year and the recruitment is done.

(*i*) ***Labour Contractors:***Recruitment may also take place through intermediaries known as agents or labour contractors. In some cases, the labour contractors are themselves the members of the companies concerned. This system of recruitment has lost popularity these days. If a labour contractor leaves enterprise, the workers employed through him will also leave the organisation. Thus, it is not a reliable method of employment.

(*j*) ***Telecasting:*** The practice of telecasting of vacant posts over T.V. (Doordarshan and other channels) is gaining importance these days. The detailed requirements of the job and the qualities required to do it are publicised along with the profile of the organisation where the vacancy exists.

(*k*) ***Web Publishing:*** Internet is becoming a popular source of recruitment these days. There are certain websites specifically dedicated for the purpose of providing information about job seekers and job opening both. In fact, websites such as www.naukri.com, www.monster.com, www.jobstreet.com,www. careerAge.com etc. are very commonly visited both by the perspective employees and the organisations searching for suitable people.

Advantages/Merit of External Source of Recruitment

1. External sources provide a large number of applicants. This permits the enterprise to have a free hand is making the right choice of candidates.
2. The enterprise can expect to get fresh, talented candidates from outside. This means infusion of new blood and new ideas into the enterprise.
3. Internal candidate have to compete with outside candidates for the higher jobs. It serves as inducement for the existing employees to show better performance.

Limitations of External Recruitment

1. Recruitment from outside may cause dissatisfaction and frustration among the existing employees who aspire for promotion.
2. External recruitment takes more time than the internal recruitment since the enterprise has to publicise the vacancies, and wait for the response of prospective candidates.
3. The prospective candidates from outside may or may not be good for the enterprise. There is no guarantee that the enterprise will be able to attract suitable applicants even after advertisement and other steps.
4. It is very costly to recruit staff from external sources. A lot of money has to be spent on advertisement and processing of applications.

5.3 MEANING OF SELECTION

Selection is the process of choosing the best person for particular job. It leads to employment of workers. Selection is the negative process as it involves rejection of unsuitable candidates. More candidates are rejected than are selected.

Selection involves several steps to weed out the unsuitable candidates for the job under consideration. Criteria are laid down at each stage. Those who do not fulfil these criteria are rejected.

"Selection is the process of choosing from among the candidates from within the organisation or from the outside, the most suitable person for the current position or for the future position".

– *Dale Yoder*

"Selection is a managerial decision making process as to predict which job applicants will be successful if hired".

– *S.P. Robins*

5.3.1 Recruitment Vs Selection

Recruitment is a positive process searching the prospective employees where as selection is a negative process because it involves rejecting of unsuitable candidates. However, both recruitment and selection are the two phases of the employment process. Recruitment being the first phase envisage taking decision on the choice of taping the sources of labour supply.

Selection is the second phase which involves giving various types of tests to candidates and interviewing them in order to select the suitable candidates.

Recruitment	*Selection*
1. It aims at searching for sources of labour and stimulating people to apply for jobs.	1. It aims at selection of right kind of people for various jobs.
2. It is a positive process.	2. It is negative process as it involves rejection of unsuitable candidates.
3. It creates a large pool of applications for jobs.	3. It leads to screening out unsuitable candidates.
4. It is simple process. It involves contacting the various sources of labour.	4. It is complex and time consuming process. The candidates have to clear a number of hurdles before they are selected for job.

5.3.2 Significance of Selection

The benefits of selecting right kind of people for various jobs are:

1. Proper selection and placement of personnel go a long way towards building up a suitable workforce. It will keep the rates of absenteeism and labour turnover low.

2. Component employees will show higher efficiency and enable the organisation to achieve its objective effectively.
3. The rate of industrial accidents will be considerably low if suitable employees are placed on various jobs.
4. When people get jobs of their taste and choice, they get higher job satisfaction. This will build up a contended workforce for the organisation.
5. The morale of the employees who are satisfied with their jobs is often high.

5.3.3 Steps in the Selection Procedure

1. **Preliminary Interview:** It is generally taken for the executive jobs. It is brief, about three to four minutes and has the job of eliminating the totally unsuitable candidates. Lack of certain requirements in education, training or experience may determine unsuitability. The candidates who clear the interview are asked to fill in the application blank. It is mainly for the management jobs.
2. **Screening of Application Forms:** When a candidate comes for an interview, he is given an application form on request to fill. It asks for a written records of qualifications, Experience and other qualities of the candidate. The application form should be as simple as possible and incorporate questions having effect on the suitability of the applicant for the job. Application forms are processed by the screening committee and properly screened to select the candidates who are to be given the employment tests and called for an interview. The purpose of preliminary screening is to reject the totally unsuitable candidates.

   This will save a lot of time and money.

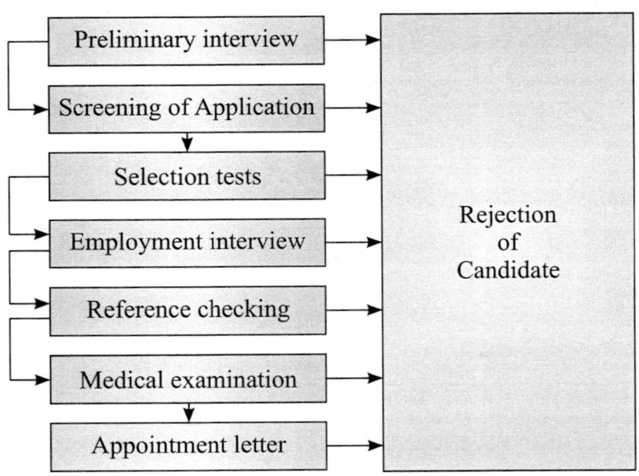

Fig. 5.2 Selection Procedure

3. **Selection or Employment Test:** Employment tests are being widely used to select persons for various jobs. These tests help in matching the characteristics of individuals with the vacant jobs so as to employ right type of personnel. The following types of tests are given to the candidates:

(*a*) Intelligence test

(*b*) Aptitude test

(*c*) Occupational or Professional test

(d) Personality test.

4. **Employment Interview:** After an applicant clears the various employment tests. He is then called for final interview. The interview is taken by the board consisting of the personnel manager and the representatives of other departments. The interview is held to determine the suitability of the candidate and to provide them a complete picture of the responsibilities associated with the job.

   It is of great importance to the applicant and the employer. There should be proper physical arrangement for the interview. It will enhance the reputation of the employers and give confidence to the candidates. Privacy and comfort should not be forgotten to ensure that the objectives of interview are achieved.

   If this is not done, there would be wastage of money on interviewing and right type of candidates may not be selected. The interviewers should also come fully prepared for the interview and should give proper attention to each candidate.

5. **Reference Checking:** Many employers request names, addresses and telephone number of reference for the purpose of verifying information and gaining additional information on an applicant. Previous employers known persons, teachers and university professors, can act as references.

6. **Medical Examination:** It is given to the selected candidates to ascertain their physical capability and protect the employer from employing physically unsuitable candidates. Medical test is given by the medical expert appointed by the employer.

7. **Appointment Letter or Job Offer:** When a candidate has cleared the above listed hurdles in the selection procedure, he is given an appointment letter and is requested to join the organisation. When the employee reports for duty, he is given some sort of orientation and is placed on the job for which he is appointed.

5.4 PSYCHOLOGICAL TESTS AND INTERVIEWING

5.4.1 Psychological/Employment Tests

Individuals differ with respect to physical characteristics, capacity, level of mental ability, likes, dislikes as well as personality traits. The differences among the individual candidates can be analysed with the help of various psychological and trade tests. These tests can provide important information about the candidates as regards his/her intelligence, aptitude, interest, personality etc.

Various types of tests available to match the characteristics of the candidates with the requirements of the job advertised are as follows:

1. **Intelligence Tests:** It is used to judge the mental ability and capacity of an applicant. It measures the individuals learning ability, *i.e.*, ability to catch

or understand instructions and also ability to make decisions. There are many verbal and non-verbal intelligence tests constructed by psychologists for different kind of jobs.

2. **Aptitude Tests:** Aptitude means the potential which an individual has for learning the skills required to do a job efficiently. Aptitude tests measure an applicant's capacity and his potential for development. Aptitude tests are the most promising indices for predicting worker's success.

   There are two types of aptitude tests:

   (*a*) ***Cognitive Tests:*** Which measure mental and intellectual aptitude.

   (*b*) ***Motor Tests:*** Which measure physical dimensions.

3. **Achievement or Proficiency Tests:** Proficiency tests are those tests which are designed to measure the skills which are already acquired by the individuals. They are also known as occupational or trade tests. The level of knowledge and proficiency of an applicant to do a specific job is used to test by the achievement tests.

4. **Interest or Motivation Tests:** Interest tests identify patterns of interest, *i.e.*, areas in which the individual shows special concern, fascination and involvement. These tests would suggest what types of jobs may be satisfying to the employees.

5. **Situational or Personality Tests:** These type of tests are designed to observe how the job applicants react to real life situations. It sees the quality of the person as a whole, *i.e.*, the combination of aptitude, interest, usual mood and temperament. They are concerned with discovering clues to an individuals value system, his emotional reactions, maturity etc. For example- a candidate in a managerial job may be given a typical problem which requires on the spot decision-making.

Importance of Employment Tests

It has following advantages:

1. A test is an objective and standardized sample of certain qualities. It tends to eliminate business in selection of personnel.
2. Tests can identify talents of individuals which might otherwise be overlooked.
3. Tests reduce the costs of selection and placement because large number of applicants can be evaluated within the least possible time.
4. Psychological test can measure the aptitude of candidates and predict their success.
5. Test provide a healthy basis for evaluating applicant's performance. They compel the interviewers to think through their evaluation more carefully.

5.4.2 Employment Interview

According to Jucius Michael, "An interview is a face-to-face, oral observational and personal appraisal method". It is used as a means of getting information from the candidate. It also includes giving information that will help the applicant make his mind about the company.

Interviewing the candidates is an important aspect of selection procedure. The final selection of an applicant is partly based on the performance of the candidate in different tests and partly on his performance in the final interview. In the interview, the candidate has to appear before the interviewer or a group of interviewers. The candidates overall is judged in the interview. The interview may last for ten to twenty minutes or even more. Various questions are asked from the candidate in order to judge his ability, knowledge, capacity and so on.

Interviewing technique is used in all companies and in the case of all categories of staff to be recruited.

Importance of Interviewing

For the selection of right type of people, employment interview is very important. The advantage of employment interview is as follow:

1. The employer can assess the personality traits of the candidate as there is a face-to face contact between the employer and the candidate.
2. The candidate can seek more information about the employer and the job. This creates better understanding in the mind of the candidate.
3. Candidate's communication skills, his way of thinking can be judged in the interview.
4. Interview is very important where the candidate has not to go through employment tests. The informations contained in the application form can be checked during the interview.
5. Many companies do not follow the elaborate selection procedure as it is costly and time consuming. They can rely on interview if it is properly planned and administered.

5.4.3 Kinds of Interview

Depending on the purpose, the interview programme of an enterprise may take any one or more than one of the following forms:

1. **Preliminary Interview:** This interview is conducted to see the worth of a candidate which is of few minutes duration. It must be conducted very carefully by some expert as many desirable applicants may be eliminated at the outset.
2. **Stress Interview:** In this type of interview, stress environment is created in an interview hall. The applicant is intentionally put under stress by interrupting him at various points, criticising his view points and sometimes keeping silence after he has finished speaking. It requires the interviewer to be tactful as well as skillful. It is used for higher posts.
3. **Background Information Interview:** This type of interview is conducted to get the background information or we can say that when the history of the applicant has to be known in terms of his experience, education, health, insurance, likes dislikes etc. It is more or less like the preliminary interview.
4. **Discussion Interview:** It is a non-directive interview conducted informally by the interviewers by giving an opportunity to the interviewer to speak his mind freely.

5. **Group Interview:** It is an interview of certain applicants in a group. Reaction of the applicants with each other can be seen by the interviewers. It is used where it is essential to know how the candidates behave in groups.
6. **Non-directed Interview:** The direct and specific questions are not asked by the interviewer in this type of interview. An atmosphere is created in which the candidate is free to talk, on the subject selected by the interviewer. Here, the interviewer is basically a passive listener as he avoids making value judgements interrupting the candidates and revealing his personal opinion.
7. **Patterned or Structured Interview:** In this type of interview, series of questions are phrased in advance which put light on the strategic parts of the applicants background. Answers to these questions are compared with a critical score and used in determining one who is to be selected. In this interview process, these standard questions are asked as they are written, the order may be varied but not the phrasing of the questions.

5.5 PLACEMENT

5.5.1 Meaning/Concept of Placement

The last step in the procurement of the employees is placing the selected candidates on the right jobs and introducing them to the organisation.

According to Pigors and Myers, "Placement may be defined as the determination of the job to which a selected candidate is to be assigned, and his assignment to the job." It is matching of what the supervisor has reason to think he can do, with the job requirements (job demands); it is matching of what he imposes (in strain, working conditions) and what he offers in the form of payroll, with other promotional possibilities etc. It is the assignment or reassignment of an employee to a new job. It includes the initial assignment of new entrants and the transfer and promotion of the existing employees.

5.5.2 Significance/Importance of Placement

It is important for both the organisation as well as staff that each employee should be placed on a suitable job. While doing so, the factor to be considered is not only the suitability of the job to the individual. But, initial placement is always a problem because there is little knowledge about the new employee. Therefore, placement should be done after due considerations of the demands of the job, as well as the social, psychological needs of the individual. An enterprise which has spent a lot of time and money in making a very careful selection may lose due to wrong placement.

Right placement of workers can have the following advantages:

1. Reduced labour turnover rate.
2. Reduced absenteeism rate.
3. Increased safety of workers and lower accidents.
4. Increased morale of workers.
5. Better human relations in the organisations.

Placement is not an easy process. It is very difficult to adjust for a new employee who is quite unknown to the job and the environment. For this reason, the employee is generally put on a probation period ranging from one year to two years. At the end of this probation period, if the employee shows good performance, he is confirmed as a regular employee of the organisation. Thus the probation period or trial period is the transition period at the end of which management takes a decision whether to make the employee regular or discharge him from the job.

5.5.3 Placement of New Employees

Placement decisions are generally taken by line executives. It is the supervisor who in consultation with higher levels decide the placement of each employee. The personnel department's role is to advise the line managers about the human resource policies of the company and to provide counselling to the employees.

While taking the placement decision, the following consideration or principles must be kept in mind:

1. **Job Requirements:** An employee should be placed on the job according to the requirements of the job such as physical and mental ability, eyesight, hearing, stress etc. The job shouldn't be adjusted according to the qualification and abilities of the employees. Job placement profile charts can be used to match the worker's physical and mental abilities with the job requirements. This profile chart displays an evaluation of both job requirements and worker abilities for a key features of the job so that the management can easily determine how well worker fits a job.
2. **Suitable Qualifications:** The job should be offered to only that person who is suitably qualified. Over qualified and under qualified persons might create problems for the organisation in the long run.
3. **Adequate Information to the Job Incumbent:** The employee should be provided with the complete information and facts relating to the job, including the working conditions prevailing in the firm. He should also be made known to the rewards associated with the performance levels.
4. **Commitment and Loyalty:** While placing the new employee, an effort should be made to develop a sense of commitment, loyalty and cooperation in his mind so that he may realise his responsibilities better towards the job, the organisation and his associates.
5. **Flexibility:** The placement in the initial period may be temporary as changes are likely after the completion of training. The employee may be later transferred to the job where he can do better justice.

Proper placement helps to improve employee morale. The capacity of the employee can be utilised fully if he is placed on the job for which he is most suitable. Right placement also help to reduce labour turnover, absenteeism and accident rates. If a candidate adjusts himself to the job and continues to perform as per expectation, it might mean that the candidate is properly placed.

5.6 INTRODUCTION OR ORIENTATION OF EMPLOYEES

According to Edwin B.Flippo, "Induction is concerned with introducing or orienting a new employee to the organisation. It is the welcoming process to make the new employee feel at home and generate in him a feeling of belongingness to the organisation". It is the first step in a proper communication policy which seeks to build a two-way channels of information between the management and employees.

The new employee, on his joining the organisation, must be helped to get adjusted and acquainted with the fellow employees and the work environment. Rather than leaving him to make his own way through the organisation. It is much better to properly and systematically introduce him to the company, its philosophy, its place in the industry or economy, its major policies etc. This phase of induction is properly handled by the personnel department. Some large organisations show pictures telling about their history and operation to give a better idea of the company to the new employees who have joined. The new employee is also given the necessary information about canteen, lunch period, rest hours, convenience facilities etc.

A complete induction programme also consists of a follow-up after few weeks to ascertain how well the new employee has adjusted himself with his job and environment. This provides an opportunity to clarify or remove any misunderstanding in him. But then also, if an employee is not found adjusted after that, it may be the case of wrong selection or wrong placement. It thus, gives an early warning of a later trouble. In any case, the follow-up interview enhances the employee's satisfaction by demonstrating to him the company's interest in his welfare.

5.6.1 Induction Procedure/Techniques

Induction may be informal in case of small firms and formal may be of duration of two or three weeks in case of large organisations. Induction training should not be too lengthy. It need not necessarily be given on the day when the new employee joins the work, but if a formal course is to be attended after two or three weeks on the job, the initial introduction and the immediately needed information may be given by the superior of the department.

Orientation procedure consist of the following steps:

1. The new employee should be given a definite time and place to report.
2. The immediate boss or the supervisor should meet and welcome the new employee to the organisation when the latter reports for the duty.
3. Complete information about the vacations, probationary period, working hours, medical leave, suggestion styles etc. should be conveyed to the employee.
4. Departmental orientation should be conducted which include introduction to the department, explanation of the functions of the department, job instructions and to whom he should look for help and guidance when he has any problem.
5. Verbal explanations may also include a wide variety of printed materials, employee handbook, employee manuals, house journals pamplets etc.

6. The induction programme should be handled by the persons who are through with the course contents. The success of the induction course also depend upon the quality of the trainers and their ability to draw out the interest of their listeners.

5.6.2 Contents of Induction Programme

Topics to be covered in induction

1. Company's history, mission, vision and philosophy.
2. Products and services of the company.
3. Company's organisation structure.
4. Location of departments and employee services.
5. Employees activities like clubs, credit society.
6. Personnel policies and procedures.
7. Standing orders.
8. Rules and Regulations.
9. Terms and conditions of services.
10. Grievance procedures.
11. Safety measures
12. Benefits and services for employee's.
13. Training, promotions and transfer facility.
14. Career advancement schemes.
15. Counselling facility.

5.7 TRAINING AND DEVELOPMENT

5.7.1 Introduction

Training is concerned with increasing the knowledge and skills of the employees for doing specific work where as development is concerned with the growth of the employees in all aspects. Training enhances job skills, where as development shapes attitudes of the employees.

5.7.2 Concepts of Training and Development

5.7.2.1 Meaning of Training

According to Edwin B. Flippo, "Training is an act of increasing the knowledge and skills of an employee for doing a particular job". It is an organised activity for enhancing the technical skills of the employees to enable them to do a particular job effectively and efficiently. In other words, it provides the workers with facility to gain technical knowledge and to learn new skills to do specific jobs. Training is equally important for the existing as well as new employees. It enables the new employees to get acquainted with their jobs and also increase the job related knowledge and skills.

5.7.2.2 Meaning of Development

Development involves growth of an employee in all respects. It is the process by which the employees acquire skills and competence to do their present jobs and increase their capabilities for handling higher jobs in the future. Thus, development has a wider perspective. It involves acquiring not only technical skills but also problem solving and decision-making skills.

That is why, development is more relevant for the managers or executives.

It is a continuous process of building competencies of the employees. This term is generally used in relation to managers or executives and is described as management development or executive development.

5.7.2.3 Comparison of Training and Development

Training	*Development*
1. Training means imparting skills and knowledge for doing a particular job.	1. Development means the growth of an employee in all respects.
2. It is concerned with main training and improving current job performance. Thus, it has a short-term perspective.	2. Executive development seeks to develop competence and skills for future performance. Thus, it has a long-term perspective.
3. It imparts job related technical skills.	3. It shapes attitudes and imparts technical, human and conceptual skills.
4. It is job centred in nature. It seeks to make the workers proficient in their existing jobs.	4. It is career-centred in nature. It seeks to prepare the employees for handling higher level jobs.
5. The term 'training' is used in the sense of imparting specific skills among operative workers and employees.	5. The term 'development' is associated with the overall growth of the executives. Depth of knowledge gained is very wide.
6. The boss takes initiative for imparting training to his subordinates. The role of the boss or trainer is very important in training.	6. The individual takes the initiative himself for his growth and development. The executive has to be internally motivated for self development.
7. It is imparted through on-the job methods like coaching, vestibule training, job rotation etc.	7. It is based more on off the job methods like lectures, conferences, senstivity training etc.

5.7.3 Need for Training

The need for training the employees arises due to the following factors:

1. **Higher Productivity:** It is essential to increase productivity and reduce cost of production for meeting competition in the market. Effective training can help increase productivity of the worker by imparting the required skills.
2. **Improvement in Quality:** Now-a-days, customer's requirement keep on changing with the changing world and customer have become more quality

conscious. So, as to satisfy the customers, quality of products must be continuously improved through training of workers.

3. **Reduction in Learning Time:** If the employees learn through trial and error, they will take longer time and even may not be able to learn right methods of doing work. So, systematic training with the help of trained instructors is essential to reduce the learning period.
4. **Reduction of Absenteeism and Turnover:** Training creates a feeling of confidence in the minds of the workers. It gives them security at the workplace. As a result, labour turnover and absenteeism rates are reduced.
5. **Industrial Safety:** The workers who are trained know the use of various safety devices in the factory and can handle the machine safely. Thus, they are less prone to industrial accidents.
6. **Technology Update:** Technology is changing at a fast pace. The workers must learn new techniques to make use of advanced technology. Thus, training should be treated as a continuous process to update the employees in the new methods and procedures.
7. **Effective Management:** Training can be used as an effective tool of planning and control. It develop skills among workers and prepare them for handling present as well as future jobs. It helps in increasing productivity and quality which are cherished goals of any modern organisation. It also helps in reducing costs of supervision, wastages and industrial accidents.

5.7.4 Importance of Training to the Organisation

Training is beneficial for the organisation as discussed below:

1. **Economy in Operations:** Trained workers will be able to make better and economic use of materials and equipments, wastage will be low. In addition, the rate of accidents and damage to machinery and equipment will be kept to minimum by the well trained employees. This will lead to less cost of production per unit.
2. **Greater Productivity:** A well trained employee usually shows higher quality of work-output and greater productivity than an untrained employee. Training increases the skills of the employees in the performance of a particular job. An increase in the skills usually helps to increases both quantity and quality of output.
3. **Uniformity of Procedures:** With the help of training, the best available methods of work can be standardised and made available to all employees. Standardisation will make high level of performance rule rather than the exception.
4. **Systematic Imparting of Skills:** The employees do not learn by trial and error method or by observing others and waste their time if the formal training programme exists in the organisation. Thus, a systematic training programme helps to reduce the learning time to reach the acceptable level of performance.
5. **Less Supervision:** If the employees are given proper training, then the responsibility of the supervision is lessened. So, training does not eliminate

the need for supervision, but it reduces the need for detailed and constant supervision.

6. **Creation of Inventory of Skills:** When the organisation require totally new skills, then it has to face great difficulties in the employment. Training can be useful in finding out promising men and in removing the defects in the selection procedure. It is better to select and train the employees from within the organisation rather than seek the skilled employees from the outside sources.
7. **Higher Morale:** A good training programme will mould the employees attitude to achieve support for organisational activities and to obtain greater cooperation and loyalty. With the help of training, dissatisfaction, complaints, absenteeism and turnover rate will be reduced among the employees. The morale of the employees is increased if they are given proper training.

5.7.5 Importance of Training to the Workers

Training is beneficial for the employees in the following ways:

1. **Acquisition of New Skills:** Employees acquire new skills and knowledge which will help them in their career prospects.
2. **Industrial Safety:** Trained workers are less prone to accidents as they can handle the machines carefully. They also know the use of various safety devices in case of any emergency.
3. **Preparation for Promotion:** The efficient workers sharpen their skills during training. This helps them to prepare for promotion to higher jobs.
4. **Higher Wages:** With the help of training, productivity of the workers is increased. Due to this, they earn higher wages and bonus.
5. **Versatile Utility:** By this term it is meant that the trained employees can shift from one job to another easily if there is no work at the present job.
6. **Higher Morale:** Trained employees know their jobs well and are less prone to accidents. They are satisfied with their jobs which increases their morale.

5.7.6. Types of Training

Depending upon the purpose of training the following kind of training programme are used in the industry:

1. **Induction Orientation Training:** This training has been explained before in detail in 6.6.
2. **Apprenticeship Training:** It involves imparting knowledge and skills in a particular trade or craft such as printing, tool making etc. The government of various countries have passed laws which make it compulsary on certain employers to provide apprenticeship training to the young people. It is desirable in the industries which require a constant flow of new employees expected to become all-round craftsmen.

It is very much prevelant in printing trade, building and construction, and vocations (profession entered) like mechanists, electricians, welders etc. It is similar to on the job training.

Under apprenticeship training, the trainee is placed under the supervision of an experienced person who imparts him the necessary skills and regulates his performance. The advantages of apprenticeship training to the trainers are that they receive stipend while learning and acquiring valuable skills which command high wage in the labour market. In India, there are so many "earn when you learn" schemes, both in the private as well as public section undertaking. This is also advantageous to the employers. Some employers look upon apprentices as a source of cheap labour.

5.7.7 Methods of Training

The important methods of imparting training to the operative employees are as follows:

On-the-job-Management
(at the place of work)

- Coaching
- Understudy
- Job Rotation
- Vestibule training

Off-the-job Management
(away from the place of work)

- Special lectures cum discussion
- Conference
- Seminar
- Role playing

Fig. 5.3

1. On the Job Training: Under this method, the worker is given training by his immediate supervisor itself at the workplace. In other words, the worker learns in the actual work environment. It is based on the principle of "learning by doing". On the job training is considered to be the most effective method of training the operative personnel.

On the job training is suitable for imparting skills that can be learnt in a relatively short time. It has the chief advantage of strongly motivating the trainee to learn. It permits the trainee to learn on the equipment and in the work environment. It is not located in an artificial situation. This method is relatively cheaper and less time consuming. Here supervisors play an important part in training subordinates.

There are four methods of on-the-job training:

(*a*) **Coaching:** Under this method, the supervisor imparts job knowledge and skills to the subordinate. The emphasis is coaching or instructing the subordinate is on learning by doing. This method is very effective if the superior has sufficient time to provide coaching to his subordinates.

(*b*) **Understudy:** The superior gives training to his subordinate as his understudy or assistant. The subordinate learns through experience and observation. This technique prepares the subordinates to assume the responsibilities of the superiors job in case the superior is absent or if he leaves the organisation.

(*c*) **Job Rotation:** The trainee is systematically transferred from one job to another so that he may get experience of different jobs. Rotation of an employee from one job to another should not be done frequently. He should be allowed to stay on a job for sufficient time so that he may acquire the full knowledge of the job. This broaden the horizon and capacity to do a variety of jobs.

Job rotation is used by many organisations to develop all-round workers. The employees learn new skills and gain experience in handling different kind of jobs. They also come to know the interrelationship between different jobs. Job rotation is also used to place workers on the right job and prepare them to handle other jobs in case of need.

(*d*) **Vestibule Training:** Vestibule training is an attempt to duplicate as nearly as possible the actual conditions of the workplace. The learning conditions are carefully controlled. The trainees can concentrate on training because they are not under any pressure of work. Their activities do not interfere with the regular process of production. Thus, vestibule training is very much suitable where a large number of persons are to be trained and where mistakes are likely to occur which will disturb the production schedules. The main emphasis is on learning rather than on production. It is also known as simulation or artificial training. Best examples of vestibule training is the training given to the pilots. They are given training in the similar environment so that they can do good practice and become expert.

2. Off-the-Job Training: Under this method, worker is required to undergo training for a specific period away from the workplace. Off-the-job methods are concerned with both knowledge and skills in doing certain jobs. The workers are free from tension of work when they are learning. The prominent off- the-job methods of training and development are described below.

(*a*) **Special Lectures and Discussions:** Training through special lectures is also known as "class-room training". It is more associated with knowledge than with skills. The special lectures may be delivered by some executives of the organisation or specialist from professional and vocational institutes. Many firms follow the practice of inviting experts for special lectures on matters like health, safety, productivity, quality etc.

These days, lectures also include discussions, films, demonstrations etc. The main emphasis of the special lectures is to enrich the workers with advanced knowledge in specific areas related to the effective performance of the job. Such courses can be more effective if they ensure two-way traffic between the trainer and the trainee. Discussions are more effective than the simple lectures because of participation by the trainees or listeners.

(*b*) **Conferences:** A conference is a group meeting conducted according to an organised plan in which the members seek to develop knowledge and understanding of a topic by oral participation.

It is an effective training device for all the members and the leader present in the conference. Member present in the conference can learn from other members by comparing his/her opinion with those of others. He learn to respect the view points of other and also realises that there is more than one workable approach to any problem.

The literal meaning of conference is "consultation". But, conference implies sharing some information with the audience consisting of large number of people. It is conducted in a big hall where the participants are allowed to exchange their views and solve their queries. The proceedings of the conference are conducted by the chairman who is also responsible for summing up the proceedings of the conference. These days video conferencing is also gaining popularity under which people can participate in the conference through link via satellite.

(*c*) **Seminar:** A seminar is organised like a conference, but it is comparatively on a smaller scale. It is generally placed around a single topic such as "Emerging issues in Human Resource Management or Information Technology in 2011". The topic or theme is examined in great detail by different experts in the concerned fields. The experts make their presentations and answer the question raised by the participants. In practice, the terms conference and seminar are used interchangeably.

(*d*) **Role Playing:** This technique is used for human relations and leadership training. Its purpose is to increase the trainee's skills in dealing with others. In this method, two trainees are assigned different role of a sales manager and the other of a customer. Both will interact with each other and play their respective roles. This will also help the trainees in learning how to behave in a conflict situation. This will also learn to appreciate the view points of each other.

5.8 MANAGEMENT DEVELOPMENT

"Executive or management development is a planned process of learning and growth designed to bring behavioral change among the executives."

It is continuous process of learning. It implies that there will be a change in knowledge and behavior of the individuals undergoing development programme. The employee will be able to perform his present job better and will increase his potential for future work.

Managers develop themselves by participating in formal training courses organised by the organisation. They also make use of actual job experience in learning new behavior and the organisation must provide opportunities for development of its managers. But an equal, but more important, counterpart to the efforts of the organisation are those of the individuals. Self development is an important concept in the whole programme of executive development.

The characteristics of management development are as follows:

1. It is an organised process of learning rather than a haphazard or trial and error approach.
2. It is a long term process as managerial skills cannot be developed overnight.
3. It is an ongoing exercise rather than a "oneshot" affair. It continues throughout an executive's entire professional career because there is no end to learning.
4. Management development aims at preparing managers for better performance and helping them to realise their full potential.

5. Executive development is guided self development. An executive can provide opportunities for development of its present and potential managers. But the image for learning has to come from the executive itself. Executive development is possible only when the individual has the desire to learn and practice what he learns.

The organisation can provide the environment for development or advancement but the initiative for development has to be taken by the individual himself. Unless the person has the required potential and urge, he cannot be considered fit for development. He can also be given the necessary guidance to help him face the present job's problems or to prepare him for the higher jobs. Training is also necessary to prevent obsolescence of knowledge and skills of individuals.

Executive development programmes seek to develop the overall personality of the executives and not only the skills necessary to do the job efficiently. The job of executive is a very challenging one. So, the executive development programmes should be more concerned with conceptual and human skills of the executives. A comprehensive development programme uses various learning aids to increase the knowledge, decision making ability, skill in dealing with people and maturity of the managers.

5.8.1 Purpose of Management Development

The purpose or objective of management development programmes are discussed below:

1. To sustain better performance of managers throughout their careers.
2. To improve the existing performance of managers at all levels.
3. To encourage existing managers to increase their capacity to assume and handle greater responsibility.
4. To enable the organisation to have the availability of required number of managers with the required skills to meet the present and anticipated (future) needs of the organisation.
5. To replace elderly executives who have risen from the ranks by highly competent and academically qualified professionals.
6. To provide opportunities to the executives to fulfill their career aspirations.
7. To ensure that the managerial resources of the organisation are utilised optimally.

5.8.2 Methods of Management Development

A large number of methods of management development have been designed. Some of them are discussed below:

1. **Special Courses:** The executive may be required to attend special courses which are formally organised by the enterprise with the help of experts. The executives may also be sponsored to attend the courses conducted by the managerial institutions. This method is gaining popularity these days. It may be noted that only the big enterprise can send their executives to the management development courses run by the management institutes because the fee of these courses is very high.

2. **Special Project:** Under this method, a trainee may be assigned a project that is closely related to the objectives of his department. For example, trainee may be asked to develop a system of cost control in the execution of an order. The trainee will study the problem and make recommendations upon it. This project will also help in educating the trainee about the importance of cost and to understand the organisational relationships with the accounting and other departments. Thus, the trainee acquires the knowledge of related subjects also.
3. **Committee Assignment:** Under this method, an ad-hoc committee of junior executives is constituted. It is assigned a subject to discuss and make recommendations. The committee has assigned objectives and responsibilities related to the work of the organisation. It will make the study of the problem and present its suggestion to the departmental managers. Committee assignments can provide the necessary general background to the trainees because every member of the committee gets a chance to learn from others. It is an important device of educating the executives and of changing their behavior towards the selected problem.
4. **Case Study:** Case method is a means of simulating experience in the classroom. Under this method, the trainees may be given a problem to discuss which is more or less related to the principles already taught. This method gives the trainee an opportunity to apply his knowledge to the solution of realistic problems. Here, the trainee is expected to study the information given in the case and make decisions or give conclusion based on the situation. If the student is provided a case of an actual company, he is expected to research the firm to gain a better appreciation of its financial condition and corporate culture. It is also necessary to come up with good case material based on real life situations and present the same before trainees in an interesting manner. When cases are meaningful, and are similar to work related situations, trainees can improve their decision making skills and problem solving abilities.
5. **Conference Training:** A conference is a group meeting conducted according to an organised plan in which the members seek to develop knowledge and understanding by oral participation. It is an effective training device for the persons in the positions of the conference member and conference leader. As a member, person can learn from others by comparing his opinions with those of other members. He also learns to respect the view point of others and to realise that there is more than one workable approach to motivate people through his direction of discussion. He learns the effect of closely controlling and dominating the discussion as compared to adopting a more permissive type of direction.
6. **Management Games:** A variety of business and management games have been devised and are being used with varying degrees of success in development programmes. A management game is a classroom exercise in which teams of students compete against each other to achieve common objectives. The game is designed to be a close representation of real-life conditions. The trainees are used to make decisions about production cost,

research and development, inventories sales etc. for a simulated organisation. Since they are often divided into teams as competing groups, experience is also obtained in team-work.

7. **Sensitivity Training:** Sensitivity training or T-group training means the development of awareness and sensitivity to behavioral patterns of oneself and others. It is an experience in interpersonal relationships which results in a change in feelings and attitudes towards oneself and others. T-group are helpful in unlearning and learning certain things. They help the participants to understand how groups actually work and give them a chance to discover how they are interpreted by others. It also aims at increasing tolerance power of the individual and his ability to understand others. The sensitivity training programmes are generally conducted under controlled laboratory conditions.

5.9 HRD AND OD

Organization Development (OD) is broadly defined as the way organizations change and evolve. The process of planned change is best achieved through organizational strategy. The elements involved in an OD functional processes need to be carefully analyzed to create a plan for change. How the plan is to be implemented, evaluated and measured must be well established within the OD strategy. The following example of a major OD function that may be affected through strategic planning is Innovation. The concept of innovation as it relates to creative thinking to improve processes, research and development, and marketing all serve as possible competitive advantages that are of particular importance in OD. Being the first to market with a new product or service and being able to deliver in a timely manner for instance directly involve innovative planning. OD strategy should involve a systemic plan to monitor the entire organization as a whole. The plan should also include ways to make changes and corrections as needed to achieve the intended result. Changes to a single process usually will affect other process. These secondary process should also be analyzed avoid to unintended results.

Organizations strive to plan ways to reduce costs and maximize production and profits. This ideal concept involves many different concepts and process that are implemented by personnel. The Human Resources Development (HRD) department should be an interracial part of the overall OD strategy by directing personnel toward the desired outcomes. Since HRD is a part of OD the strategic planning will involve similar and or the same concepts.

Human Resource Development entails many concepts that center on the economic value of personnel to an organization. In recent years, one particular concept of increased interest is the "measuring" of the processes and affects of HRD. These HRD measurements are seen as a leading indicators rather than lagging indicators such as financial reports. The HRD measurements are documented to track how well each process is contributing to the overall OD strategy. HRD must define and identify the performance drivers and enablers of the organization that fit into the OD strategy. A performance driver that is identified to increase customer satisfaction may stem from training sales associates to ask customers certain questions that will better describe the customer's needs and desires. These measurements also allow for the verification of how HRD contributes to the strategy and bottom line of the organization.

A quality HRD department is a strategic partner and asset to the organization.

A major concept in HRD is the process of defining and training personnel in specific competencies. One of the functions of training is developing behavioral changes in personnel to the betterment of both the person and the organization. These are valued changes such as increasing competencies that improve process skills that further lead to the creation of opportunities for recognition, promotions for the employee and increased productivity for the organization. The process of matching proper personnel to the position or task is also an important concept in HRD. Spending the right amount of time and money in recruiting quality personnel is a trade off that, done properly, will result in lower turnover rates, higher aptitude toward learning the required competencies of the positions and innovations of improving the positions and processes.

CASE STUDY

Organization Development in a Voluntary Organization

We present here our experience of an Organizational Development (OD) effort in an unusual setting. Conceptually, OD is expected to cover the entire system. However, this is a normative statement. Instances of OD activities encompassing the entire organization are rare; in actuality, OD directly involves the upper management and hopefully generates systemic changes which permeate through the rank and file. Presented here is a case in which every member of the organization was directly involved in the programme.

Most OD experiences derive from enterprise organizations. In these, as also in various non-enterprise organizations, bonds between members are essentially based on economic relationships. In this case, however, the bonds were ideological.

The total span of the OD programme was two years. The organization involved was a voluntary service agency, which we shall call International Services (IS). This was an international organization, providing services in only one specialized field. It operated six geographical sectors across the world.

This paper is based on an OD project conducted by the Behavioural Science Centre (India). The project planning group consisted of Somnath Chattopadhyay, Manohar S. Nadkarni, and Udai Pareek. The facilitators in the basic laboratories were Abad Ahmed, Sujit Bhattacharjee, Somnath Chattopadhyay, N.K. Jaiswal, Gordhan Kathuria, Prayag Mehta, Ed McGrath, Manohar S. Nadkarni, Udai Pareek, and Pandit Patankar. The facilitators for the internal resource development laboratories were Ken Benne, Somnath Chattopadhyay, Manohar S. Nadkarni, and Udai Pareek. The project director was Somnath Chattopadhyay.

The Setting

The governance of the organization was conducted structurally by the Central Coordinating Authority (CCA), with a chief. The CCA coordinated the activities of the six sectors. Each sector also had a chief, who was guided by the Board of Sector Representatives (BSR). Each sector was divided into divisions. Each division had a divisional chief with a Board of Divisional Representatives (BDR). Each division was further subdivided into local units (LUs). Any member of the organization was really a member of an LU. The members of an LU elected their chief. All these chiefs were members of the BDR, all divisional chiefs were members of the BSR, and all sector chiefs were members of the CCA.

The organization was committed to its ideology and served the ideology by working for the people. All over the world, this organization was known for its service to humanity in distress. It also ran a number of specialized institutions. Initially, they concentrated upon individualized care in the institutional framework. The activities of all these institutions were not performed by members of this organization only. Many other people of different backgrounds were employed in the institutions.

The sources of income for the organization were charity, gifts, en-dowments, and the like. Institutions managed by them generally had two systems: free service and paid. The payment charged was linked to the paying capacity of the person. These were non-profit institutions; money collected from payments was spent on the running of the institutions.

The CCA had a clear structure and authority pattern. Its administration was guided by a constitution. At the end of every six years, the constitution of the CCA was reviewed to see whether any changes were necessary. We had the impression that the organization chose bureaucracy as their method of administration. Power was extremely centralized in the hands of the chiefs at each level, the communication flow was almost one-way from the top down, decisions were taken at the top in each level, and members were expected to follow, implement, and work according to them. There were a number of implications of this system which could be easily visualized. A few years ago, there was a special review of the CCA constitution and drastic changes were brought about. Essentially, the changes centred around a more democratic, participative management. At that time, the CCA took a close look at the organization and redefined its objectives, mainly in terms of greater emphasis on total development orientation.

Following the new provisions in the constitution and the proposed management system, a number of controversies arose. A feel for some of the controversies could be had from the excerpts of two letters which were published for international circulation.

> *'It's like ... "My, my. Doesn't the organization have such a lot of problems, and so many members are troubled and insecure! We really must do something about it—we must get expert help"... We have strong faith indeed in the almighty expert! The differences that divide us, that cause endless misunderstandings, that make it virtually impossible to implement the new constitutions are, in my opinion, irreconcilable.'*

The other point of view was:

> *'With regard to the experts, I admit that we have a number of our own in the organization ... We often reject them together with their message ... Outside experts also have the advantage of objectivity. I know that what problems we have are no greater or less than those being experienced by many other ... committees today. However, I feel that to just live with the problems without doing our best to find solutions is shirking our responsibility.'*

Controversies and differences of viewpoints, often blended with reasons and intense feelings, filled up every LU meeting during the following two years. Some of the differences were voiced loud and clear; some were muffled and whispered. To some, the new trend was the reviving breeze that energized and liberated; to some it was the dreaded storm that uprooted everything that came in its way and left behind only destruction. There was utter confusion. The confusion, however, did not lead to any sharp, organized polarization.

During this period, the organization realized that changing the constitution *per se* was not enough. Operationalizing the new constitution needed a series of activities. Accordingly, a concerted effort in terms of a new scheme (we shall call it plan of renovation, or PR) was thought up and discussed. The spirit of and the expectations from PR were necessarily vague. Its unstructuredness and ambiguity matched the uncertainty that followed the new constitution. One of the tasks of PR, we understood, was to remove the fear and apprehension of rootlessness, supportlessness, and shifting anchors, that were inadvertently generated as a consequence of the new constitution; to reduce the level of anxiety and channellize it for creative problem-solving; and, most importantly, to find out ways and means not only to operationalize the new constitution but stimulate the desire to live with all changes. No one was very clear about the objectives of PR, the ways and means of achieving them, the barriers and blocks it would face, or its results. One decision was, however, taken—instead of taking directions from the central authority, each sector should find its own answers.

Around this time, we were invited by one sector to help it formulate and implement PR. Before we describe our entry, it would be worthwhile to consider a few details about our client sector.

The Client System

The sector geographically comprised two countries and had two divisions, each division having several local units. Numerically, this sector had the largest membership in the organization.

The number of members and consequently the strength of the LUs was evidently not equal. The numbers depended upon convenience, availability of regular work, or special work assignments. Migration from one LU to another, dictated by the requirements of work, was quite common.

Life in the LU was marked by austerity and plain living—this also meant group living. Domestic activities were shared by members on the basis of competence and accommodation. Every morning, after breakfast and after completing domestic chores, they went to their workplace. At the end of the day's work, they came back to the house again for tea, relaxation, cooking, dinner, studies, meditation, and sleep.

Guests came quite often—members from other countries or from other LUs—who might stay for a few days. Callers from outside communities visited occasionally in the afternoons or evenings. The members were not socially insulated from the external community but, at the same time, had set up for themselves an implicit boundary on social mixing at home. Their involvement in their ideological work and their work schedules made them economical with time. Besides food and shelter, the LU provided them with a small allowance for personal necessities, which included the simplest of clothes and sandals, toothpaste, and books. One could buy anything one liked within the budget limit. We did not see anybody complaining about the modesty of the allowance—far from it. Their acceptance of plain living had developed ease, grace, and simplicity.

To become a member of the organization, one had to apply to the chief of the LU. If one was a minor, explicit permission of the parents was necessary. The candidate was not immediately accepted as a member and was required to spend quite some time in learning and training. If the applicant chose a particular vocation with which the society agreed and if that vocation required the person to pursue academic studies, the member was free to do so. After the candidate had passed through the final phase of training and had had a taste of what membership of the organization would mean, he or she had to make promises and pledges at three stages. After the third stage, which was the stage of final commitment, the person became a full member. A member had the freedom to leave the organization, although psychologically it was a difficult option to exercise.

The requirements of membership and the constitution of the society were what one abided by and upheld respectively. One left the parental home and no longer identified oneself with the old familiar relationships but found a new home with the other members in a bond of comradeship. The members were required to lead a communal life, live simply, be loyal and obedient, not marry, and observe celibacy.

The sector had, at the time of this exercise, 188 members, mostly below the age of 50. There were only 11 members above the age of 50. The majority were in their 30s (79), followed by those in their 20s (51) and 40s (47). It was obvious that younger members were more in number than the older. But if one divided the range into three—young (below 30), middle-aged (between 30 and 39), and older (above 40)—one got a real feel of the forces behind the composition. The number of those in these three categories were 51 (young), 79 (middle age), and 58 (older), respectively. The implication was that none of these three groups was a 'minority': each had sufficient strength. Only if one group joined with another did the third became a minority. The organization could thus be divided into two age camps of almost equal strength with 35 as the median age.

The members ranged from one who had just joined the organization to one who had spent 38 years with it. The quartiles fell on five years, 10 years, and 15 years, that is, about one-fourth of the members had spent five years or less with the organization, about half the members had spent 10 years or less with the organization, and about one fourth had spent 15 years or more.

In terms of vocations, the nursing profession had the highest numbers, followed by midwifery. Many of the members were also in education, public health, or administration at various levels. Some were also students.

Members came from 13 different geographical backgrounds: eight countries and five different parts of one country.

To gain some insight into their interpersonal orientation styles, we subsequently administered the FIRO-B (Schutz, 1958). The scores showed that the members of this organization did not have extremely high (7, 8, 9) or low (0, 1, 2) scores except on control dimensions. It seemed that they had a tendency not to express control. Their desire to be controlled by (to receive instructions, ideas, and decisions from others) was higher than their need to exercise control. By and large, the data showed that the community as a whole was a psychologically mature group.

Entry of OD

One fine morning, N came up the narrow stairs of the Behavioural Science Centre (India), or BSC(I), office. She discussed the purpose of this visit with the executive director of the BSC(I) with one of the authors present. N enquired about our willingness to help her organization.

We wanted to know more about the organization and what the organization expected us to do. Thus followed a series of discussions. N had already contacted the other author, who was not living in Delhi at that time. In the beginning, he was not too keen to enter into this contract. While the idea of effecting OD in a voluntary service-oriented organization was attractive, the hesitation over helping an ideological group (and the possibility of consequent conflict in values) was dominant.

Our client solved the dilemma beautifully. N said, 'We are asking you to help us not in our ideology. We seek your help in the areas of your competence and expertise. When a patient goes to a doctor he does not enquire about the religion of the doctor or his values or attitudes towards life.' So the dilemma was resolved. We were struck by N's openness and we must acknowledge how much we enjoyed a relationship which was both professionally and personally stimulating and rewarding. We learnt a great deal ourselves and our respect for the organization—and more than that, the individuals whom we came to know so well—was greatly enhanced.

When these discussions were going on, one of the authors was invited by the apex body of the organization to participate in a three-day international meeting in Europe to work out the dimensions of PR. At the same time, the director of BSC(I) returned to India from his foreign consultancy assignments and joined the authors in the initial explorations.

For the sake of convenience, let us call ourselves A, B, and C. A and B had several meetings with N, the contact person for the organization. The main purpose of meeting with N, and with a few other members, was to understand the organization. There were a few meetings between A, B, and C to discuss the planning of the project.

On the basis of these meetings and discussions, B prepared a working paper on the project, which was then analysed and discussed by A, B, and C.

The Working Paper

The working paper consisted of four parts dealing with:

- Our understanding, at that time, of some of the key issues involved.
- A conceptual framework developed to formulate interrelations.
- Certain hypotheses, assumptions, and propositions that were to be the basis of the OD programme design.
- An outline of the intervention plans.

This working paper was designed for a hypothetical local unit (LU) where a hypothetical member (M) lived. Edited extracts from the working paper are presented below:

Some Key Issues

Member M does not have any problem as such. M is secure, adequate, carrying out his or her duties honestly to the best of his or her abilities, and fulfilling the organizational mission. The main issue, however, is for M to deal effectively with the changes occurring in the environment and shifting his or her three anchors—ideological life, pro-fessional life, and missionary life.

The role constitutes a crucial issue in the renovation plan. The behaviour that M expects from his or her own self and what others expect from him or her—that M fulfil himself or herself in belonging to an ideology, and to a particular order and as a professional—will not be the same as it used to be with the changes occurring in the environment. The changing environment also demands a critical outlook, a questioning spirit. The significance, meaning, and impact of these issues have to be seen in the person, in the organization, and in the community for all the areas of functions.

The Conceptual Framework

Member M is seen as an individual (self), a person committed to an ideology, a member of a task force (work organization), and a member of a community. His or her observed behaviour is a result of forces operating on him or her from within and without. It is not one value, or even one set of values, but different sets of values that are operating in M's forcefield at any given point of time. Also operating are different sets of expectations and need dispositions. These act, interact, and react until the individual finds some kind of a balance, an equilibrium, which determines his or her behaviour.

But with the introduction of 'change', the 'peaceful', 'stable' equilibrium will be disturbed. Change will affect all these—values, expectations, needs, and the overall climate—simultaneously but in different magnitudes. In order to accommodate and adjust to the new associative network, resulting in a newer equilibrium, different behaviour will be required. The individual may face barriers to changing his or her behaviour, to pass on from his or her observed behaviour of today to his or her required behaviour of tomorrow according to him or her. The barriers are likely to bring forth at least four major concerns. These concerns may be of

- essentiality,
- relevance,
- congruence, and
- authenticity.

In each of these areas of concern, the member may ask himself or herself, 'Now that times have changed, how essential am I to myself?', 'How essential am I now to my organization, to my community?', and 'How relevant are these new demands?' He or she will have a number of such questions. When these questions are handled properly and the blocks and barriers removed, the gates are opened for a movement towards self-actualization.

Some Hypotheses

From the perspective of the above conceptual frame-work, intervention designs were developed. In developing the designs, the following hypotheses were used:

- It is possible to examine knowledge and its linkage with behaviour.
- Such an examination can be made at both the rational as well as emotional levels.
- Suitable hypotheses for the personal, organizational, and community levels can be framed by an individual, who can gather data regarding his or her behaviour in each dimension. He or she can analyse these data and accept or reject the hypotheses he or she has framed with regard to his or her behaviour patterns.
- Confrontation of issues leads to development of one's understanding of the latent issues involved. This develops his or her insight, which may provide him or her with added strength and adequacy in dealing with his or her problems.
- When the latent issues—anxiety, fears, apprehensions, etc.—are brought to the surface and shared with others, perceptions of issues change and become more realistic. It is then possible to sort out how far the fears are real or unreal and how far they are personal or general, individual or systemic. Such sorting out may help in energy mobilization.
- The sharing of feelings of anxiety and inadequacy reduces the levels of anxiety and inadequacy.

Some Assumptions

The next phase in the design for the process facili-tation was to consider a set of assumptions. It was assumed that:

- With the changes occurring (as seen above) in the framework, the organization, and more particularly, any 'local', is willing to critically consider a strategy for dealing with changes.
- There is a certain degree of eagerness to move from the curative to the preventive perspective in a new professional area.
- There is a certain degree of eagerness to develop a critical and yet constructive outlook.
- Each individual has enormous potential for growth and unlimited resources for securing strength.

There was another set of assumptions regarding the methodology of intervention:

- Sensitivity training will be used only when required.
- The programme should be based on laboratory training (a separate note on laboratory training was given).
- The learning will be experience based. An attempt will be made to create an atmosphere in which the participants can generate and share their experiences. Experimental learning will have greater pay-off than direct teaching.
- The individual himself or herself is the main actor in the learning situation; the trainer will take the role of a facilitator only.
- The training strategy will be one of unfreezing, movement, and refreezing.

Some Propositions

The following propositions were made for the training:

- Changes are more likely to occur when the setting dramatizes the importance of working intensely on certain issues and lifts them out of the routine of everyday life, thereby creating an in-group feeling among the participants.
- The more reasons an individual has to believe that he or she can, will, or should change, the more the organized attempts to change are likely to succeed.
- The more thoroughly an individual explores his or her personal orien-tations, styles, apprehensions, etc., and relates them to self-actualization, the more likely he or she is to experiment and change.
- The more an individual can confront his or her usual style of interpersonal interaction, the more he or she is likely to change.
- The more an individual can perceive and experience the new attempts at change as consistent with the ideal self-image, the more likely that changes will be sustained.
- The more the self-confrontation in a face-to-face group and the more the climate supports such confrontation, the more the individual is likely to develop insight into his or her behaviour.
- The more a group develops a climate of support with a norm of unhesitatingly giving feedback to each other, the more likely it is that the learning will be deep and sustained.
- The more intense and persistent the interactional meetings are, the more likely it is that the learning about the self will be deep and enduring.
- The more an individual perceives that change in his or her behaviour (*e.g.*, being more confronting) is required by the demands of his or her career and life situation, the more he or she is likely to experiment and accept the change.
- The more an individual can perceive and experience the changes attempted as consistent with the prevailing cultural values and norms, the more likely it is for the change to be sustained.
- The more an individual can examine the dynamics of events in everyday life, the more likely is the internalization of his or her learning.
- The more an individual can link his or her newly developed insight into personal and interpersonal dynamics to action, the more likely that change in both behaviour and action will occur and endure.
- The more an individual commits himself or herself to achieving concrete goals in life consistent with the new insights, the more likely the change is to get stabilized.
- The more an individual records his or her progress towards achieving goals to which he or she is committed, the more the accepted change is likely to influence his or her future thoughts and actions.
- Changes are likely to occur and persist if they are a sign of membership in a new and continuing reference group.

Within a few days of our presenting the working paper to S, the sector chief, there happened to be a divisional meeting. S thought that it would be a good idea if the working paper was discussed by the members of the divisional committee. To help them go through the working papers as well as to have a dialogue with the committee, S invited us to attend the meeting. Half a day in the two-day meeting agenda was reserved for the purpose. A and B attended this meeting—they presented the paper afresh to the members and then answered questions raised. The main stand taken was that of tentativeness both in terms of the contract as well as the proposal.

Incubation Period and Contracting

The committee requested us to prepare a shorter version of the working paper which they wanted to circulate to all the members of the organization before a decision was taken. So a simpler version of the working paper was prepared and submitted.

There was complete silence for almost six months. It was decided by the organization that the decision to implement an OD programme would have to be taken not from above but by each member and that the decision of each member would be communicated upwards. This then became an issue by itself. A system that usually followed the norm of decision-making from the top was now required to reverse the norm and initiate a new practice. The result was indecision.

At the same time, an important process began: A good deal of questioning started. Questions were raised not only on whether to accept or reject the working paper but also on the various norms and practices associated with the governance of the system. Such questions were raised by the members individually, in small groups, and in the LUs. All the LUs were buzzing with these questions. It may be noted here that the consultants did not actively generate the process, but the task performance (of discussing the working paper) made this process almost inevitable.

We were satisfied with the process. We clearly indicated that we were not at all eager for a hasty decision. We could wait. During this phase, S showed a great deal of leadership, courage, determination, and above all, a sense of positive patience. She was under strong pressures. Some people were eager to start OD; others were reluctant to have anything to do with it. To some it would be a process of rejuvenation; some thought it would be a sure road to chaos and confusion. So we waited.

After a period of six months, we thought it would be necessary to churn things up a little. First of all, it was necessary that a final decision—yes or no—emerge and that the members should have the success experience of being able to arrive at a decision. In order to facilitate the process, we offered to meet the representatives of the houses for a couple of days. So a two-day programme was arranged.

A and B conducted this programme. The programme was attended by representatives of all LUs. In this programme, we evolved a situation where the members could discuss the new constitution, the emergent issues facing them since the introduction of the constitution, the meeting, the implications of PR and societal changes, adaptability, etc. We presented various aspects of change: how we face issues of change, what change means to us personally and professionally, and how we cope or fail to cope with the demands of change. We gave a demonstration of laboratory training in micro form and presented a possible outline for the OD programme and its financial implications. We made it clear that the members should not feel obligated to call us in for OD work and should have a wider choice of consultants.

After this programme, the representatives went back to their LUs and vigorous discussion ensued. We did not hear anything, one way or the other, for two months until most of the LUs became ready to participate in the OD programme and it was sensed that the entire community of members might, by and by, choose to participate in the programme. Finally, a draft contract was prepared, which was readily accepted. We waited for another three months, by which time a few more members expressed their willingness. Very few remained undecided; those who remained undecided said that they wanted to wait and see what happened in a few labs before they made up their minds. But as a policy they had no objection if the community went in for OD. At this stage we thought our entry phase was over and we started planning the interventions.

Initial Diagnosis

Meaningful planning of interventions could be done on the basis of our understanding of the organization. The main task in understanding an organization is to sort reality from illusion. Several methods are available for organizational diagnosis. However, we did not follow any of these standard methods. As stated earlier, based on our preliminary interviews with a few key persons, we prepared a working paper. We clearly mentioned to our client that the working paper was based to some extent on our limited understanding of the organization and mostly on our general understanding of OD.

To gain a thorough understanding, we did the following:

- Visited a few LUs and saw the normal way of life there.
- Observed some of the formal meetings, both at the LU level as well as the division level.
- Acquainted ourselves intensively with the organizational process in the members workplaces.

Visits to the LUs were informal. Sometimes these were in response to their requests, sometimes on our own initiative. During the visits we talked freely about a number of things. We had no particular question in mind, yet we had so many questions! We were aware of our ignorance of the organization. So we sat there for long, lazy hours with cups of coffee in hand, talking and listening. We imbibed, slowly and quietly, the life that easily flowed in the house. And we did not take any notes!

We went to their formal meetings on invitation. Our role was that of a learner rather than that of an observer. We were interested in the content of their discussion as well as in the processes of discussion and decision-making. Sometimes we also took part in the discussions, despite our limitations.

We observed their functioning at work quite thoroughly. We undertook an organizational diagnosis of the workplace and analysed the role of the members of the BDR. With all this data and understanding, we agreed amongst ourselves that we had developed enough insight to help us go ahead with planning the interventions.

The process of initial diagnosis we undertook was rather unorthodox. The nearest standard method to our approach was the anthropological method of observation. Our view is that too much reliance on instrument-generated data tends to shadow the holistic nature of the client system. Attention tends to be given to the quantitative picture and the consequent interpretation of data.

Intervention Planning

Some of the parameters of intervention planning had been presented earlier in the working paper that we discussed with our clients and these remained unchanged even after the initial diagnosis, although we were ready to change them if needed.

In the actual planning of the interventions, the issues that we considered were:

- The unit with which the intervention was to be used
- The overall objectives and the specific objectives of interventions
- The type and nature of the interventions
- Phasing, sequencing, and linkage of the interventions.

There are usually extreme positions taken in the planning of interventions. While some consultants may use their general insights and experience in working with the client system and evolve a strategy based on their intention, some others plan out everything in detail (we do not know whether they plan the number of sugar cubes required for coffee at 11.30 a.m. sharp!). However, we took a somewhat intermediate position. Although the individual facilitators in our project group had a wide variety of styles, we considered it necessary to plan interventions only up to a point, beyond which planing became futile.

The Unit of Intervention

The basic element in intervention planning is the question whether or not it can be totally changed, or whether the change has to be viewed by delineating the various systems (or sub-systems) that together constitute the organization. These systems (or sub-systems) may be concerned with the technological process, the economic process, or the human process. At the initial stage, we decided that both the technological process and the economic process were out of bounds for us. First, there were very few so-called technological processes involved in the organizational life, though in one of their workplaces, the hospital, the technological processes played a key role. The other technical part of their life emerged from their religious orientation. We left that part aside also as it was beyond our competence. We felt that the new constitution was itself a strong intervention in the organizational process. Any further intervention—for example, structural changes—was more likely to be dysfunctional at that stage. We therefore concentrated our efforts on the human system.

One can conceive of different levels of interventions in the human system. We conceived of four levels (see Pareek et al., 1972; Pareek, 1975). We thought that interventions could be directed towards the elements of the human system: person, interperson, intragroup or team, and interteam or the total human system of the organization. By person, we meant the uniqueness of the person with his or her personal characteristics and qualities, skills, knowledge, motives, needs, values, attitudes, ways of thinking, expecta-tions, ambitions, fears, strengths, and weaknesses—all of which were his or her own. It is that aspect of human existence that allows each one of us to have a distinct identity. The interpersonal aspect was meant to focus on the transactional world created by two people who together might or might not constitute a group. The relevance of the intra-group was with reference to group life in each house of assembly. The inter-group element referred to the interrelationship amongst all the houses forming the community of the entire division. From our initial diagnosis, we were convinced that interventions had to be directed towards each one of these four elements of the human system.

The Objective

In planning the interventions, we raised several questions about each of the four elements mentioned earlier, with an eye to their present state and the future expectations from them. We realized that our objectives in the interventions would not be to provide answers to these questions (did we even know the answers!). The main overall objective of the interventions would be to help people bring these questions up to the surface; confront them; and evolve, through their own strength, the answers to the questions they formulate individually and in groups; and help them to increase their effectiveness as individuals, groups, and a community.

The specific objectives would, therefore, be to help them move towards the positions they valued and become self-actualized, to help them function as optimally as possible, and to help them use freedom and responsibility in being relevant to the world and its needs today. These three objectives included other objectives, such as:

- Developing an orientation of alertness concerning what is happening in the organization and why it is happening. This would presumably strengthen both religious motivation and concern for others.
- Fostering a tendency to help others and receive help from others (interdependence of growth).
- Instilling comfort with discussing things, including differences, openly (openness).
- Working towards continual growth.
- Developing tolerance and flexibility at all levels.
- Creating awareness and improving sensitivity to one's own strengths so that better utilization of resources for the community takes place.

The Type and Nature of the Interventions

Using two dimensions—whether the intervention is focused on the statics of the organization or on its dynamics and whether it is itself structured or unstructured—we get a typology of interventions (see Pareek, 1975). We do not prefer any one intervention to any other. We believe that one or the other intervention may be useful, depending upon the needs of the client system as well as the presence of certain conditions within the organization such as its flexibility, reward system, risk-taking capacity, ability to tolerate ambiguity, availability of internal and external facilities, and so on. Based on our understanding of these conditions in the organization, we thought that unstructured inter-ventions directed both towards the dynamics and the statics of the organization would serve our purpose best.

Phasing, Sequencing, and Linkage

The nature of the task and the objectives of the programme were such that the interventions had to target the living systems. For example, all members participated—after their participation in the basic lab (see below)—in organizational diagnosis of their houses and in preparation and implementation of action plans.

Before our withdrawal, stabilization and consolidation of the effort for change had to be planned. This would technically be an attempt to refreeze and would obviously be the last input. Accordingly, the interventions were attempted in the following order:

- Basic laboratory
- Internal resource development
- Organizational diagnosis
- Review laboratory for internal resource people
- Stabilization.

The entire OD programme was to be completed in a period of 12–15 months. The basic laboratories were to be provided within the first six months, followed by internal resource development. It was planned to have organizational diagnosis after four months and continue the practice thereafter. At the end of the year, the review laboratory was to be conducted. And after the review laboratory was over and the internal resource persons went back to their respective houses, the terminal phase was to start.

The following are brief descriptions of the interventions, notwithstanding the obvious and understandable difficulties of verbalizing unstructured interventions.

Basic Laboratory

The basic intervention was a five-day laboratory. All the members of the organization had the opportunity to participate in such a laboratory. In all, 15 basic laboratories were conducted and all of these were completed within six months.

Structure of the Laboratory

The programme for each laboratory varied, depending on the need of the members of that particular group, with certain common foci. The schedule of a typical lab is given in Table 1. The laboratories started with a micro lab session for the specific purpose of unfreezing. This was followed by a number of forms that the members filled. These included six TAT pictures, FIRO-B, and the self-anchoring ladder. Data from these were later fed back to and used in the groups. This was followed by L (Learning)-groups, or what are also called T-groups. Generally, two sessions of L-groups were followed by a short concept session on the conditions of learning. This was followed by further L-groups with concept sessions on congruence, defensive behaviour, and feedback. The first two days saw intensive L-group activity. At its height, the L-group activity usually climaxed into a marathon L-group on the second night.

On the third day, a number of structured exercises were done and data from the instruments were discussed in detail as feedback. New groups were formed and a case was used to discuss the sources of behaviour and how role play emphasized empathy. Structured exercises were conducted on perception and motivation. A panel discussion was held on PR. Towards the end of the third day, participants moved towards refreezing activity, the main item being working on action plans—new groups working on the diagnosis of the problems facing their LUs, prioritizing these problems, and developing action plans. This was followed by a fantasy exercise which was designed to bring up the gestalt of the members' existence in terms of time, that is, past, present, and future, both for the individual and the organization. This was followed by the last L-group on action plans, individual plans, and personal work commitments.

The lab usually ended with a review and evaluation of the lab. A basic lab was of four or five days' duration.

Micro Lab

This was intended to start the unfreezing process as well as to introduce the participant to the new situation. Micro labs were generally structured as a quick succession of events, each event being of three to four minutes' duration and requiring the participants to interact in dyads, triads, quadrads, or pentads. Verbal exercises required the participants to share with other member(s) in their groups their perceptions, feelings, information about themselves, etc. The movement in the micro lab was from person to group to community life. Two non-verbal exercises which were quite common were the trust walk and a presentation on the community. The last item was performed as a joint decision and action effort of about 10 participants. From time to time, activities were halted and the participants were given time to reflect upon the meaning and implications of the preceding items. The micro labs were conducted as community sessions.

Table 5.1 A Sample Schedule for Basic Laboratory

Day	*Session I* **8.00 a.m.– 10.30 a.m.**	*Session 2* **11.00 a.m.– 1.00 p.m.**	*Session 3* **2.00 p.m.– 4.30 p.m.**	*Session 4* **7.30 p.m.– 9.30 p.m.**	*Session 5* **9.30 p.m.– 11.00 p.m.**
1	Micro lab—form filling	Leadership	L-group—concept	L-group	L-group
2	L-group—concept feed-back and defensive behaviour	L-group—perception exercise	Exercise on collaboration and trust	Data feedback, L-group (marathon)	
3	N-group—PR panel	Case: Concept of conflict	Exercise on motivation and expectations, action plans	L-group—feedback	Fantasy exercise, discussion of feedback
4	L-group—action plans	Individual action plan	Life goals exercise, personal commitment, review, and evaluation closing		

Personal Structured Data

As mentioned earlier, instruments were used for the purpose of giving feedback on the data. One of the forms required the participants to state what they expected from the programme. The responses related to the self (more self-knowledge, self-acceptance, more self-confidence, reduced oversensitivity, increased listening, seeing others' strengths, and self-discipline), the community (openness and sensitivity, learning to recognize one another's strengths, increase in individual responsibility, mutual understanding and love, greater participation by all, a better climate, and mutual support and help), and solving problems (practical guidance to problems, choosing better leaders, and working out definite changes in the community).

As already stated, FIRO-B and six cards of TAT were used by the participants to write stories. The concept sessions were very short and mostly grew out of the needs of the groups. Themes included conditions of learning, feedback, defensive behaviour, sensitivity, congruence, authenticity, levelling, collaboration–competition–conflict, role–role analysis, motivation, motivating others, helping relationships, working effectively in groups and in organizations, effective membership behaviour and leadership functions, and dynamics of planned change.

The basic skills emphasized in the laboratory for practice were mainly effective speaking, effective listening, giving and receiving feedback, giving and receiving help, confrontation, and action planning.

Several exercises were incorporated in the programme to facilitate the process of movement, specially in acquiring new knowledge, and developing insights, attitudes, and skills. Structured exercises were focused on perception, trust and mutuality, self-motivation and motivating others, planning and communication, decision-making, and life and society goals.

Panel Discussion of PR

About the third day, a panel discussion used to be arranged on plan renovation. Generally, the panel was constituted of one representative from each L-group. Each member put before the community his or her own ideas as well as the ideas of other members of his or her L-group on PR. A single chair was used for any member of the community to occupy when stating a point for the consideration of the panel. On the whole, the discussions brought out in the open many hidden agendas, fears, apprehensions, as well as hopes about PR. The session was usually followed by a period of free time, in which the discussion continued in small groups of twos and threes.

Action Plan

As stated in one of the propositions in the working paper presented earlier, we strongly believe that behavioural changes are better sustained in the context of actual action. Resolutions, howsoever pious they may be, tend to remain in ether, losing their vigour and becoming mere wishes or fantasy, if they are not reinforced continually by commitments in the work sphere. Accordingly, we place a great deal of emphasis on action plans. Work on action plans was started towards the end of the third day, followed by night assignments, which was again followed by considerable time spent on this on the fourth day. Essentially, the action plan module started with a concept discussion on the dynamics of planned change. For the exercise of action plans, participants worked not in L-groups but in new groups composed on the basis of geographical proximity of their living quarters, often coming from the same house. Each participant prepared two action plans—one for the LU and one for himself or herself.

The following steps were involved in preparing action plans for the houses:

- Preparing a master list of all problems of the LU as perceived by the members of the LU present in the new group
- Prioritizing the problems on the basis of three criteria—urgency of the problem, feasibility of action, and minimal requirements in terms of external resources
- Selection and definition of a specific problem
- Dynamics of possible causes
- Forcefield analysis of the problem
- Devising alternative strategies for solutions and selecting one of the alternatives
- Planning the action steps, including the time when an action is expected to be completed
- Anticipation of difficulties
- Consideration of help needed and planning for the help.

These plans were presented by the planning module to the whole group in the community session for critical review and, later, revision. After revision, the group committed themselves to working on the action plan on their return. Strategies to involve other members of the house (who were not present) were also discussed.

Evaluation

At the end of the programme, the whole experience was evaluated. Different methods were used. Typically, the participants worked in six groups to discuss the various aspects of the programme and presented their conclusions to the community. For example, six groups were formed for the following aspects:

- Insight into self and planning for self-development: L-group, including marathon group; FIRO-B; individual plans and commitment; micro lab data feedback
- Learning to understand others: L-group; case; perception exercise
- Developing a climate of collaborative work in the houses: exercise on trust; exercise on motivation; N-group
- Understanding community problems and preparing to deal with a problem: action plans; fantasy; N-groups; panel
- Developing conceptual understanding of interpersonal, intra-personal, and group processes: exercises; concept sessions
- Management of the programme (including facilities, kind participants, and nagging trainers!)

The programme ended with songs on commitment and energizing the learning obtained with the transferring it for the good of all. Usually, it was a very intensive programme, highly packed with events both at the rational as well as the emotional level. The working day started at 8.00 a.m. and often closed in the small hours of the following day.

The programme was designed to be intensive. The intensity of the experience, though very exhausting, was highly stimulating. Never for a moment did it seem boring to any participant or trainer. The credit for this may partly be given to the designer of the programme. However, the keenness and eagerness of a group of individuals who had dedicated their life to a purpose, with their high degree of motivation and commitment to the programme, was a unique experience for any consultant. To each trainer, this programme became extremely involving and satisfying, an experience from which we believe we learnt as much as the participants did.

Organizational Health Survey

At a much later stage and as a part of an intervention to help the members develop some skill in organizational diagnosis, we undertook a survey with the help of the members. We believe that interventions themselves are not enough in the OD facilitation process. What happens during the intervening phase between two interventions is also important. To sustain the momentum gained in the basic laboratories, we thought that an intervention would be necessary at the level of each LU. This took the form of organizational diagnosis. This was one of the OD interventions.

Unit Health Survey

This intervention was named the unit health survey. This exercise required the would-be facilitator to enter into a dialogue to assess the health of the LU. What was meant by 'organizational health', the characteristics of a healthy organization, and the action steps to be followed were explained in a letter written by the coordinator N and sent to each house. In the absence of a facilitator, it was decided that any person nominated by the LU could undertake this exercise. The following are extracts from the letter sent. A list of characteristics of healthy and unhealthy organizations, adapted from Fordyce and Weil (1971) were appended to the letter.

'You may like to undertake a survey of the health of the LU you live in. (We are not using the word 'Health' in the way medicos use it often. [sic] Would we prefer WHO's definition—mental, physical, and social well-being, considering the house like an organism? Well, we mean by health here—[sic] organizational health).'

'Examine the strengths of the LU, its weakness too; its problems and prospects; what are the diseases, if any?'

'How is information shared in the unit? How does "news" about events, persons, or work flow? Who talks to whom? Are there regular patterns?'

'How are decisions made in the house? Are there regular patterns in the way of [sic] decisions regarding activities, events, etc., involving all/most/some of the members of the units?'

'Who are the influentials in the unit? Who do not have much influence? Whose words, decisions and ideas carry maximum weight? Is there any pattern in which influences are exerted in the house?'

'How do members feel about the unit? Are they proud of their LU? Do they enjoy living there—with the work, the company, the facilities? Would they like to move out to another house?'

'How are relationships? Strained? Neither good or bad? So? Is there a close bond of affinity? What are the attitudes of members towards each other and towards the house?'

'These and similar other [sic] questions may help you determine the present state of affairs in the house.'

The procedure suggested is as follows:

Step 1: Interview as many members (one at a time) as you can. Devise your own way of interviewing; for example, you may prepare a questionnaire, or a schedule, or just a plain dialogue.

Step 2: After step 1 is over and you are sure you have understood different points of view sufficiently well, arrange small group discussions to discuss the issues mentioned in the objectives above. A small group may comprise three–five [sic] persons. Have a number of small groups, if the LU is not too small. You may change the composition of the group. You may spend a few days for this part of the programme.

Step 3: Organize a conference of all (most) [sic] members of the LU. List all issues raised. Evolve a consensus of the issues. Let the house group the issues under a few broad headings. Under each heading let the LU prioritize [sic] the issues. Finally put the headings and issues under them in order of priority.

'Some LUs may spend more time in step 3 [sic] than in step 2. You may like to evolve your own method which suits your time, place, and people. You are welcome to do so.'

'Remember to keep a record from step 1 through step 3 [sic]. Keep records of individual interviews, small group meetings and the LU conference.'

The survey helped in integrating the various action plans prepared by the members for the LUs during the basic labs, as described in a later section. We hoped that the organizational diagnosis itself would throw up important issues for the LUs and motivate the members towards their management.

The following picture came out of the survey regarding the organizational climate:

- There was lack of involvement of members in making decisions in the community (or, all members did not participate in making decisions in the community).
- There was a felt absence of an atmosphere of complete trust, openness, and acceptance in which the most timid might feel free to speak.
- There was a feeling of inadequacy in some members.
- There was a lack of sufficient information.
- There was a fear of hurting and being hurt.
- There existed a feeling among members that small groups would be the answer to all the problems, while at the same time realizing that this was not physically possible.

- Communications did not always get through, though there had been improvement in this area.
- Members did not feel entirely free to express their feelings and opinions.

After describing the process work in detail, one LU wrote, '. . . in the months that followed we continued to handle conflicts openly. Our [sic] action plan being activated we have now moved into another one'. We found in our survey that the greatest attention needed was in our communi-cation. We lacked skills in communication. We are now working for improvement in [sic] ways of communicating with each other. . . .After collecting the responses, it was read to the community and priority was given then [sic]. We needed to start working. Development of mutual trust and openness was agreed upon as the first need. A weekly meeting where we could openly talk out [sic] and clarify certain issues that might cause or might have caused some tensions, was suggested as a means and the idea was heartily welcomed. This weekly gathering was carried on regularly and was found to be producing the desired effects. It was also an opportunity for learning more and understanding one another. This was adapted accordingly in daily life. Tensions were reduced as the members had [a] chance to speak out things [sic] openly. Proper participation in [the] LU was another point for improvement. A change for the better came about. Those who remained quiet usually started breaking the monopoly of the vocal ones. Evaluation of our efforts followed.'

Internal Resource Development (IRD) Internal Resource Development (IRD)

In the plan of process facilitation, it was conceived that the role of the external consultants should be self-liquidating. In order to do so, it was decided that a self-supporting system would be developed within the organization. This included:

- Developing a group of persons who might acquire the expertise to become agents of change from within the organization. They might be termed internal change agents, internal resource persons, or internal facilitators. The term, 'internal facilitators' was most acceptable to the organization.
- Developing, from amongst the internal facilitators, a few who had some knowledge of the behavioural sciences further in terms of acquiring more knowledge and expertise in application through higher education and training.
- Contracting young behavioural scientists to work full time within the organization, in close liaison with the internal resources, on problems requiring the application of the behavioural sciences. This was a suggestion made to N for putting up for consideration before the governing board.

We discuss here the programme for the development of the internal resources.

Selection of Internal Resource Persons

Selection of internal resources turned out to be a very potent intervention in itself. In almost all the LUs, many a battle was fought in the open and under cover; armistices were signed; peace was made, broken, and remade. An intense confrontation process came into operation. Many a tear was shed and much laughter rolled in, literally.

What were the issues? To answer that, let us examine the content, the process, and the implications of the intervention. Content-wise, it was a straight task—each house would select a person whom they would regard as an internal resource.

As soon as the task became apparent, a reply was sent to the consultants, however, which said 'nothing doing'. The parallels of this behaviour in laboratory situations are too frequent to require elaboration. When the consultants did not relieve members of their responsibility to do the task and the task had to be done, three courses were open: confront the situation and do the task as conscientiously as possible; avoid the situation, and express bottled-up emotions in an indirect fashion; or, instead of doing the task properly, distort it and go through the motions and rituals of doing it. All these things happened.

At a certain stage, when the organization communicated to us that they understood the position we had taken and asked us for help in selecting the resource persons, we agreed to join them in evolving certain criteria that might be used in the selection.

To understand these dynamics it would be helpful to go through the letter that caused all the furore. Some extracts from the letter are quoted below:

- 'Please keep in mind that it is important that these members who are to participate in Phase 2 (internal resource development phase) participate in Phase 1 first. That is, the internal resource persons chosen by each local unit should be chosen from amongst these members who will attend the third series of the laboratories or who have already participated in Phase 1 first and the second series of laboratories.
- 'The consultant team has made some suggestions for us to consider in choosing the participants for phase 2. They should have a knack of working with person [sic] as opposed to material things; should enjoy working with people; should be interested in examining the dynamics of situations; should enjoy tinkering with ideas, and what would happen if the elements of the experience were varied? [sic] They should, of course, above all, be willing. The team emphasises that we should keep in mind that we are not looking for a leader type, but for someone who is sensitive to others, is often heard in the house, matters [a] great deal in the house in terms of his views, and who at the same time doesn't want to be in the limelight. The community should consider what this selection means to those who are not selected and might wish to be.
- 'The community *must realise and accept* the fact that in terms of the future group process, this member will be expected to be a 'facilitator' of discussions and interpersonal relationships, and their selections [sic] means a commitment on the part of the community to support him in this role and not to resent his interjecting himself when he feels he can be helpful. So, be prepared to select the one amongst you whom you feel you can best accept in this role.
- 'The one who accepts the responsibility must realise that for a period of approximately six months he will be expected to have time (not away from the work situation) for readings and exercises assigned by the consultant team. He must be prepared to participate with the consultant team in one of

the Phase 1 sessions scheduled in the near future. He should be able occasionally to free himself to gain experience by working with the consultant team with groups outside the society. 'He should be free to place his priorities in the community and its life and plan on sticking with his local assembly for a minimum of one year. He need *not* have any background at all in the social science [sic].

- 'We'd appreciate knowing the names of those chosen for Phase 2 just as soon as possible—the consultant team wishes to send them some material for study before the June experience. Plan renovation needs prayerful support—especially at this crucial stage. Let's join together in one huge *Hear, Oh Lord, the sound of our call.*'

One of the latent concerns was whether the resource persons selected would become too powerful—by the fact of their selection by the assembly and by the advanced expertise generated out of specialized training—and replace, in effect, the chief of the LU. This did disturb the existing power equations. The entire issue of power distribution came under scrutiny. Backed up by experience from the basic laboratory and the members' newfound energy for verbalizing, sharing, opening up, and owning, there was vigour and strength enough to deal with the issue. However, the task became very involved and difficult indeed.

The list ultimately did arrive, but only just before the start of Phase 2. The list contained 21 names, which included one facilitator from each house and a few more members who could look after the process facilitation for special tasks at the corporate level.

Structure of the Internal Resource Development Programme

The basic programme was a 10-day laboratory. The schedule of the laboratory, which appears in Table 5.2, indicates the programme structure.

Table 5.2 Schedule for the Internal Resource Development Laboratory

Day	*Session1* 8.30 a.m.– 10.00 a.m.	*Session 2* 10.30 a.m.– 12.00 p.m.	*Session 3* 12.00 p.m.– 1.00 p.m.	*Session 4* 3.00 p.m.– 4.00 p.m.	*Session 5* 4.30 p.m.– 5.30 p.m.	*Session 6* 8.00 p.m. onwards
1						Micro lab
2	L-group	L-group	Concept session on self-aware-ness	Plan of the lab	YCT	L-group
3	L-group	Planning the module	PCS—Feedback	Interper-sonal communi-cation	YCT	L-group
4	C-group	Diagnosis	PCS—team building	Collabo-ration and helping	YCT	L-group

5	C-group	Module 1	PCS—decision-making	Observation skills	YCT	L-group
6	L-group	C-group human values	Change and	Free	Free	Free
7	C-group—hidden agenda	Module 2	PCS—data for inter-vention	Intervention for change	YCT	L-group
8	C-group, PCS—authority and leader-ship	Module 3 planning	PCS—and commit-ment	Facilitating	YCT	L-group
9	C-group	Module 4	PCS—managing conflicts	Facilitator's role	YCT	L-group
10	C-group	Action plans	Open	Open	Review and closing	

Some of the concepts are explained as follows:

Consultant–client Group (C-group)

The structure of the C-group was that of a group of 21 participants, further divided into three small groups of seven persons each. The objective was to provide an opportunity to develop skills of giving and receiving help and other associated skills such as listening, communicating, interviewing, understanding, and building empathy.

One person (X) from the group would be the help receiver and another person (Y) the help giver. A third person was the process observer. Each member played each role in turn. The role of the trainer was to intervene at an appropriate time to focus attention on the process. The content was a real-life problem which X confronted.

The session would end with a critique by all the members of the sub-group.

Participants Concept Session (PCS)

In this session, participants were required to present different concepts. The objective of the session was: (*a*) to create the confidence that they could present concepts and (*b*) to develop knowledge and understanding of the theoretical constructs necessary for any meaningful role in a concept session. Generally, three members worked together for this session. The trainer provided them with some help. But they themselves designed their session plan, studied the content to be presented, made thorough and elaborate notes, and prepared the necessary audio-visuals to present the concepts. The presentation was often shared by the three-member group and the questions after the presentation were taken up interdependently by the group members. The various concepts the members presented were: feedback, team-building, decision-making, hidden agendas, data for intervention, authority and leadership, planning and commitment, and managing conflicts.

Your Choice Time (YCT)

This part of the schedule provided for any activity that the participants or the trainer wished to do. Surprisingly, to none of the participants did it mean free time. It was mostly used for study, consultation, writing the lab diary, small group (dyad/triad) meetings, or resolving some issue that bothered participants in the 'here and now'.

Module

This was a small group working from planning to implementation of a small intervention project. The group was subdivided into four modules. Each module was like a group of consultants working on a problem. Each group diagnosed the need for intervention, collected data for it, planned the intervention, and then implemented it. After all this was over, the group's activity was reviewed in a critique where the trainer joined the other members in sharing their observations with the community.

Faculty concept session (FCS)

Besides PCS, there were also FCSs, that is, concept sessions provided by the faculty. It was considered necessary that, unlike in a straight laboratory situation where the maximum emphasis is on experiential learning, the advanced laboratory for facilitators should provide a conceptual base for the experiential learning as well. The concept sessions provided by the faculty were on self-knowledge, interpersonal communication (with role-playing), collaboration, observation skills, change, human values, facilitation, and the facilitator's role. Many other concepts were discussed in lecture forms at the appropriate time, once other activities brought about sufficient experience that needed backing up by the relevant concepts.

L-group

The last event the day was the L-group, held after dinner but continuing till the small hours of the morning. Every night, the session turned out to be a mini marathon. With the gentle breeze combined with the silently dancing leaves of the coconut trees, the chirping of crickets, the wavering lights from the lamp and the candles, and the dancing shadows on the wall, all that was sombre and melancholy in life danced with the joy and peace of being. It was a memorable experience indeed!

Reinforcement Lab

The reinforcement lab (RL) was held nine months after the internal resource development laboratory (IRDL). It was intended that all the facilitators who participated in the IRDL would also participate in the RL. All the facilitators had expressed their desire to attend the RL, however. As a matter of fact, they wrote to the coordinator (N) expressing their desire and indicating the time when it should be held. Accordingly, the programme dates were fixed. But when the laboratory started, it was seen that about five or six participants from a single district had not turned up. The reason mentioned was that a concurrent programme was being conducted at that very time. Something was amiss. The dynamics of the situation were rather involved and we will discuss this later.

The reinforcement lab was quite unstructured. The participants were familiar with the trainers, having participated in the basic lab as well as in the IRDL. They knew the norms and values of the laboratories.

The content of the programme essentially revolved around the role of the facilitators: the skills needed in performing the role effectively and in identifying and resolving their dilemmas facilitators while working in their LUs.

The first part of the programme focused on sharing their experience as facilitators in the respective houses during the preceding nine months. The following are some excerpts from what they narrated:

> '...After the facilitators' programme, I was full of enthusiasm to put into practice what we had learned. On the other hand, on my return to my community, I did not see nor feel any positive response or interest from the group. Somehow this affected me and I withdrew. Our community was very fluid. The climate was one of restlessness due to many factors. At one point when the confusion was too much, I thought the group was ready, made an intervention, applying the forcefield analysis. When I came to the specific point of what we wanted to work [for] as a group, the response was silence. I took that silence for the group not being ready and therefore, did not push. Later on, seeing all the complications I gave up the idea of introducing a planned and directed action plan. I, therefore, concentrated on being a support, a classifier, and helping the individuals and the group to become aware of certain things going on in the group. I also concentrated on providing feedback on a one-to-one basis as and when the occasion arose.
>
> 'I was operating on the principle that unless the individual was freed from his inner conflicts, there can be no group action. But suddenly events took an unexpected turn. In one of the meetings of the LU, the entire group on their own came out and expressed their discontent at a lot of things and explored as to why [sic] we could do. At this point I intervened and suggested working on something specific. We started to open up slowly. The atmosphere made us more acceptance-oriented and understanding, and problems came up to the surface, one by one. We tried to cope with tensions and conflicts. At present I feel that we are ready to start new things and venture into new horizons.
>
> 'I feel that I have not done much, nothing concrete, because I did not come as 'a mighty and gusting wind but rather a gentle breeze. The issues I see as relevant are: (*a*) the responsibility of my professional work—the attention, devotion, and the time it requires versus the responsibility of my facilitation work, (*b*) durability of roles—as a facilitator and as a member, (*c*) my inability to be free to provide feedback to certain individuals in the LU, and (*d*) skills required to deal with unpleasant situations.'
>
> **Another person:**
>
> 'I doubt whether this [facilitators and their functioning] works in small community [sic]. To be facilitators requires you to be withdrawn from active participation in the group meetings so you can observe the process and give feedback to the group—do process analysis...When I dropped the facilitator role, everyone was much happier and we could function much better. A member in authority possessing the skills of a facilitator coming into a LU on its request might be a good idea. But this "pool" idea might not work because it would be a roundabout way...Some goals I set I did achieve...Unforeseen, very deep conflicts within individuals came to the surface which we as a group could not

handle. One problem was solved (not really solved but was recognised and help is being offered); but another was there half recognised by the person who is not yet ready to face it; we are not afraid, we shall give him the needed help as he gradually comes to face himself. But how can we acquire greater expertise?'

'...When I was elected as the internal resource person, I felt that I was chosen, against some strong opposition, in a hurry because there was nobody else and because I pushed myself in. This feeling of mine has, in my opinion, slowed down the process of change. When I returned, after the IRD programme, I had the LU action plan ready, based on the data of the survey. The first meeting with the community after my return was to clarify my roles and to discuss and take decision [sic] on the action plan. We ended this meeting with what seemed very negative results. The roles remained as vague as they were before, and it was decided to put up all action plans for further study and thought. I felt further rejected. Again, at a subsequent meeting I brought up the issue of the action plan once again. This led to a lot of emotions. I expressed that I felt I was not accepted and supported by the community. Support came forth then. About a month later I sought clarification of my role and presented my understanding of my role. The LU felt that they could not spell out the role until they knew the action plan. However, I was asked to do process observations at meetings. As for the action plan, it was decided to do a LU climate survey once more. Subsequently we did conduct the survey again and on the basis of the survey, we made some structural changes in the house. Things are moving on, but now I have to move out of the LU on some specific assignment to another sector...'

The reinforcement laboratory spent most of its time in discussing the issues brought out thus by the facilitators from their experience in working as facilitators. They themselves formulated the schedule of the programme. The issues were listed out and prioritized, and then taken up one after another. The faculty raised questions about the issues and helped the participants in exploring ways of increasing their own effectiveness. Some conceptual and strategic issues were also raised and explored. The issues included questions such as: Should the facilitator work for his or her own visibility? Should the facilitator work with authority? Is backing by the authority necessary? Does the 'facilitator' mean a 'leader'?

Every day, the facilitators had one extended session on skill practice. Work was done on skills of communication, collaboration, conflict manage-ment, consensus-building, commitment and involvement, help giving, help receiving, and trusting. The duration of the laboratory was four days.

Stabilization and Termination

One of the thrusts of the reinforcement lab was to bring in maturity in place of an exuberance of enthusiasm resulting in over-ambitious planning of action. Notwithstanding the shortcomings in the original selection of internal facilitators and the subsequent struggle, after the reinforcement lab we had renewed faith in the strength of the internal facilitators and their self-confidence was reinforced. We thought it was time for us to plan our withdrawal. Before we withdrew finally, we thought it was necessary to attend to a few issues. One of these issues was helping to resolve some conflicts that arose at the management level of plan renovation. One of us spent quite some time in an LU to counsel on this issue and arrange confrontation meetings between the conflicting parties. The confrontation proved very useful and resolution was achieved through the hard work of the conflicting parties.

At almost the same time, the members of the LU had reinforcement and psychological rewards from other sectors. There was praise and appreciation (and a little bit of jealousy also!) that this sector had moved ahead of the other sectors in their work. In the meantime, an international meeting of representatives from all over the world was held. One of us agreed to provide process help to the meeting and four internal resource persons were selected for help. We spent time in planning the process feedback strategy, and a few days on back-up help to the internal facilitators, who took on the main process work. Almost everyone in the meeting was highly impressed by the superb professional work the internal facilitator did. And we were naturally very proud of it.

The central authority was also keeping itself informed about the activities of this sector and they had witnessed their work and professional competence. Happily, they did give positive reinforcement. A professionally competent person visited the sector several times, met the members almost in every LU, and entered into detailed dialogue in each LU. His visits had a stimulating effect on the members in their new pursuit.

In our gradual withdrawal process, we made it clear to the organization that we did not think any other direct intervention was needed now, but that we would like to visit and help any LUs and the coordinator if it was needed. However, we made it clear that such requests for help must be channelled through their coordinator and they should approach the external consultants only when they had done all that they could do to help themselves. These criteria were set forth to underscore the need for self-help and self-reliance. We are glad to say they did not need much outside help. They really depended on their own skills and abilities. A year later, however, we received a request for help in training a group of members on certain specific aspects of management. But that is another story.

OD in Voluntary Organizations: Towards Organizational Maturation

We have described one long-term attempt to implement OD in a voluntary organization. Our experience has been very rewarding. We learnt a great deal from this association. Voluntary organizations have some special features which distinguish them from enterprise. While these can be regarded as their main strengths, they also pose some problems. In our experience, these problems can be resolved through effective OD interventions and their special features can be turned into their greatest strengths. We would like to comment on three main features.

Commitment

In the first place, voluntary organizations work in a climate of commitment. In most voluntary organizations, the members come together on the basis of some ideology and a high level of commitment to the goals which they share in common. This level of commitment is not found in other organizations. While this may be a great strength of these organizations, this may also create some problems. Since commitment is the main distinguishing feature of these organizations, individuals may become manipulative through their high level of commitment to the cause and thereby may control the organization. If this happens, a great deal of dependency may be generated in the organizational culture. It often happens that those who start as sources of inspiration feel that they have—and they may, in fact, have—a much higher commitment to the goals of the organization. This may give them a sense of ownership of the organization, with the right to direct and control. This may be done in good faith and with noble intentions. However, this does cause a problem when more people begin to join the organization and a differential strata of psychological membership begins to emerge. OD in such organizations, then, has to deal with the problem of dependency.

Extension Motivation

Another feature of these organizations is that the level of extension motivation—concern for others and a desire to serve others (Pareek, 1968)—is quite high. Compared to other organizations, the stress on service and working for others is a great strength of these organizations. However, the stress on service may lead to the tendency to use oneself for others rather than to work for others without necessarily sacrificing one's own self.

We make a distinction between self-sacrifice (giving precedence to larger goals over immediate individual goals and integrating the self with the goals) and sacrificing or using the self (denying the self in the name of serving the goals by splitting the two and taking the role at the cost of the self). While in the concept of self-sacrifice, an integration of the self and the higher goals is involved, in using the self, a split and denial of the self is implied.

This is likely to be the problem in India where denying oneself is a prime value. There may be two consequences if the self is used for, rather than integrated with, service to others. On the one hand, it may lead to self-rejection and an escape into the role of a helper. This may disturb mental health and the individuals working in voluntary organizations with this tendency are likely to relish self-rejection and escape the problem of confronting their personal problems with the members with whom they work. On the other hand, excessive stress on serving others may lead to lack of affection and personal relationships amongst members of the organization. As a result of these, there may be a lack of mutuality and collaboration. One problem, therefore, which OD has to deal with in volun-tary organizations is that of working towards mutuality and collaboration.

Empathy

A corollary of a high level of extension motivation is that members of these organizations may have a high level of empathy. This in itself is a great strength. However, it may create problems if empathy makes people over-sensitive and may result in the tendency to avoid hurting the feelings of others. If this happens, people are likely to try to be 'good' and would generally avoid confrontation. This may create issues which OD may have to deal with.

As discussed earlier, some of the strong points that these organizations may have are likely to become problems if the organization does not mature as an organization. Organizational maturation (and not organizational maturity)—a state of becoming and not of being—is likely to be an issue in voluntary organizations, since their main emphasis is on service and commit-ment to goals rather than on organizational effectiveness. Over-emphasis on such goals may make them insensitive to organizational issues. It may be important for OD to work towards organizational maturation—a process of developing a rational and empirical approach to the understanding of the problems of the organization and its functioning and finding a systematic way of dealing with these problems, working towards a high commitment and positive confronting culture on the organization. A combination of these two is likely to result in the maturation of an organization.

For working towards organizational maturation, it is necessary that the consultants help the organization deal with the problems of dependency, independence, and counter-dependence and enable it to move towards mutuality and collaboration.

EXERCISES

Questions

Read and reflect on the case in the light of your readings of Part Five of the book. Answer the following questions:

1. Describe the culture and climate of the organization, giving their indicators.
2. How was diagnosis made in the organization?
3. Which interventions were used to introduce change, and with what effects?
4. Can you call this organization a learning organization? Give reasons.
5. What was the role of the external consultants and how were internal resources developed?

Short Questions Answers

1. Name the most important factors of production?
2. What is human resource?
3. What is human resource planning?
4. Explain the nature and significance of human resource planning?
5. Discuss the process of human resource planning?
6. What is job analysis?
7. What is recruitment and selection?
8. Give the difference between the recruitment and selection?
9. Discuss the sources of recruitment?
10. Explain the steps in the selection procedure?
11. Define psychological tests and interviewing?
12. Discuss the importance of employment tests?
13. Discuss the various kind of interview programme held in an enterprise?
14. What is placement?
15. Discuss the significance or importance of placement?
16. While taking the placement decision, some consideration or principles are kept in mind in an organization, elaborate?

17. What is induction of employees?
18. Discuss the induction procedure?
19. What is training and development?
20. Give the difference between training and development?
21. Why is there a need for training the employees?
22. What is the importance of training to the organization?
23. Explain the importance of training to the workers?
24. Which kinds of training programme are used in the industry?
25. Discuss the methods of training?
26. What is management development?
27. What is the purpose of management development programme?
28. Discuss some of the methods of management development?

Long Questions Answers

1. Explain the following:
   (*a*) Human resource planning
   (*b*) Recruitment and selection
   (*c*) Training and development
   (*d*) Management development.
   (*e*) Psychological test and interviewing
   (*f*) Placement and induction.
2. Compare and contrast training and development. Also discuss different training methods?
3. What is the need of human resource planning? Discuss the training and development of human resource for detail?
4. Why is there need for human resource planning and explain its process in detail?
5. Compare and contrast recruitment and selection. Also discuss the technique of induction?

■■■

CHAPTER

6

Organizational Behavior

6.0 CHAPTER OVERVIEW

Managers need to develop their interpersonal or people skills if they are going to be effective in their jobs. Organizational behavior (OB) is a field of study that investigates the impact that individuals, groups, and structure have on behavior within an organization, and then applies that knowledge to make organizations work more effectively. Specifically, OB focuses on how to improve productivity, reduce absenteeism and turnover, and increase employee citizenship and job satisfaction.

We all hold generalizations about the behavior of people. Some of our generalizations may provide valid insights into human behavior, but many are erroneous. Organizational behavior uses systematic study to improve predictions of behavior that would be made from intuition alone. Yet, because people are different, we need to look at OB in a contingency framework, using situational variables to moderate cause-effect relationships.

Organizational behavior offers both challenges and opportunities for managers. It recognizes differences and helps managers to see the value of workforce diversity and practices that may need to be changed when managing in different countries. It can help improve quality and employee productivity by showing managers how to empower their people as well as how to design and implement change programs. It offers specific insights to improve a manager's people skills. In times of rapid and ongoing change, faced by most managers today, OB can help managers cope in a world of "temporariness" and learn ways to stimulate innovation. Finally, OB can offer managers guidance in creating an ethically healthy work climate.

6.1 INTRODUCTION

The study of Organizational Behavior (OB) is very interesting and challenging too. It is related to individuals, group of people working together in teams. The study becomes more challenging when situational factors interact. The study of organizational behavior relates to the expected behavior of an individual in the organization. No two individuals are likely to behave in the same manner in a particular work situation. It is the predictability of a manager about the expected behavior of an individual. There are no absolutes in human behavior. It is the human factor that is contributory to the productivity hence the study of human behavior is important. Great importance therefore must be attached to the study.

Researchers, management practitioners, psychologists, and social scientists must understand the very credentials of an individual, his background, social framework, educational update, impact of social groups and other situational factors on behavior. Managers under whom an individual is working should be able to explain, predict, evaluate and modify human behavior that will largely depend upon knowledge, skill and experience of the manager in handling large group of people in diverse situations. Preemptive actions need to be taken for human behavior forecasting. The value system, emotional intelligence, organizational culture, job design and the work environment are important causal agents in determining human behavior. Cause and effect relationship plays an important role in how an individual is likely to behave in a particular situation and its impact on productivity. An appropriate organizational culture can modify individual behavior. Recent trends exist in laying greater stress on organizational development and imbibing a favorable organizational culture in each individual. It also involves fostering a team spirit and motivation so that the organizational objectives are achieved.

There is a need for commitment on the part of the management that should be continuous and incremental in nature.

The scope of the organizational behavior is as under:

(*a*) Impact of personality on performance
(*b*) Employee motivation
(*c*) Leadership
(*d*) How to create effective teams and groups
(*e*) Study of different organizational structures
(*f*) Individual behavior, attitude and learning
(*g*) Perception
(*h*) Design and development of effective organization
(*i*) Job design
(*j*) Impact of culture on organizational behavior
(*k*) Management of change
(*l*) Management of conflict and stress
(*m*) Organizational development
(*n*) Organizational culture
(*o*) Transactional analysis
(*p*) Group behavior, power and politics
(*q*) Job design
(*r*) Study of emotions

The field of the organizational behavior does not depend upon deductions based on gut feelings but attempts to gather information regarding an issue in a scientific manner under controlled conditions. It uses information and interprets the findings so that the behavior of an individual and group can be canalized as desired. Large number of psychologists, social scientists and academicians has carried out research on various issues related to organization behavior. Employee performance and job satisfaction are determinants of accomplishment of individual and organizational goals.

Organizations have been set up to fulfill needs of the people. In today's competitive world, the organizations have to be growth-oriented. This is possible when productivity is ensured with respect to quantity of product to be produced with zero error quality. Employee absenteeism and turnover has a negative impact on productivity. Employee who absents frequently cannot contribute towards productivity and growth of the organization. In the same manner, employee turnover causes increased cost of production. Job satisfaction is a major factor to analyze performance of an individual towards his work. Satisfied workers are productive workers who contribute towards building an appropriate work culture in an organization. Organizations are composed of number of individuals working independently or collectively in teams, and number of such teams makes a department and number of such departments makes an organization. It is a formal structure and all departments have to function in a coordinated manner to achieve the organizational objective. It is therefore important for all employees to possess a positive attitude towards work. They need to function in congenial atmosphere and accomplish assigned goals. It is also important for managers to develop an appropriate work culture. Use of authority, delegation of certain powers to subordinates, division of labor, efficient communication,

Benchmarking, re-engineering, job re-design and empowerment are some of the important factors so that an organization can function as well-oiled machine. This is not only applicable to manufacturing organizations but also to service and social organizations.

6.2 DEFINITION

Organizational Behavior: OB is a field of study that investigates the impact that individuals, groups, and structure have on behavior within organizations for the purpose of applying such knowledge toward improving an organization's effectiveness.

Organizational behavior is a field of study:

- OB studies three determinants of behavior in organizations: individuals, groups, and structure.
- OB applies the knowledge gained about individuals, groups, and the effect of structure on behavior in order to make organizations work more effectively.
- OB is concerned with the study of what people do in an organization and how that behavior affects the performance of the organization.
- There is increasing agreement as to the components of OB, but there is still considerable debate as to the relative importance of each: motivation, leader behavior and power, interpersonal communication, group structure and

processes, learning, attitude development and perception, change processes, conflict, work design, and work stress.

Each of us is a student of behavior:

- A casual or common sense approach to reading others can often lead to erroneous predictions.
- You can improve your predictive ability by replacing your intuitive opinions with a more systematic approach.
- The systematic approach used in this chapter will uncover important facts and relationships and will provide a base from which more accurate predictions of behavior can be made.
- Behavior generally is predictable if we know how the person perceived the situation and what is important to him or her.
- While people's behavior may not appear to be rational to an outsider, there is reason to believe it usually is intended to be rational by the individual and that they see their behavior as rational.

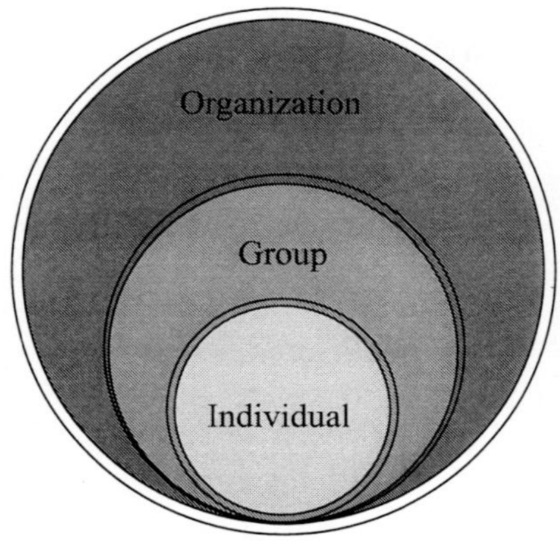

Organizational behavior can be defined as – **"the study and application of knowledge about human behavior related to other elements of an organization such as structure, technology and social systems** (LM Prasad). Stephen P Robins defines **"Organizational behavior as a systematic study of the actions and attitudes that people exhibit within organizations."** It has been observed that we generally form our opinion based on the symptoms of an issue and do not really go to the root cause of the happening. Science of organizational behavior is applied in nature. Disciplines like psychology, anthropology and political science have contributed in terms of various studies and theories to the field of organizational behavior. A leader should be able to communicate with his subordinate and keep them in picture as to the happenings in the organization. People promote organizational culture for mutual benefit. Politics is often used to create conflict with the aim of enlarging self-power base to the detrimental of organizational growth. Politics, in Indian context has made in roads based on religion, caste system in the decision making process which has led to formation of informal groups in the organization that often exploit the organization for fulfillment of personal goals at the cost of organizational goals. Conflict and manipulating power bases need to be handled in an appropriate manner to modify human behavior and stimulate various individuals towards achieving higher productivity. Power dynamics plays a significant role in organization situations in different environment.

6.3 CONTRIBUTING DISCIPLINES TO THE OB FIELD

Organizational behavior is an applied behavioral science that is built upon contributions from a number of behavioral disciplines. The predominant areas are psychology, sociology, social psychology, anthropology, and political science.

Psychology

- Psychology is the science that seeks to measure, explain, and sometimes change the behavior of humans and other animals.
- Early industrial/organizational psychologists concerned themselves with problems of fatigue, boredom, and other factors relevant to working conditions that could impede efficient work performance.
- More recently, their contributions have been expanded to include learning, perception, personality, emotions, training, leadership effectiveness, needs and motivational forces, job satisfaction, decision making processes, performance appraisals, attitude measurement, employee selection techniques, work design, and job stress.

Sociology

- Sociologists study the social system in which individuals fill their roles; that is, sociology studies people in relation to their fellow human beings.
- Their greatest contribution to OB is through their study of group behavior in organizations, particularly formal and complex organizations.

Social Psychology

- Social psychology blends the concepts of psychology and sociology.
- It focuses on the influence of people on one another.

Major area—how to implement it and how to reduce barriers to its acceptance

Anthropology

- Anthropology is the study of societies to learn about human beings and their activities.
- Anthropologists work on cultures and environments; for instance, they have helped us understand differences in fundamental values, attitudes, and behavior among people in different countries and within different organizations.

Political Science

- Frequently overlooked as a contributing discipline.
- Political science studies the behavior of individuals and groups within a political environment.
- Political science has contributed to the field of Organizational behavior. Stability of government at national level is one major factor for promotion of international business, financial investments, expansion and employment. Various government rules and regulations play a very decisive role in growth of the organization.

6.4 FEW ABSOLUTES IN OB

There are few, if any, simple and universal principles that explain organizational behavior. Human beings are complex. Because they are not alike, our ability to make simple, accurate, and sweeping generalizations is limited. That does not mean, of course, that we cannot offer reasonably accurate explanations of human behavior or make valid predictions. It does mean, however, that OB concepts must reflect situational, or contingency, conditions. Contingency variables—situational factors are variables that moderate the relationship between the independent and dependent variables. Using general concepts and then altering their application to the particular situation developed the science of OB. Organizational behavior theories mirror the subject matter with which they deal.

There are many challenges and opportunities today for managers to use OB concepts.

1. *Responding to Globalization*
   (*a*) Organizations are no longer constrained by national borders.
   (*a*) Globalization affects a manager's people skills in at least two ways.
   - First, if you are a manager, you are increasingly likely to find yourself in a foreign assignment.
   - Second, even in your own country, you are going to find yourself working with bosses, peers, and other employees who were born and raised in different cultures.
2. *Managing Workforce Diversity*
   (*a*) Workforce diversity is one of the most important and broad-based challenges currently facing organizations.
   (*b*) While globalization focuses on differences between people from different countries, workforce diversity addresses differences among people within given countries.
   (*c*) Workforce diversity means that organizations are becoming more heterogeneous in terms of gender, race, and ethnicity. It is an issue in Canada, Australia, South Africa, Japan, and Europe as well as the United States.
   (*d*) A melting-pot approach assumed people who were different would automatically assimilate.
   (*e*) Employees do not set aside their cultural values and lifestyle preferences when they cometo work.
   (*f*) The melting pot assumption is replaced by one that recognizes and values differences.
   (*g*) Members of diverse groups were a small percentage of the workforce and were, for the most part, ignored by large organizations (pe-1980s); now:
   - 47 percent of the U.S. labor force are women
   - Minorities and immigrants make up 23 percent

- More workers than ever are unmarried with no children.

(*h*) Workforce diversity has important implications for management practice.

- Shift to recognizing differences and responding to those differences
- Providing diversity training and revamping benefit programs to accommodate the different needs of employees

3. *Improving Quality and Productivity*

(*a*) Total quality management (TQM) is a philosophy of management that is driven by the constant attainment of customer satisfaction through the continuous improvement of all organizational processes.

(*b*) Implementing quality programs requires extensive employee involvement

(*c*) Process reengineering asks the question: "How would we do things around here if we were starting over from scratch?"

- Every process is evaluated in terms of contribution to goals
- Rather than make incremental changes, often old systems are eliminated entirely and replaced with new systems

(*d*) To improve productivity and quality, managers must include employees.

4. *Responding to the Labor Shortage*

(*a*) If trends continue as expected, the U.S. will have a labor shortage for the next 10-15 years (particularly in skilled positions).

(*b*) The labor shortage is a function of low birth rates and labor participation rates (immigration does little to solve the problem).

(*c*) Wages and benefits are not enough to keep talented workers. Managers must understand human behavior and respond accordingly.

5. *Improving Customer Service and People Skills*

(*a*) The majority of employees in developed countries work in service jobs—jobs that require substantive interaction with the firm's customers. For example, 80 percent of U.S. workers are employed in service industries.

(*b*) Employee attitudes and behavior are directly related to customer satisfaction requiring management to create a customer responsive culture.

(*c*) People skills are essential to managerial effectiveness.

(*d*) OB provides the concepts and theories that allow managers to predict employee behavior in given situations.

6. *Empowering People*

(*a*) Today managers are being called coaches, advisers, sponsors, or facilitators, and in many organizations, employees are now called associates.

(*b*) There is a blurring between the roles of managers and workers; decision making is being pushed down to the operating level, where workers are being given the freedom to make choices about schedules and procedures and to solve work-related problems.

(*c*) Managers are empowering employees.

- They are putting employees in charge of what they do.
- Managers have to learn how to give up control.

- Employees have to learn how to take responsibility for their work and make appropriate decisions.

7. *Coping with "Temporariness"*

(*a*) Managers have always been concerned with change:

- What is different today is the length of time between changes
- Change is an ongoing activity for most managers. The concept of continuous improvement, for instance, implies constant change
- In the past, managing could be characterized by long periods of stability, interrupted occasionally by short periods of change.
- Today, long periods of ongoing change are interrupted occasionally by short periods of stability!

(*b*) Permanent "temporariness":

- Both managers and employees must learn to live with flexibility, spontaneity, and unpredictability
- The jobs that workers perform are in a permanent state of flux, so workers need to continually update their knowledge and skills to perform new job requirements.

(*c*) Work groups are also increasingly in a state of flux.

- Predictability has been replaced by temporary work groups, teams that include members from different departments and whose members change all the time, and the increased use of employee rotation to fill constantly changing work assignments.

(*d*) Organizations themselves are in a state of flux.

- They reorganize their various divisions, sell off poor-performing businesses, downsize operations, subcontract non-critical services and operations to other organizations, and replace permanent employees with temporaries.

8. *Stimulating Innovation and Change*

(*a*) Successful organizations must foster innovation and the art of change.

(*b*) Companies that maintain flexibility, continually improve quality, and beat their competition to the marketplace with innovative products and services will be tomorrow's winners.

(*c*) Employees are critical to an organization's ability to change and innovate.

9. *Helping Employees Balance Work-Life Conflicts*

(*a*) The creation of the global workforce means work no longer sleeps. Workers are on-call 24-hours day or working non-traditional shifts.

(*b*) Communication technology has provided a vehicle for working at any time or any place.

(*c*) Employees are working longer hours per week—from 43 to 47 hours per week since 1977.

(*d*) The lifestyles of families have changes creating conflict: more dual career couples and single parents find it hard to fulfill commitments to home, children, spouse, parents, and friends.

(*e*) Employees want jobs that allow flexibility and provide time for a "life."

6.5 BEHAVIOR MODEL FOR ORGANIZATIONAL EFFICIENCY

Organizational behavior is a study and application of managerial skills and knowledge to people in the organization to investigate individual and group behavior. Various concepts and models in the field of organizational behavior attempt to identify, not only the human behavior but also modify their attitude and promote skills so that they can act more effectively. This is done scientifically; therefore, organizational behavior field is a scientific discipline. The knowledge and models are practically applied to workers, groups and organizational structure that provide tools for improved behavior and dynamics of relationship.

The field of organizational behavior also provides various systems and models for international relationship that are applied to organizations. Leaders must look for indicators (effects) of individual behavior and of groups in any organization. Indicators have a root cause beneath. As a leader, it is that symptom, which must be evaluated, and cause of human behavior established so that if the behavior is good, the manager can establish the norms of behavior. If the behavior is not conducive to achieve the organizational objective then suitable alternative model can be applied to channelize individual behavior towards an appropriate organizational value system and thus individual behavior modified. An organization has three basic elements namely, people, structure, and technology. An organization must have suitable organizational structure, with appropriate number of tier and reporting system properly explained. Principle of unity of command, delegation of authority and responsibility, formulation of objectives and its allotment to various groups is very important so that workers achieve a required level of job satisfaction. They must be trained to handle sophisticated machines and equipment. It is the people, their value system, and faith in the leadership that make an organization. Leader must be able to describe, understand, predict and control individual behavior in the organization. This is explained in the succeeding paragraphs.

(*a*) **Describe:** Study of organizational behavior is based on scientific methods, which have been applied on human beings. It is a science, that analyses as to how people behave in different situations in the organization. A manager should be able describe the behavior of each of the individuals under his command, identify attitude, and be able to pinpoint his behavior so that the situation in the organization is under control.

(*b*) **Understand:** Leaders must understand human behavior as to why people behave in particular manner and try to identify reasons so that corrective actions can be taken.

(*c*) **Predict:** By frequent closer interaction, a leader is in a position to identify the nature of workers. Some are more productive while the others are tardy and disruptive. In such situation, a leader should be able to handle each individual differently so that his or her actions can be channelized to higher productivity.

(*d*) **Control:** Managers in the organizations should train their subordinates continuously; aim being development of skills, promotion of productivity and improvement of individual behavior. It is a continuous process on the part of manager. He must lay down control measures so that the energy of workers is diverted towards organizational objectives. Communication should be used to ensure that the behavior of individual is controlled. Environment has a great impact on human behavior. Appropriate internal environment would help organizations to build favorable work environment that will help individuals and groups within organizations to work effectively towards higher productivity.

6.6 ORGANIZATIONAL COMPONENTS THAT NEED TO BE MANAGED

People

People are the main component of any organization that has to be managed. Every individual has a personal goal to be achieved. Organizations must identify the **need spectrum** of individuals and take suitable steps for its fulfillment to enable them to perform effectively so that they complete their allotted task in time. **Relationship** between the workers, with subordinates and superiors should be established based on full understanding and complete faith based on mutual trust so that it is easy to communicate and understand each other's views. **Work teams** and **Groups** play a vital role in the organization. Individual may have to keep his personal interest aside if it conflicts with team or group goals. It is the team goals, accomplishment of which contributes towards achieving organizational goals. Apart from managing internal workforce, it is also important to manage customers who are the end persons using organization's products or services. Utmost interest of stakeholders, government, employees, social groups and non-governmental organizations (NGOs) must be kept in mind as they play a dominant role in the society. Apart from the above, adequate consideration should also be given to competitors, regulatory agencies, labor force, suppliers and resource persons.

Structure

There are two types of organizations, formal and informal. Informal organizations do not have a specified structure. Formal organizations are building based upon the objective set for it. Organizational structure in such organization is hierarchical in nature, with people at each level having their own objectives, which contributes towards fulfillment of overall organizational objectives. In such organization people at lower levels report to higher level managers. The tier system has the principle of unity of command inbuilt in it. The organization structure may depend upon the size, number of products/services produced, skill and experience of the employees, managerial staff and geographical location of the organization. An organization may have several levels and pyramid like organizational structure or flat structure. The efficiency of the organization will depend upon the free flow of the information, efficient communication system prevailing in the organization, well-defined authority and responsibility supported by detailed policies, rules and regulations.

The organization must have well laid out systems, which are understood by workers, supervisors and managers. The leader must keep open mind while dealing with subordinates and exercise full control over various systems, levels and ensure planned productivity and achieve high level of job satisfaction.

Technology

Managing technology is an important job of any management. It is an important element of any unit. Selection of technology, procurement, installation, operation and maintenance is important and no compromise should be made in procuring **latest or advanced technology.**

Various systems and sub-systems should support technology that exists in an organization. Based on the technology, an organization should formulate job structure and resultant procurement of human resource so that they are complimentary to each other. Adequate attention is also be paid to service industry. For example an appropriate drill, procedures are installed in hospital industry to ensure that the patients' record is maintained properly. On line operations of all systems relating to admission record, past treatment, drugs, availability of beds, schedule of operations maintained so that the level of patients satisfaction is raised. In minimum number of days, maximum numbers of patients should be treated. Various processes required to regulate these functions form the important part of service industry.

Jobs

Job is an assignment assigned to an individual. It encompasses various tasks within it. For example, Personnel manager wants to fill up twelve vacancies in production department within three months. Job will have various tasks inbuilt in it like designing of job specification, selection of media, advertising vacancies, scheduling of selection and recruiting process. Manager, therefore have to manage various tasks to accomplish a particular job. This may form a part of managerial functions. Adequate delegation, supervision, application of various control techniques makes the job simpler for the manager. Introduction of computers have made managerial functions simpler, as required information is available for decision making.

Processes

Management of processes and its inter-dependence is very crucial to high productivity and higher job satisfaction. What is important for a manager is to ensure high morale of the work force. To ensure this, he must identify various managerial dictums. Select appropriate subordinates to carry out a job based on aptitude, personality traits, mental build up and attitude. He should also involve himself and lead subordinates by personal example. In defense services, it is the quality of leadership that motivates troops to achieve near impossible task where everything appears to be going wrong. Various role models assist leaders in identifying as to which process, method or approach would be suitable to mold subordinates in suitable frame that may be required by any organization. Nothing motivates workers better if you give them their entitlements in full and train them to take up higher jobs. By doing so, manager must develop and build an organizational culture that will bind employees to a common cultural bond. During day-to-day functions, managers must be transparent and maintain a high degree of value system and display ethical behavior. There are no short cuts to this and will pay rich dividends in times to come.

External Environment

What we have so far discussed is various components of an organization that should be managed properly. External environment also plays an important role in managing the points discussed above. When we talk about managing people in the organization, what we have to study and manage is the influence of culture and its impact on the individual. A manager should examine as to how he is going to cope up with the changes. Study of external environment is very wide and encompasses economic, cultural, social, government rules and regulations, legal aspects, political climate, demographics and its impact. If one scans the external environment that is prevailing in Indian context, one will find that individuals are racing to catch up the upper classes it relates to standards of living, material possession, higher education, attempt to copy western culture, food habits, dressing pattern and the like. Beauty parlors, pubs and cyber cafes around each corner are an ample evidence of the impact of external environment. This trend has an impact on what products or services are on priority in the society and indicates the behavior of an individual. If the above factors are evaluated appropriately, a manager will be able to examine and predict human behavior in the organization. It is therefore important to evaluate market situation, competitors, and availability of raw material, technology, availability of skilled, semi-skilled and non-skilled personnel. In addition, evaluate prevailing culture and how individuals are likely to respond to the call of the organization. Some factors like government rules, and political stability keep changing, the organizations must cater for such contingencies. Manager must therefore keep in mind the internal and external factors and make the best amalgam and work to achieve organizational effectiveness.

6.7 ORGANIZATIONAL THEORY

Organizational theory is the sociological study of formal social organizations, such as businesses and bureaucracies, and their interrelationship with the environment in which they operate. It complements the studies of organizational behavior and human resource studies.

Organizations, which are defined as "social units of people that are structured and managed to meet a need or to pursue collective goals," are said to have risen in the United States within a variety of social and historical contexts. Several of those factors are credited with making organizations viable and necessary options for citizens, and they built on one another to bring organizations to the level of importance that they are at today.

In 1820, about 20% of the United States population was dependent on a wage income. That number increased to 90% by 1950. Generally, farmers and craftsmen were the only ones by 1950 who were not dependent on working for someone else; prior to that, most people were able to survive by hunting and farming their own food, making their own supplies, and remaining almost fully self-sufficient. As transportation became more efficient and technologies were further developed, self-sufficiency became an economically poor choice. As in the Lowell Textile Mills, various machines and processes were developed for each step of the production process, thus making mass production a cheaper and faster alternative to individual control. In addition, as the population grew and transportation progressed, the pre-organizational system struggled to support the needs of the market. These conditions made for a wage dependent population that sought out jobs in growing organizations, leading to a shift from individual and family production.

In addition to a shift to wage dependence, externalities from industrialization also created a perfect opportunity for the rise of organizations. Various negative effects such as pollution, workplace accidents, crowded cities, and unemployment became rising concerns. Rather than small groups such as families and churches being able to control these problems as they had in the past, new organizations and systems were required in order to keep their heightened effects down. The smaller associations that had contained various social issues in the past were no longer viable, and instead were collapsed into larger formal organizations. These organizations were less personal, more distant, and more centralized; but, what they lacked in locality, they made up for in efficiency. Along with wage dependency and externalities, growth of industry also played a large role in the development of organizations. Markets that were quickly growing and expanding needed employees right away – because of that, a need developed for organizational structures that would help guide and support these new employees. Some of the first New England factories relied on daughters of farmers at their onset; later, as the economy changed, they began to gain work from the farmers, and finally, European immigrants. Many Europeans left their homes for the promises of US industry, and about 60% of those immigrants stayed in the country. They became a permanent class of workers in the economy, which allowed factories to increase production and produce more than they had before. With this large growth came the need for organizations and leadership that was not previously needed in small businesses and firms.

Overall, the historical and social context in which organizations rose in the United States allowed for not only the development of organizations, but also for their spread and growth. Wage dependency, externalities, and growth of industries all played into the change from individual, family, and small-group production and regulation to large organizations and structure.

Even though the decline in small business might not seem to substantiate how the development in organizations leads to increased aggregate economic return, it exemplifies the cut-throat nature of capitalism. As organizations develop, they devour the smaller organizations that cannot keep up, but also allow for the evolution of innovative management and production techniques for other larger companies. The development of organizations demands a higher level of skillset from workers as it continues to grow. It also builds precautionary measures on cutting edge technology. It amplifies the need for specialization and accounts of functionalism in various organizations and their respective societies. Through much advancement in the interaction of capitalistic bureaucracies, the development of organizations is what has driven contemporary firms to thrive in its modern day society.

Competing Theories of Organization

As organizations are implemented over time, many people experimented as to which one was best. These theories of organizations include Bureaucracy, Rationalization

(Scientific Management), and the Division of Labor. Each theory provides distinct advantages and disadvantages when implemented.

6.7.1 Weber's Ideal of Bureaucracy

Official Jurisdiction on all areas are ordered by rules or laws already implemented. There is an office hierarchy; a system of super- and subordination in which there is supervision of lower office by higher ones.

The management of the modern office is based upon written rule, which are preserved in original form. Office management requires that of training or specialization. When the office is developed/established it requires the full working capacity of individuals. Rules are stable and can be learned. Knowledge of these rules can be viewed as expertise within the bureaucracy (these allow for the management of society)

When a bureaucracy is implemented, they can provide accountability, responsibility, control, and consistency. The hiring of employees will be an impersonal and equal system. Although the classical perspective encourages efficiency, it is often criticized as ignoring human needs. Also, it rarely takes into consideration human error or the variability of work performances (each worker is different).

In the case of the Space Shuttle Challenger disaster, NASA managers overlooked the possibility of human error.

Rational system perspective

In a rational organization system, there are two significant parts: Specificity of Goals and Formalization. Goal specification provides guidelines for specific tasks to be completed along with a regulated way for resources to be allocated. Formalization is a way to standardize organizational behavior. As a result, there will be stable expectations, which create the rational organizational system.

Scientific Management: Taylor analyzed how to maximize the amount of output with the least amount of input. This was Taylor's attempt to rationalize the individual worker.

- Divide work between managers and workers
- Provide incentive system (based on performance)
- Scientifically trained workers
- Create a science for each individual's responsibilities
- Make sure work is done on time/efficiently

There are problems that arose out of scientific management. One is that the standardization leads workers to rebel against mundaneness. Another is that workers may reject the incentive system because they are required to constantly work at their optimum level, an expectation that may be unrealistic.

Division of Labor

The division of labor is the specialization of individual labor roles. It is often associated with increasing output and trade. According to Adam Smith, the division of labor is efficient due to three reasons: occupational specialization, saving from not changing tasks, and machines taking the place of human labor. Occupational specialization leads to increased productivity and distinct skill. Also, Smith argued that human and physical capital must be similar or matched; if the skill of workers were matched with technological improvements, there would be a major increase in productivity.

Although the division of labor is often viewed as inevitable in a capitalistic society, there are several specific problems that may arise. They include a lack of creativity, monotony, and lack of mobility. Creativity will naturally suffer due the monotonous atmosphere that the division of labor creates. Doing the same routines may not be for everyone. Also, employees aren't familiar with other parts of the job. They cannot assist employers of different parts of the system.

6.7.2 Modernization Theory

Modernization "began when a nation's rural population started moving from the countryside to cities". It deals with the cessation of traditional methods in order to pursue more contemporary effective methods of organization. Urbanization is an inevitable characteristic of society because the formation of industries and factories induces profit maximization. It is fair to assume that along with the increase in population, as a result of the subsequent urbanization, is the demand for an intelligent and educated labor force. Following the 1950s, Western culture utilized the effects of mass media coverage to communicate their good fortune attributed to modernization. The coverage promoted "psychic mobility" among the social class and increased the aspirations of many hopefuls in developing economic countries. Under this theory, any country could modernize by using Western civilization as a template.

Although this theory of modernization seemed to pride itself on only the benefits, countries in the Middle East saw this movement in a new light. Middle Eastern countries believed that the media coverage of modernization implied that the more "traditional" societies have not "risen to a higher level of technological development". Consequently, they believed a movement that benefits those who have the monetary resources to modernize technological development would discriminate against the minorities and poor masses. Thus, they were reluctant to modernize because of the economic gap it would create between the rich and the poor.

The growth of modernization took place beginning in the 1950s. For the ensuing decade, people analyzed the diffusion of technological innovations within Western society and the communication that helped it disperse globally. This first "wave" as it became known had some significant ramifications. First, economic development was enhanced from the spread of new technological techniques. And second, modernization supported a more educated society (as mentioned above), and thus a more qualified labor force. The second wave took place between the years 1960 and 1970. This period was labeled anti-modernization, because it saw the push of innovations of Western society onto developing countries as an exertion of dominance ("modernization theory"). It refuted the concept of relying heavily on mass media for the betterment of society. The last wave of modernization theory, which took place in the 1990s, depicts impersonality. As uses of newspapers, TVs, and radios become more prevalent, the need for direct contact, a concept traditional organizations took pride in, diminishes. Thus, organizational interactions become more distant ("Modernization Theory").

According to Frank Dobbin, the modern worldview is the idea that "modern institutions are transparently purposive and that we are in the midst an evolutionary progression towards more efficient forms". This phrase epitomizes the goal of modern firms, bureaucracies, and organizations to maximize efficiency. The key to achieving this goal is through scientific discoveries and innovations. Dobbin discusses the outdated role of culture in organizations. "New Institutionalists" explored the significance of culture in the modern organization. However, the rationalist worldview counters the use of cultural values in organizations, stating, "transcendental economic laws exist, that existing organizational structures must be functional under the parameters of those laws, [and] that the environment will eliminate organizations that adopt non-efficient solutions". These laws govern the modern organizations and lead them in the direction that will maximize profits efficiently. Thus, the modernity of organizations is to generate maximum profit, through the uses of mass media, technological innovations, and social innovations in order to effectively allocate resources for the betterment of the global economy.

Classical perspective

The classical perspective emerges from the Industrial Revolution and centers on theories of efficiency. There are two subtopics under the classical perspective: the scientific management and bureaucracy theory.

Efficiency and teleological arguments in Weberian bureaucracy

Max Weber believed that an ideal bureaucracy consists of six specific characteristics: hierarchy of authority, impersonality, written rules of conduct, promotion based on achievement, specialized division of labor, and efficiency. This ultimate characteristic of Weberian bureaucracy, which states that bureaucracies are very efficient, is controversial and by no means accepted by all sociologists. There are certainly both positive and negative consequences to bureaucracy, and strong arguments for both the efficiency and inefficiency of bureaucracies.

While Max Weber's work was published in the late 1800s and early 1900s, before his death in 1920, his work is still referenced today in the field of sociology. Weber's theory of bureaucracy claims that it is extremely efficient, and even goes as far as to claim that bureaucracy is the most efficient form of organization. Weber claimed that bureaucracies are necessary to ensure the continued functioning of society, which has become drastically more modern and complex in the past century. Furthermore, he claimed that without the structured organization of bureaucracy, our complex society would be much worse off, due to the fact that society would act in an inefficient and wasteful way. He saw bureaucracies as organizations driven towards certain goals, which they could carry out efficiently. In addition, within an organization that operates under bureaucratic standards, the members will be better off due to the heavy regulation and detailed structure. Not only does bureaucracy make it much more difficult for arbitrary and unfair personal favors to be carried out, it also means that promotions and hiring will generally be done completely by merit.

Weber most definitely saw bureaucracies as goal-driven, efficient organizations, but one must not come to the quick and incorrect conclusion that he saw no downfalls to bureaucracy. He recognized that there are constraints within the bureaucratic system. First of all, he realized that bureaucracies were ruled by very few people with very large amounts of unregulated power. This tends to lead to a situation of oligarchy, whereby a limited number of officials become the political and economic power. Furthermore, Weber considered further bureaucratization to be an "inescapable fate," due to the fact that it is supposedly superior to and more efficient than other forms of organization. Weber's analysis of bureaucracies led him to believe that they are too inherently limiting to individual human freedom and he feared that people would begin to be too controlled by bureaucracies. His rationale comes from the knowledge that the strict methods of administration and legitimate forms of authority associated with bureaucracy act to eliminate human freedom.

Regardless of whether or not bureaucracies should be considered positively efficient or too efficient to the extent that they become negative, Weberian bureaucracy tends to offer a teleological argument. A theory, in this case bureaucracy, is considered to be teleological if it involves aiming at specific goals. Weber claimed that bureaucracies are goal-oriented organizations, which use their efficiency and rational principles to reach their goals. A teleological analysis of businesses leads to the inclusion of all involved stakeholders in decision-making. The teleological view of Weberian bureaucracy postulates that all actors in an organization have various ends or goals, and attempt to find the most efficient way to achieve these goals.

6.7.3 Scientific Management Theory

The scientific management theory was introduced by Frederick Winslow Taylor to encourage production efficiency and productivity. Taylor argues that inefficiencies could be controlled through managing production as a science. Taylor defines scientific management as "concerned with knowing exactly what you want men to do and then see in that they do it in the best and cheapest way." According to Taylor, scientific management affects both workers and employers, and stresses the control of the labor force by management.

The Principles of Scientific Management

Taylor identifies four inherent principles of the scientific management theory.

- The creation of a scientific method of measurement that replaces the "rule-of-thumb" method
- Emphasis placed on the training of workers by management
- Co-operation between manager and workers to ensure the principles are being met
- Equal Division of labor between managers and workers.

6.7.4 Bureaucratic Theory

The scholar most closely associated with Bureaucratic theory is Max Weber. In Economy and Society, his seminal book published in 1922, Weber articulates the necessary conditions and descriptive features of bureaucracy. An organization governed under Weber's conception of bureaucracy is characterized by the presence of impersonal positions that are earned and not inherited, rule-governed decision-making, professionalism, chain of command, defined responsibility, and bounded authority.

Weber begins his discussion of bureaucracy by introducing the concept of 'jurisdictional areas': institutions governed by a specific set of rules or laws. In a 'jurisdictional area' regular activities are assigned as official duties, the authority to assign these duties is distributed through a set of rules, and duties are fulfilled continuously by qualified individuals. These elements make up a bureaucratic agency in the case of the state and a bureaucratic enterprise in the private economy.

There are several additional features that comprise a Weberian bureaucracy:

- It is possible to find the utilization of hierarchical subordination in all bureaucratic structures. This means that higher-level offices supervise lower level offices.
- In bureaucracies, personal possessions are kept separate from the monies of the agency or the enterprise.
- People who work within a bureaucracy are usually trained in the appropriate field of specialization.
- Bureaucratic officials are expected to contribute their full working capacity to the organization.
- Positions within a bureaucratic organization must follow a specific set of general rules.

Weber argued that in bureaucracy, taking on a position or office signifies an assumption of a specific duty necessary for the organization. This conception is distinct from historical working relationships in which a worker served a specific ruler, not an institution.

The hierarchical nature of bureaucracies allows employees to demonstrate achieved social status. When an office holder is elected instead of appointed, that person is no longer a purely bureaucratic figure. He derives his power 'from below' instead of 'from above.' When a high-ranking officer selects officials, they are more likely to be chosen for reasons related to the benefit of the superior than the competency of the new hire. When high-skilled employees are necessary for the bureaucracy and public opinion shapes decision-making, competent officers are more likely to be selected.

According to Weber, if 'tenure for life' is legally guaranteed, an office becomes perceived as less prestigious than a position that can be replaced at any time. If 'tenure for life' or a 'right to the office' develops, there is a decrease in career opportunities for ambitious new hires and overall technical efficiency becomes less guaranteed. In a bureaucracy, salaries are provided to officials. The amount is determined on the basis of rank and helps to signify the desirability of a position. Bureaucratic positions also exist as part of stable career tracks that reward office-holders for seniority.

Weber argues that the development of a 'money economy' is the "normal precondition for the unchanged survival, if not the establishment, of pure bureaucratic administrations". Since bureaucracy requires sustained revenues from taxation or private profits in order to be maintained, a money economy is the most rational way to ensure its continued existence.

Weber posits that officials in a bureaucracy have a property right to their office and attempts at exploitation by a superior means the abandonment of bureaucratic principles. He articulates that providing a status incentive to inferior officers helps them to maintain self-respect and fully participate in hierarchical frameworks. Michel Crozier reexamined Weber's theory in 1964. He determined that bureaucracy is flawed because hierarchy causes officers to engage in selfish power struggles that damage the efficiency of the organization.

Criticism of the Weber's Theory of Bureaucracy

Weber's theories were purposed to set a stage for other organizations to follow, and the characteristics are so ideal that they may be impossible for any actual organization to succeed. He wanted to come up with a set of guidelines that would favor both efficiency and, most importantly, conditions that would make the workers top priority. It was common for earlier theorists to distort Weber's views, and today, people still make the same mistakes as they did when Weber's views first came into play. He has always been critiqued for the branches of his ideas that don't work in reality, but the point of his theory was not to actually create an organization, but to create an ideal model for other organizations to follow.

One big misconception that people have had in the past is a question of Weber's morality due to their oversimplification of his characteristics of a pure bureaucracy. "There is dangerous risk of oversimplification in making Weber seem cold and heartless to such a degree that an efficiently-run Nazi death camp might appear admirable" (Bureaucracy Theory). In reality, Weber believed that by using human logic in his system, we could achieve improvement of human condition in various workplaces. Complexity in an organization yields the highest success, therefore simplifying it leads to the illusions of over-authority and intense hierarchical power that are inaccurate of Weber's beliefs.

Another critique of Weber's theory is the argument of efficiency. Highest efficiency, in theory, can be attained through pure work with no regard for the workers (for example, long hours with little pay), which is why over simplification can be dangerous. If we were to take one characteristic focusing on efficiency, it would seem like Weber is promoting unhealthy work conditions, when in fact, he wanted the complete opposite. Put all of them together, and we have the ideal organization, but since a pure bureaucracy is nearly impossible to obtain, efficiency takes the back seat in his beliefs. Though his theories include characteristics of a highly efficient organization, we must remember that these characteristics are only meant to set a model for other organizations to follow, and if all the other conditions are not perfect, the organization is not pure. Is it really a bad thing that Weber's priorities were for the people rather than the company itself?

With this said, the characteristics of Weber's theory have to all be perfect for a bureaucracy to function at its highest potential. "Think of the concept as a bureau or desk with drawers in it, which seems to call out to you, demanding that everything must fit in its place" (Bureaucracy Theory). If one object in the drawer does not fit properly, the entire drawer becomes untidy, which is exactly the case in Weber's theory; if one characteristic is not fulfilled the rest of them are unable to work in unison, leaving the organization performing below its full potential.

One characteristic that was meant to better workplace conditions was his rule that "Organization follows hierarchical principle — subordinates follow orders or superiors, but have right of appeal (in contrast to more diffuse structure in traditional authority)" (Bureaucracy (Weber)). In other words, everyone in a company or any sort of work environment has the opportunity and right to disagree or to speak up if they are unhappy with something rather than not voice their opinion in fear of losing their job. Open communication is a very important part of Weber's ideal bureaucracy, and is practiced today. Because of the communication it may not be the most efficient, but Weber would argue that improved human conditions are more important than efficiency.

It is hard to critique Weber's theories strictly because of the fact that they are theories; they are nearly impossible to perform in real life, therefore how can we know if they work or not? They are merely a set of guidelines that make up bureaucracy, which today many believe is the best way to run organizations in all aspects.

Neoclassical Perspective

The Neoclassical perspective began with the Hawthorne studies in the 1920s. This approach gave emphasis to "affective and socio-psychological aspects of human behaviors in organizations." The human relations movement was a movement which had the primary concerns of concentrating on topics such as morale, leadership, and mainly factors that aid in the cooperation in Organizational behavior.

6.7.5 Hawthorne Study

A number of sociologists and psychologists made major contributions to the study of the neoclassical perspective, which is also known as the human relations school of thought. Elton Mayo and his colleagues were the most important contributors to this study because of their famous Hawthorne study from the "Hawthorne plant of the Western Electric Company between 1927 and 1932."

The Hawthorne study suggested that employees have social and psychological needs along with economic needs in order to be motivated to complete their assigned tasks. This theory of management was a product of the strong opposition against "the Scientific and universal management process theory of Taylor and Fayol." This theory was a response to the way employees were treated in companies and how they were deprived of their needs and ambitions.

In November 1924, a team of researcher – professors from the renowned Harvard Business school of USA began investigating into the human aspects of work and working conditions at the Hawthorne plant of Western Electric Company, Chicago. The company was producing bells and other electric equipment's for the telephone industry. Prominent Professors included in the research team were Elton Mayo (Psychologist), Roethlisberger and Whilehead (Sociologist), and William Dickson (Company representative). The team conducted four separate experimental and behavioral studies over a seven-year period. These were:

- 'Illumination Experiments (1924–27) to find out the effect of illumination on worker's productivity.'
- 'Relay Assembly Test Room experiment (1927–28) to find out the effect of changes in number of work hour and related working condition on worker productivity.'

'Experiment in interviewing Working: In 1928, a number of researchers went directly to workers, kept the variables of previous experiment aside, and talked about what was, in their opinion, important to them. Around 20,000 workers were interviewed over a period of two years. The interviews enabled the researchers to discover a rich and intriguing world that previously remained undiscovered and unexamined within the Hawthorne studies undertaken so far. The discovery of the informal organization and its relationship to the formal organization was the landmark of experiments in interviewing workers. These experiments led to a richer understanding of the social, interpersonal dynamics of people at work.'

Results from the Hawthorne studies

The Hawthorne studies helped conclude that "a human/social element operated in the workplace and that productivity increases were as much an out growth of group dynamics as of managerial demands and physical factors." The Hawthorne studies also concluded that although financial motives were important, social factors are just as important in defining the worker-productivity.

Hawthorne Effect was the improvement of productivity between the employees, it was characterized by:

- The satisfactory interrelationships between the coworkers.
- It classifies personnel as social beings and proposes that sense of belonging in the workplace is important to increase productivity levels in the workforce.
- An effective management understood the way people interacted and behaved within the group.
- The management attempts to improve the interpersonal skills through motivations, leading, communication and counseling.
- This study encourages managers to acquire minimal knowledge of behavioral sciences to be able to understand and improve the interactions between employees

Criticism of the Hawthorne Study

Critics believed that Mayo gave a lot of importance to the social side of the study rather than addressing the needs of an organization. Also, they believed that the study takes advantage of employees because it influences their emotions by making it seem as if they are satisfied and content, however it is merely a tool that is being used to further advance the productivity of the organization.

6.7.6 Contingency Theory

The Contingency Theory is a class of the behavioral theory that claims that there is no best way to organize a corporation, to lead a company, or to make decisions. An organizational, leadership, or decision making style that is effective in some situations, may not be successful in other situations. The optimal organization, leadership, or decision making style depends upon various internal and external constraints (factors).

Contingency Theory Factors

Some examples of such constraints (factors) include:

- The size of the organization
- How the firm adapts itself to its environment
- Differences among resources and operations activities

1. Contingency on the organization: In the Contingency Theory on the Organization, it states that there is no universal or one best way to manage an organization. Secondly, the organizational design and its subsystems must "fit" with the environment and lastly, effective organizations must not only have a proper "fit" with the environment, but also between its subsystems.

2. Contingency theory of leadership: In the Contingency Theory of Leadership, the success of the leader is a function of various factors in the form of subordinate, task, and/ or group variables. The following theories stress using different styles of leadership appropriate to the needs created by different organizational situations. Some of these theories are:

The contingency theory: The contingency model theory, developed by Fred Fiedler, explains that group performance is a result of interaction between the style of the leader and the characteristics of the environment in which the leader works.

The Hersey–Blanchard situational theory: This theory is an extension of Blake and Mouton's Managerial Grid and Reddin's 3-D Management style theory. This model expanded the notion of relationship and task dimensions to leadership, and readiness dimension.

3. Contingency theory of decision-making: The effectiveness of a decision procedure depends upon a number of aspects of the situation:

- The importance of the decision quality and acceptance.
- The amount of relevant information possessed by the leader and subordinates.
- The amount of disagreement among subordinates with respect to their alternatives.

Criticism of the Contingency Theory

It has been argued that the contingency theory implies that a leader switch is the only method to correct any problems facing leadership styles in certain organizational structures. In addition, the contingency model itself has been questioned in its credibility.

6.8 LEADERSHIP AND ORGANIZATIONAL BEHAVIOR

Organizational Behavior (OB) is the study and application of knowledge about how people, individuals, and groups act in organizations. It does this by taking a system approach. That is, it interprets people-organization relationships in terms of the whole person, whole group, whole organization, and whole social system. Its purpose is to build better relationships by achieving human objectives, organizational objectives, and social objectives.

As you can see from the definition above, organizational behavior encompasses a wide range of topics, such as human behavior, change, leadership, teams, etc. Since many of these topics are covered elsewhere in the leadership guide, this paper will focus on a few parts of OB: elements, models, social systems, OD, work life, action learning, and change.

Elements of Organizational Behavior

The organization's base rests on management's philosophy, values, vision and goals. This in turn drives the organizational culture which is composed of the formal organization, informal organization, and the social environment. The culture determines the type of leadership, communication, and group dynamics within the organization. The workers perceive this as the quality of work life which directs their degree of motivation. The final outcomes are performance, individual satisfaction, and personal growth and development. All these elements combine to build the model or framework that the organization operates from.

6.9 MODELS OF ORGANIZATIONAL BEHAVIOR

There are four major models or frameworks that organizations operate out of, Autocratic, Custodial, Supportive, and Collegial.

Autocratic—The basis of this model is power with a managerial orientation of authority. The employees in turn are oriented towards obedience and dependence on the boss. The employee need that is met is subsistence. The performance result is minimal.

Custodial—The basis of this model is economic resources with a managerial orientation of money. The employees in turn are oriented towards security and benefits and dependence on the organization. The employee need that is met is security. The performance result is passive cooperation.

Supportive—The basis of this model is leadership with a managerial orientation of support. The employees in turn are oriented towards job performance and participation. The employee need that is met is status and recognition. The performance result is awakened drives.

Collegial—The basis of this model is partnership with a managerial orientation of teamwork. The employees in turn are oriented towards responsible behavior and self-discipline. The employee need that is met is self-actualization. The performance result is moderate enthusiasm.

Although there are four separate models, almost no organization operates exclusively in one. There will usually be a predominate one, with one or more areas over-lapping in the other models.

The first model, autocratic, has its roots in the industrial revolution. The managers of this type of organization operate mostly out of McGregor's Theory X. The next three models begin to build on McGregor's Theory Y. They have each evolved over a period of time and there is no one best model. In addition, the collegial model should not be thought as the last or best model, but the beginning of a new model or paradigm.

Social Systems, Culture, and Individualization

A social system is a complex set of human relationships interacting in many ways. Within an organization, the social system includes all the people in it and their relationships to each other and to the outside world. The behavior of one member can have an impact, either directly or indirectly, on the behavior of others. Also, the social system does not have boundaries... it exchanges goods, ideas, culture, etc. with the environment around it.

Culture is the conventional behavior of a society that encompasses beliefs, customs, knowledge, and practices. It influences human behavior, even though it seldom enters into their conscious thought. People depend on culture as it gives them stability, security, understanding, and the ability to respond to a given situation. This is why people fear change. They fear the system will become unstable, their security will be lost, they will not understand the new process, and they will not know how to respond to the new situations.

Individualization is when employees successfully exert influence on the social system by challenging the culture.

The quadrant has shown below shows how individualization affects different organizations (Schein, 1968):

Quadrant A: Too little socialization and too little individualization creates isolation.

Quadrant B: Too little socialization and too high individualization creates rebellion.

Quadrant C: Too high socialization and too little individualization creates conformity.

Quadrant D: while the match that organization wants to create is high socialization and high individualization for a creative environment. This is what it takes to survive in a very competitive environment having people grow with the organization, but doing the right thing when others want to follow the easy path.

This can become quite a balancing act. Individualism favors individual rights, loosely knit social networks, self-respect, and personal rewards and careers—it may become look out for Number One! Socialization or collectivism favors the group, harmony, and asks "What is best for the organization?" Organizations need people to challenge, question, and experiment while still maintaining the culture that binds them into a social system.

Organization Development

Organization Development (OD) is the systematic application of behavioral science knowledge at various levels, such as group, inter-group, organization, etc., to bring about planned change (Newstrom, Davis, 1993). Their objective is a higher quality of work-life, productivity, adaptability, and effectiveness. It accomplishes this by changing attitudes, behaviors, values, strategies, procedures, and structures so that the organization can adapt to competitive actions, technological advances, and the fast pace of change within the environment.

6.10 CHARACTERISTICS OF OD (NEWSTROM, DAVIS, 1993)

Humanistic Values: Positive beliefs about the potential of employees (McGregor's Theory Y).

Systems Orientation: All parts of the organization, to include structure, technology, and people, must work together.

Experiential Learning: The learners' experiences in the training environment should be the kind of human problems they encounter at work. The training should NOT be all theory and lecture.

Problem Solving: Problems are identified, data is gathered, corrective action is taken, progress is assessed, and adjustments in the problem solving process are made as needed. This process is known as Action Research.

Contingency Orientation: Actions are selected and adapted to fit the need. Change Agent: Stimulate, facilitate, and coordinate change. Levels of Interventions: Problems can occur at one or more level in the organization so the strategy will require one or more interventions.

Quality of Work Life

Quality of Work Life (QWL) is the favorableness or unfavorableness of the job environment. Its purpose is to develop jobs and working conditions that are excellent for both the employees and the organization. One of the ways of accomplishing QWL is through job design. Some of the options available for improving job design are:

Leave the job as is but employ only people who like the rigid environment or routine work. Some people do enjoy the security and task support of these kinds of jobs. Leave the job as is, but pay the employees more. Mechanize and automate the routine jobs.

And the Area that OD Loves—Redesigns the Job

When redesigning jobs there are two spectrums to follow—job enlargement and job enrichment. Job enlargement adds a more variety of tasks and duties to the job so that it is not as monotonous. This takes in the breadth of the job. That is, the number of different tasks that an employee performs. This can also be accomplished by job rotation.

Job enrichment, on the other hand, adds additional motivators. It adds depth to the job—more control, responsibility, and discretion to how the job is performed. This gives higher order needs to the employee, as opposed to job enlargement which simply gives more variety. The chart below illustrates the differences (Cunningham & Eberle, 1990):

The benefits of enriching jobs include:

- Growth of the individual
- Individuals have better job satisfaction
- Self-actualization of the individual
- Better employee performance for the organization
- Organization gets intrinsically motivated employees

- Less absenteeism, turnover, and grievances for the organization
- Full use of human resources for society
- Society gains more effective organizations

There are a variety of methods for improving job enrichment (Hackman and Oldham, 1975):

Skill Variety: Perform different tasks that require different skill. This differs from job enlargement which might require the employee to perform more tasks, but require the same set of skills.

Task Identity: Create or perform a complete piece of work. This gives a sense of completion and responsibility for the product.

Task Significant: This is the amount of impact that the work has on other people as the employee perceives.

Autonomy: This gives employees discretion and control over job related decisions.

Feedback: Information that tells workers how well they are performing. It can come directly from the job (task feedback) or verbally form someone else.

For a survey activity, Hackman & Oldham's have given Five Dimensions of Motivating Potential.

Action Learning

An unheralded British academic was invited to try out his theories in Belgium—it led to an upturn in the Belgian economy. "Unless your ideas are ridiculed by experts they are worth nothing," says the British academic Reg Revans, creator of action learning.

Action Learning can be viewed as a formula: [L = P + Q]:

Learning (L) occurs through a combination of

Programmed knowledge (P) and

The ability to ask insightful questions (Q).

Action learning has been widely used in Europe for combining formal management training with learning from experience. A typical program is conducted over a period of 6 to 9 months. Teams of learners with diverse backgrounds conduct field projects on complex organizational problems that require the use of skills learned in formal training sessions. The learning teams then meet periodically with a skilled instructor to discuss, analyze, and learn from their experiences.

Revans basis his learning method on a theory called System Beta, in that the learning process should closely approximate the scientific method. The model is cyclical—you proceed through the steps and when you reach the last step you relate the analysis to the original hypothesis and if need be, start the process again. The six steps are:

- Formulate Hypothesis (an idea or concept)
- Design Experiment (consider ways of testing truth or validity of idea or concept)
- Apply in Practice (put into effect, test of validity or truth)
- Observe Results (collect and process data on outcomes of test)

- Analyze Results (make sense of data)
- Compare Analysis (relate analysis to original hypothesis)

Note that you do not always have to enter this process at step 1, but you do have to complete the process.

Revans suggest that all human learning at the individual level occurs through this process. Note that it covers what Jim Stewart (1991) calls the levels of existence:

- We think — cognitive domain
- We feel — affective domain
- We do — action domain

All three levels are interconnected, i.e., what we think influences us and is influenced by what we do and feel.

6.11 CHANGE

In its simplest form, discontinuity in the work place is change. - Knoster& Villa, 2000

Our prefrontal cortex is a fast and agile computational device that is able to hold multiple threads of logic at once so that we can perform fast calculations. However, it has its limits with working memory in that it can only hold a handful of concepts at once, similar to the RAM in a PC. In addition, it burns lots of high energy glucose (blood sugar), which is expensive for the body to produce. Thus when given lots of information, such as when a change is required, it has a tendency to overload and being directly linked to the amygdala (the emotional center of the brain) that controls our fight-or-flight response, it can cause severe physical and psychological discomfort. (Koch, 2006)

Our prefrontal cortex is marvelous for insight when not overloaded. But for normal everyday use, our brain prefers to run off its hard-drive—the basal ganglia, which has a much larger storage area and stores memories and our habits. In addition, it sips rather than gulps food (glucose).

When we do something familiar and predictable, our brain is mainly using the basal ganglia, which is quite comforting to us. When we use our prefrontal cortex, then we are looking for fight, flight, or insight. Too much change produces fight or flight syndromes. As change agents we want to produce insight into our learners so that they are able to apply their knowledge and skills not just in the classroom, but also on the job.

And the way to help people come to insight is to allow them to come to their own resolution. These moments of insight or resolutions are called epiphanies—sudden intuitive leap of understanding that are quite pleasurable to us and act as rewards. Thus you have to resist the urge to fill in the entire picture of change; rather you have to leave enough gaps so that the learners are allowed to make connections of their own. Doing too much for the learners can be just as bad, if not worse, than not doing enough.

Doing all the thinking for learners takes their brains out of action, which means they will not invest the energy to make new connections.

6.12 ORGANIZATIONAL CHANGE AND DEVELOPMENT

6.12.1 Introduction

Change is a constant, a thread woven into the fabric of our personal and professional lives. Change occurs within our world and beyond — in national and international events, in the physical environment, in the way organizations are structured and conduct their business, in political and socioeconomic problems and solutions, and in societal norms and values. As the world becomes more complex and increasingly interrelated, changes seemingly far away affect us. Thus, change may sometimes appear to occur frequently and randomly. We are slowly becoming aware of how connected we are to one another and to our world. Organizations must also be cognizant of their holistic nature and of the ways their members affect one another. The incredible amount of change has forced individuals and organizations to see "the big picture" and to be aware of how events affect them and vice versa.

Organizational development (OD) is a field of study that addresses change and how it affects organizations and the individuals within those organizations. Effective organizational development can assist organizations and individuals to cope with change. Strategies can be developed to introduce planned change, such as team-building efforts, to improve organizational functioning. While change is a "given," there are a number of ways to deal with change — some useful, some not. Organizational development assists organizations in coping with the turbulent environment, both internally and externally, frequently doing so by introducing planned change efforts.

Organizational development is a relatively new area of interest for business and the professions.

While the professional development of individuals has been accepted and fostered by a number of organizations for some time, there is still ambiguity surrounding the term organizational development. The basic concept of both professional development and organizational development is the same, however, with an essential difference in focus. Professional development attempts to improve an individual's effectiveness in practice, while organizational development focuses on ways to improve an organization's overall productivity, human fulfillment, and responsiveness to the environment. These goals are accomplished through a variety of interventions aimed at dealing with specific issues, as well as through ongoing processes.

6.12.2 Change in Organizations

Organizational development efforts, whether facilitated by an outside expert or institutionalized and conducted on an ongoing basis, bring about planned change within organizations and teams.

However, they are but one type of change that occurs in organizations, for change can be both planned and unplanned and can occur in every dimension of the universe. A change in chief justice, appropriations, or staff support can dramatically alter the character of a judicial education organization.

Institutional alignment of the state bar, local law schools, area colleges and universities, and judicial professional associations may yield similar impacts. Planned change takes conscious and diligent effort on the part of the educator or manager. Kanter (1983) originated the concept of the change master: a person or organization adept at the art of anticipating the need for and of leading productive change. As a way to reinforce the judicial educator's role in the change process, this term will be used to refer to educators and managers who are interested in effecting change in their organizations or work teams.

Change will not occur unless the need for change is critical. Because individuals and organizations usually resist change, they typically do not embrace change unless they must. One OD consultant describes how "pain" drives change (Conner, 1990). Pain occurs when people pay the price for being in a dangerous situation or for missing a key opportunity. As such, change is needed to relieve the pain. According to this perspective, change will not occur just because "it's a good idea." It will only occur when the pain of an individual or an organization is sufficiently high to justify the difficulties of assimilating change. Therefore, a change master must focus on the absolute need of the organization to change, rather than simply on the benefits of the anticipated change. Effective change masters understand this, and they then assist others in recognizing that the organization has no choice but to change. The organization cannot afford to maintain the status quo; change is simply that critical. The Ohio Judicial College's movement to full funding emerged from such a catharsis. In other states, mandatory training saved the office of justice of the peace. Court administration emerged as an independent vocation out of exigent circumstances. These examples of change in state judicial education provide evidence that effective change masters have perceived a critical need for change and then helped to make that change happen.

6.12.3 Planning for Change

Before embarking on an organizational change initiative, it is wise to carefully plan strategies and anticipate potential problems. One useful method of planning comes from an early researcher on change, Kurt Lewin (1947), who developed the concept of force-field analysis. The term describes analysis that is deceptively simple and can be used to help plan and manage organizational change. Lewin believed that behavior within an organization was a result of the dynamic balance of two opposing forces. Change would only occur when the balance shifted between these forces. Driving forces are those forces which positively affect and enhance the desired change. They may be persons, trends, resources, or information. Opposing them are the restraining forces, which represent the obstacles to the desired change. As these two sets of forces exist within an organization, they create certain equilibrium. That is, if the weights of the driving and restraining forces are relatively equal, then the organization will remain static. As changes occur and affect the weight of either one of the forces, a new balance will occur, and the organization will return to what Lewin called **"quasi-stationary equilibrium."** Individuals practicing their vocation in the context of a political organization may intuitively employ these concepts in defining and redefining what change is possible. Judicial educators operate in such a context.

6.12.4 The Process of Change

A method such as force-field analysis is the beginning step of any planned change. There are many different models for the change process in the literature; the following is a simple, straightforward one proposed by Egan. He delineates three steps:

- The assessment of the current scenario.
- The creation of a preferred scenario.
- Designing a plan that moves the system from the current to the preferred scenario.

It is evident Egan has been influenced by Lewin, in the emphasis on both planning and assessment. Additionally, Egan argues that planning must lead to an action that produces valued outcomes or results for the organization. Thus, both planning and change must be directed toward a specific goal. Once the need for change has been determined, one follows the steps of the model in sequence. While these steps could each be examined in detail, only step three will be discussed in an in-depth manner here. The first step, "assessing the current scenario," can be accomplished through a mechanism such as force-field analysis. It provides the necessary information on the forces that can facilitate the desired change and the forces that will resist and deter the change. Step two, "creating a preferred scenario," is often accomplished through team effort in brainstorming and developing alternative futures. While the need that precipitates the change is clearly compelling, there may be several ways in which the change could actually occur within the organization. It is important to examine the various alternatives thoroughly.

The third step of the process, "devising a plan for moving from the current to the preferred scenario," includes the strategies and plans that educator and managers must develop to overcome the restraining forces in an organization. This is a political process, requiring individuals to harness and utilize power. Power is necessary for change to occur. It is neither inherently good nor bad; it simply assists individuals in accomplishing their goals. In his recent book mastering the Politics of Planning, Benveniste (1989) notes that even well-thought-out plans for change can be derailed when the politics of implementation are not considered. Change masters must gather support for the desired change throughout the organization, using both formal and informal networks. The multiplier or "bandwagon" effect, he notes, is often necessary to rally enough support for the change.

Key Roles in the Change Process

During this stage of planning, it is useful to distinguish the different roles associated with the change process. These roles must remain distinctive in order to implement planned change effectively.

However, within different settings or systems, a judicial educator may play more than one role. The various roles that individuals can play, as described by Conner (1990), are:

- Change Sponsor: Individual or group who legitimizes the change.
- Change Advocate: Individual or group who wants to achieve a change but does not possess legitimization power.

- Change Agent: Individual or group who is responsible for implementing the change.
- Change Target: Individual or group who must actually change.

One of the most critical tasks for the educator in implementing change is to harness the support of an effective change sponsor. The sponsor is in a position to legitimize the change. Sponsorship is critical to implementing the desired change. Directly or indirectly, pain can motivate the sponsor to foster the planned change. Within the state judicial system, this sponsor may be the chief justice, the head of the education committee, or the state court administrator. Conner (1990) argues that weak sponsors should be educated or replaced, even by someone at a lower level in the organization, or, he emphasizes, failure will be inevitable.

Educators and managers are often in the position of change advocates, who perceive the need for change and desire and advocate the change, but who do not have the necessary organizational power to implement the change. Alternatively, these individuals may function as the change agent, with the responsibility (but again, not the power) to implement change. And, of course, in an organizational change effort, educators and managers may be part of the group affected by the change, or the change target. It is useful to consider each of these roles in planning strategies not only for implementation, but for gathering support for the change effort.

Strategies for Implementing Change

In order to move an OD effort from the idea stage into implementation, educators and managers must also rally the resources and support of the organization. Kanter (1983) describes how the following three sets of "basic commodities" or "power tools" can be acquired by members of an organization to gain power:

- Information (data, technical knowledge, political intelligence, expertise).
- Resources (funds, materials, staff, time).
- Support (endorsement, backing, approval, legitimacy).

The first strategy in implementing a change would be to collect as many of these power tools as possible. As this occurs, individuals can "plant seeds of support" for the planned change. This is particularly important in helping others see the critical need for the planned change. It may be possible to plant these seeds before sponsorship of the change is sought so that the sponsor feels he or she is proactively responding to a critical need.

Another strategy is to "package" the change in a way that makes it less threatening and, therefore, easier to sell. For instance, it is easier to implement change of a product or a project when it is:

(*a*) Conducted on a trial basis;

(*b*) Reversible, if it doesn't succeed;

(*c*) Done in small steps;

(*d*) Familiar and consistent with past experience;

(*e*) A fit with the organization's current direction; or

(*f*) Built on the prior commitments or projects of the organization (Kanter, 1983).

This packaging should be completed prior to submitting the OD effort to the designated change sponsor, although that person needs to be involved in further assisting in the packaging and selling of the planned change. Building coalitions is a strategy that often occurs throughout the entire phase of implementing the change. Support must be gathered from all areas which will be affected by the desired change, across different levels of the organization. It is always advisable to get the support of an immediate supervisor early on, although this may not always be possible. In such instances, other support could be gathered across the organization to influence the supervisor to reconsider lending support to the change efforts.

Effective change masters use their informal networks and deal with any concerns or questions of supporters individually rather than in formal meetings. "Pre-meetings" can provide a safer environment for airing concerns about implementing change. In such settings, an individual may have the opportunity to "trade" some of the power tools that he or she has acquired in order to generate support. Additionally, some individuals will support a project or change effort for reasons that are fairly reactive:

"If so-and-so supports it, then I will, too," or "If such-and-such state is moving in that direction, then we should, too." Obviously, the more change masters know about how particular individuals may react, the better able they are to plan for ways to garner support.

Resistance to Change

With every major and minor change, resistance typically occurs. Every judicial educator has seen this tendency, whether the change involves a certain speaker or a particular topic, a customary time of year or time of day to meet, a favorite location for conferences or planning sessions, or a given style for delivering a speech or organizing a paper. This resistance should be accepted as a "given," so that the educator can predict and plan for effectively dealing with inevitable resistance. This section will address some of the causes and types of resistance to change, particularly at the organizational level.

Individual Resistance

Why does resistance to change occur? The primary reason is that people fear change. They are not usually eager to forego the familiar, safe, routine ways of conducting their business in favor of unknown and possibly unsafe territory. As humans, we tend to prefer routines and accumulate habits easily; however, fear of change may be attributed to more than a tendency toward regularity. Change represents the unknown. It could mean the possibility of failure, the relinquishing or diminishing of one's span of control and authority, or the possibility of success creating further change. It might be that the planned change has little or no effect on the organization whatsoever. Any one of these possibilities can cause doubt and thus fear, understandably causing resistance to the change efforts.

Additionally, the transition between the present state and the changed state is difficult for both individuals and organizations. On an individual level, people must be reminded that every transition or change effort begins with an ending — the end of the current state. The first step toward change is going through the process of ending. Endings must be accepted and managed before individuals can fully embrace the change. Even if the impending change is desired, a sense of loss will occur. Because our sense of self is defined by our roles, our responsibilities, and our context, change forces us to redefine ourselves and our world. This process is not easy. William Bridges does an excellent job of discussing the process of individual change in his book Transitions (1980). In describing the process of ending,

Bridges presents the following four stages that individuals must pass through in order to move into the transition state and effectively change:

- *Disengagement.* The individual must make a break with the "old" and with his or her current definition of self.
- *Disidentification.* After making this break, individuals should loosen their sense of self, so that they recognize that they aren't who they were before.
- *Disenchantment.* In this stage, individuals further clear away the "old," challenging assumptions and creating a deeper sense of reality for themselves. They perceive that the oldway or old state was just a temporary condition, not an immutable fact of life.
- *Disorientation.* In this final state, individuals feel lost and confused. It's not a comfortable state, but a necessary one so that they can then move into the transition state and to a new beginning.

In this process, it is important to recognize how the change was initiated. While all change is stressful, it is easier to go through the process of ending and into the transition if the change was internally driven, rather than if it was initiated by an external source. When we make changes in ourlives (*e.g.*, marriage, a new home, or new career), there is a greater sense of control over the change; therefore, we feel more capable of coping with the unpleasant aspects of transition. This is not necessarily true when the change is driven or mandated by an external source (*e.g.,* new job responsibilities, unexpected changes in finances, or job relocation). In externally driven change, the process of transition is more difficult, and there is more resistance as individuals refuse to begin the ending process and make the initial break with the past. This almost innate resistance to externally driven change makes a convincing argument for the change master to facilitate "ownership" of the change efforts by both the change sponsor and the change target. In other words, we should assist others in perceiving the change as desirable and internally driven, rather than an externally driven mandate that is thrust upon them.

Organizational Resistance

With even this very brief discussion on the difficulties involved in individual change, it should be apparent that this phenomenon occurs at the organizational level as well. Organizations, regardless of size, are composed of individuals. The extent to which individuals within the organization can appropriately manage change represents the overall organizational capacity for change. However, there are other factors peculiar to the organizational setting that can act as barriers to implementing change. These include:

- **Inertia.** One of the most powerful forces that can affect individuals and organizations is inertia.

The day-to-day demands of work diminish the urgency of implementing the change effort until it slowly vanishes within the organization.

- **Lack of Clear Communication.** If information concerning the change is not communicated clearly throughout the organization, individuals will have differing perceptions and expectations of the change.
- **Low-Risk Environment**. In an organization that does not promote change and tends to punish mistakes, individuals develop a resistance to change, preferring instead to continue in safe, low risk behaviors.
- **Lack of Sufficient Resources.** If the organization does not have sufficient time, staff, funds, or other resources to fully implement the change, the change efforts will be sabotaged. These factors, combined with others characteristic to the specific organization, can undermine the change effort and create resistance. A wise change agent will spend the necessary time to anticipate and plan for ways to manage resistance. Techniques such as force-field analysis, discussed earlier, are useful tools to assist in developing strategies for overcoming organizational resistance to change.

6.12.5 Organizational Development: A Neglected Need

Judicial educators may be in the best position to sense the need for organizational change and development efforts within their committees, boards, or work teams. While they may not have the organizational power to ensure that appropriate efforts are implemented (the role of a change sponsor), they can certainly "raise the consciousness" of the organization and advocate that OD needs be addressed. Healthy organizations are willing to commit time and energy toward improving both the individual and the organization as a whole.

A synergistic effect can be created when professional development and organizational development efforts are simultaneously put into place. However, many organizations and teams unfortunately overlook the need for organizational development, often because of their unfamiliarity with the concept or their emphasis on professional development. While "professionals" are generally considered to be independently-based practitioners, the majority of professionals function within an established organizational setting. Judicial personnel are no exception. While members of a judicial office may make decisions in an independent fashion, they are still part of the larger system in which they work (*e.g.*, the state justice system) and are related to the organizational context in which they function (*e.g.*, Administrative Office of the Courts). Judges in particular work in a highly independent fashion. Yet, they too are part of an over arching system. The system itself must have the benefit of development as well as the individual professionals who are a part of the system. However, judicial educators may unwittingly neglect the needs of the organization or system in favor of professional development alone. Indeed, some observers have argued that this is a fundamental part of conflict between state judicial educators and state court administrators.

With this focus on improving the practice of individual professionals, educators and managers may overlook the equally important needs of the organization to support its institutional mission as well as individual vocational growth. Individuals can improve only as much as their organizations "allow" them to grow. Nowlen, among others, has discussed this challenge in A New Approach to Continuing Education for Business and the Professions (1988). He believes that the most effective professional development is holistic in nature, involving not only the practitioner, but the individual's organization as well, so that new behavior and learning can be integrated into the work setting. Nothing is more frustrating than continuing education that is stopped at the point of entry of the practitioner's work setting. Judicial educators may often feel the limitations of working with professionals whose practice setting does not encourage their professional growth. Organizations, as well as individuals, must be committed to the concept of lifelong learning and continual growth and development. Again, we see this holistic perspective of not only individuals and organizations, but of the learning process itself.

Learning is not a discrete process that occurs only through individual effort. Implementation of the learning in practice depends upon a receptive work environment, as well as effective teaching and interaction in the actual learning environment.

Organization development (OD) is a deliberately planned, organization-wide effort to increase an organization's effectiveness and/or efficiency. OD theorists and practitioners define it in various ways. Its multiplicity of definition reflects the complexity of the discipline and is responsible for its lack of understanding. For example, Vasudevan has referred to OD being about promoting organizational readiness to meet change, and it has been said that OD is a systemic learning and development strategy intended to change the basics of beliefs, attitudes and relevance of values, and structure of the current organization to better absorb disruptive technologies, shrinking or exploding market opportunities and ensuing challenges and chaos. It is worth understanding what OD is not. It is not training, personal development, team development, HRD (human resource development), L&D (learning and development) or a part of HR although it is often mistakenly understood as some or all of these. OD interventions are about change so involve people - but OD also develops processes, systems and structures. The primary purpose of OD is to develop the organization, not to train or develop the staff.

According to Middlemist and Hitt (1988, p. 493), organizational development is: A systematic means for planned change that involves the entire organization and is intended to increase organizational effectiveness. Cummings and Huse (1988, p. 1) define OD in broader terms:

A system wide application of behavioral science knowledge to the planned development and reinforcement of organizational strategies, structure, and processes for improving an organization's effectiveness.

Several parts of the above definitions are particularly worth emphasizing. The first is that organizational development is a systematic activity, an ongoing process that can help organizations deal with current and anticipated problems, putting leaders in a proactive, rather than reactive, stance. This stance differs from the "putting-out-fires" approach that so many groups and organizations have relied upon historically.

The second noteworthy item is that OD involves planned change within an entire organization or work team. This means that a proactive stance is absolutely necessary for change to be effective; otherwise, the planned change effort will lag too far behind the need that it is intended to address. For example, judicial employees who have recently come under the jurisdiction of the State Court.

Administrator's Office may find difficulty in adjusting to new procedures, new channels of communication and authority, and new roles. Concurrently, existing staff within the Office will experience their own difficulties in assimilating the new group, sharing information, and performing roles that also may have changed. Is it then time for a "team-building" session? While it's often "better late than never," the real effort should have already occurred before the organizational changes were enacted. Staff from all areas ideally should have been brought together and the job restructuring discussed, procedures and policies formulated, and team-building efforts begun. After changes have taken place, it's often too late to begin organizational development efforts. This fact makes planning even more critical and highlights the advantage of institutionalizing the concept of team-building so that staff are empowered and able to cope with changes in the work team.

The third part of an OD definition to keep in mind is the rationale for organizational development—to improve organizational effectiveness. Organizations and work teams must be effective and efficient, particularly in the current environment of limited resources. In thinking about effectiveness, every organization and team would be well-advised to adopt a "quality" mentality regarding OD efforts. Ongoing, institutionalized OD strategies are akin to continuous quality improvement. Quality concepts originated in the manufacturing sector, but today, the quality movement is being incorporated into service organizations as well. While not as technically oriented, work teams providing services can effectively determine quality standards and work proactively as a team to insure that standards are met. Concepts of quality can be applied in the educational arena.

For instance, the National Association of State Judicial Educators (NASJE) has recently developed a set of "Principles and Standards for Judicial Education." This endeavor signifies commitment not only to quality, but also to the professionalization of the field. Ensuring quality, like strengthening an organization's effectiveness, is an ongoing task. Organizational development can, of course, occur in groups or teams, as well as within an entire organization. Effective OD is ongoing and systematic, strengthening both the individual and the group.

Many organizations today accomplish their tasks through a project management approach that brings together teams for a short period of time. Members of the team may be brought from various parts of the organization, representing different constituencies and levels within the organization. These work teams do not always have a unified focus or vision of the task to be accomplished; therefore, they can alter in their responsibilities because of conflicting perspectives, communication difficulties, or lack of clear objectives. An ongoing system of organizational development strategies is useful in this situation. Team-building in particular is effective when incorporated and supported by the organization.

6.12.6 A Model for Organizational Development

Now that we have examined the major emphases associated with organizational development, abroad picture of the actual process is needed. While there are variations on any model,

- Recognition of need for change
- Diagnosis of cause(s)
- Development of change alternatives
- Implementation of change
- Reinforcement of change
- Evaluation of change
- Further change action taken, if required
- Feedback

The first step of the model is critical. Often, educators and managers will recognize the need for change before others who have decision-making power do. In such instances, these decision-makers, or change sponsors, will need to be convinced of the need for change. Ongoing diagnosis and systematic examination of the environment can provide a rationale for planned OD efforts. An environmental scanning program, strategic planning, employee or "customer" feedback survey, or similar methods can detect changes in the internal or external environment that will have impact on the organization.

Organizations can also examine the quality of their products and/or processes to determine whether OD efforts need to be directed toward specific areas within the organization or work team. It is often advisable for an organizational development expert to assist in the implementation of change efforts. For many organizations, this consultation is not always an option due to limited resources, the unavailability of consultants, or political reasons, among others. Various strategies are available for use within organizations. Some can be used in an ongoing fashion, while others are "oneshot" efforts. Two major types of strategies that can be used are:

- Process Strategies: team-building, quality circles, sensitivity training, survey feedback, career planning.
- Structural Strategies: job redesign, job enrichment, management by objectives, organization are structuring, flextime options.

The change strategy chosen should relate appropriately to the organizational or team need.

Political, technological, or legal factors may cause an organization to move in directions that require OD efforts. All too frequently, however, interpersonal change efforts are chosen as the appropriate strategy. For instance, a simple problem such as the overuse of a particular piece of office equipment may result in disgruntled employees. This problem may be solved by purchasing more equipment or restructuring the use of the equipment, rather than developing a team-building program. Interpersonal or even process strategies cannot adequately equip an organization to deal with these types of external pressures.

Now that we are more familiar with organizational change and development, we can look at one

OD strategy that is being successfully employed in many organizations. Team-building is an OD strategy that is often used in organizations to make work groups more cohesive, committed, satisfied, and productive. A variety of assessment tools and exercises are available for groups to diagnose their capacity as a team and improve their team functioning. Attachment the lists characteristics of an effective team, which can be used as a benchmark to evaluate the level of team functioning within an organization or committee. Assessment is the first step in attempting to build an effective team. However, team-building efforts are most effective when perceived as an integral concept within an organization and not merely as a sporadic intervention. These efforts then become part of the shared vision of that organization, a priority to which leaders are committed.

If institutionalized by a work group, team-building shapes the way everyone within that group thinks about his or her role in relationship to others and to the organization. Understandably, many organizations, as well as many small work teams or committees, have difficulty orienting themselves to a team approach rather than an individual approach. Individual, rather than team, effort is traditionally recognized and appreciated within organizations. Figure 2 lists several suggestions that an organization can implement to overcome this tendency toward individual effort and to encourage a team culture. The list has been adapted for judicial education organizations.

For The Development of a Team-player Culture

1. Public statements by leaders on the importance of team players — not just the
2. State court administrators, judges, and other leaders serving as models of team players.
3. Promoting people who are both technically competent and team players.
4. Giving important assignments to positive team players.
5. Incorporating team-player behaviors in the performance appraisal system.
6. Offering training workshops on the skills of effective team players.
7. Giving higher salary increases to positive team players.
8. Developing incentive systems that reward team effort.
9. Designing flexible compensation programs that pay individuals for their contributions to a team approach.

6.12.7 Leaders as Followers

In organizations, attention is typically directed toward leaders within the group through leadership training, self-assessments, style inventories, and the like. Most believe that sound leadership is the key to an effective work team. There is no doubt that the abilities of the leader can affect the team as a whole, but some of our traditional notions of leadership include outdated modes of behavior or unrealistic expectations. Even if we have moved from thinking of leaders in the John Wayne stereotype, we may still expect leaders to create an inspirational work setting by producing vision statements that motivate and unify the organization. Recently, however, some authors have discussed the concept of leadership in different terms, moving the leaders even further from the grandstand and closer to the audience. In a recent talk to educators and business leaders, Dr. Badi Foster, President of a Etna Institute for Corporate Education, discussed how the mission of the organization must be connected to all members of the team and how leaders are needed to develop this connection. Everyone needs to believe in something bigger than just "making a buck" at work; people need to feel that they are contributing to "the big picture." Their gifts or talents brought to the work setting should be connected to an organizational vision so that they feel an integral part of the team. He prefers "vision" over "mission," believing that the first implies impelling people, while the second term connotes compelling employees in their work environment.

How can organizations develop congruence between individual team members' vision and a larger organizational vision? Leaders must first encourage all employees to create their own personal vision, motivating and inspiring them in their work setting. Secondly, leaders must discern these personal visions, as well as the particular gifts that employees bring to work. Foster's belief is that leaders can help "bring forth that which is already there" within individuals. Leaders can work with team members to help make the connection between their personal visions and goals and the vision of the organization. In some instances, organizations may be restructured; job descriptions may be altered, or jobs may be redefined to provide the employee more responsibility. Empowering employees creates an environment in which they can feel more connected and committed to the organization. If there is clearly no congruence between team members and the organization, then something must change; however, these changes will ultimately benefit both the individual and the organization.

Lee (1991) takes this idea even further and argues that often the best leaders are the best followers: [Leaders] can create an environment in which followers can develop their own goals (in other words, a culture of empowerment), as well as provide the training to develop competence. It then becomes the leader's task to sense where followers want to go, align their goals with the larger goals of the organization, and invite them to follow.

In this conception of teams, the leader does not set himself or herself apart from the other members of the group but is intimately connected to them, helping them to align their goals with the organization and providing motivation through empowerment.

6.12.8 Followers as Leaders

What, then, is the role of the follower within a committee or work team? Team members are not merely waiting for inspiration and direction from leaders; rather, they are active participants in creating the vision and direction of their organization. They have "ownership" in their work setting, are involved in making it effective, and feel a connection between themselves and that vision. They speak the truth, take responsibility for themselves, and have both personal integrity and organizational loyalty. In short, they behave in ways traditionally considered behavior of "leaders."

Team members who strive for these characteristics within a supportive organization will find that their personal satisfaction in the work environment increases. Many organizations are now embracing the "self-directed work team" concept and truly dispersing leadership throughout the organization. In small groups, this concept may already be a reality. However, it takes some managers a considerable amount of time to rethink their role in the group and move from "boss" to "enabler" or "coach." At the same time, members of a team may not initially feel comfortable taking on more responsibility and accountability. New ideas and behaviors require some adjusting on everyone's part. But the result may be worth the effort — teams become more effective and productive, and team member's experience greater satisfaction with their work and feel more connected to the organization and other team members.

Team-Building Strategies

The literature on developing more cohesive teams is vast. However, in addition to the organizational strategies to encourage a team culture, there are some strategies that team leaders can use to help build effective teams. Specific team-building efforts vary depending upon a number of variables, including the nature of the team, its duration, the task at hand, and the organizational culture. Team-building efforts can range from T-groups to outdoor adventure weekends, from self-assessments to team assessments, and from one-hour team-building videotapes to multi-day workshops by OD consultants. Therefore, they may serve as guidelines to leaders for achieving a high level of team functioning. According to Parker, successful team-building strategies for team leaders are:

- Get to know the team.
- Define the team's purpose.
- Clarify roles.
- Establish norms.
- Draw up a game plan.
- Encourage questions.
- Share the limelight.
- Be participatory.
- Celebrate accomplishments.
- Assess team effectiveness.

6.13 A LEARNING ORGANIZATION

A learning organization is the term given to a company that facilitates the learning of its members and continuously transforms itself. Learning organizations develop as a result of the pressures facing modern organizations and enables them to remain competitive in the business environment. A learning organization has five main features; systems thinking, personal mastery, mental models, shared vision and team learning. The Learning organization concept was coined through the work and research of Peter Senge and his colleagues. It encourages organizations to shift to a more interconnected way of thinking.

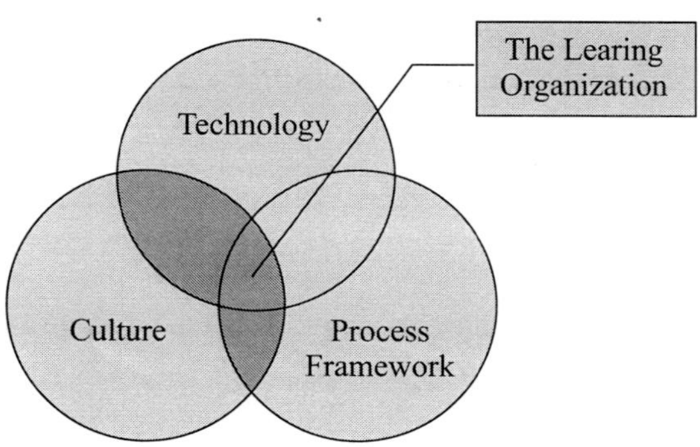

In considering these perspectives on the roles of the team leader and team members, it is helpful to bring these ideas together into a more comprehensive view of organizations. For instance, how can teams work together in an integrated whole, rather than as a group of individuals? How can the capacities of each member interact with the talents of other members to maximize the organization's productivity? Peter Senge's (1990) book The Fifth Discipline: The Art and Practice of the Learning Organization offers an inspiring look at how organizations can focus on growth and renewal, overcome threats from the external environment, and develop the capacities of individuals within the organization. He argues that changes in our "mindset" are needed to truly develop a learning organization. In the earlier discussion on change, we noted that changes in one area of the world affect others. Senge recognizes this phenomenon and emphasizes the concept of systems thinking, which he calls the "fifth discipline." This is a holistic way of viewing organizations, as well as individuals, and provides a perspective useful in planning and change efforts.

Senge writes: The tools and ideas presented are for destroying the illusion that the world is created of separate, unrelated forces. When we give up this illusion - we can then build "learning organizations," organizations where people continually expand their capacity to create the results they truly desire, where new and expansive patterns of thinking are nurtured, where collective aspiration is set free, and where people are continually learning how to learn together.

One of the disciplines that characterize the learning organization is team learning. Senge discusses how unaligned teams produce a great amount of wasted energy. While individuals within the team may work diligently, the team effort is low. He writes: By contrast, when a team becomes more aligned, a commonality of direction emerges, and individuals' energies harmonize. There is less wasted energy.... There is commonality of purpose, a shared vision, and an understanding of how to complement one another's' efforts. Individuals do not sacrifice their personal interests to the larger team vision; rather the shared vision becomes an extension of their personal visions. Again, this congruence of personal and organizational vision can only occur when visions have been developed by both the organization and the individual. A shared vision is a necessity in this type of context. Neither the organization nor the individual can shirk responsibility in developing a vision. Senge notes three critical dimensions of the concept of team learning that have implications for judicial educators, as well as other teams. First, he believes that teams need to tap the minds of individual members in order to think insightfully about complex issues. Everyone must take responsibility for contributing to the group. Secondly, there is a need for coordinated action, through which team members can trust and expect certain behaviors from one another. He likens this to jazz musicians playing together in spontaneous, yet coordinated, action.

And thirdly, team learning is dispersed throughout an organization through the actions of team members. Each member of a team will also be a member of other teams; his or her behavior can therefore encourage team learning in other areas as well. It is clear that these new ideas can radically change the way we look at teams. Changing our perspective and our expectations of teams can transform organizations and the way that work gets accomplished. Active participation is required of team leaders and followers, as well as commitment. Effective teams need a vision for the organization, but, additionally, each member must feel connected to it and perceive it as part of his or her own personal vision. Through both organizational and individual clarity of vision, organizations can expect commitment, diligence, and a passion for work from its members. Individuals and organizations who can find a passion for work will not only flourish and be productive, but will also find an excitement in carrying out even the most mundane functions in everyday work.

6.14 DECISION MAKING

The thought process of selecting a logical choice from the available options. Decision-making is an essential skill for operational team leaders. Applying a systematic method to solve problems is critical to team performance and the safety of operations. Team members share the responsibility for solving problems by contributing timely and valuable information to the team leader.

When trying to make a good decision, a person must weigh the positives and negatives of each option, and consider all the alternatives. For effective decision making, a person must be able to forecast the outcome of each option as well, and based on all these items, determine which option is the best for that particular situation.

GENERAL POINTS ABOUT DECISION MAKING

- A large fraction of decisions don't require much thought or special effort. Our primary focus in this course is on improving skills in dealing with more complex, important decisions that need special attention.
- We consider decision making in a broader sense than merely making a choice. We consider it as the often-extensive process ranging from first recognizing a problem to be solved or a decision to be made, on through framing the problem or decision, identifying and evaluating alternatives, choosing among the alternatives, implementing the chosen alternative, and stabilizing the action plus learning from the experience.
- Critical thinking and creativity are important, the latter especially in identifying alternatives. We will devote substantial attention to both topics.
- Rational decision making (as described in the *Basic Decision Making Model* web module) is viewed as a cornerstone for our work, and is considered as a basic decision making process that should be used to varying degrees of detail and thoroughness in various situations. Please study that module in detail.
- Many decision situations are messy and the outcomes of the decisions are less than optimum, for many reasons including: (*a*) multiple, interconnected problems, (*b*) incomplete and/or inaccurate information, (*c*) limited time and resources, (*d*) uncertainty in predicting outcomes of alternatives, (*e*) intelligence and thinking limitations, and (*f*) judgmental biases. We will devote considerable attention to such factors and how to reduce their negative effects on decision making.
- March & Simon (1958) coined the term "bounded rationality" to explain some of these factors that hinder rational decision making. The result is that decision makers often "satisfice" by accepting a satisfactory or reasonable decision rather than continuing the search for a better or optimum solution.

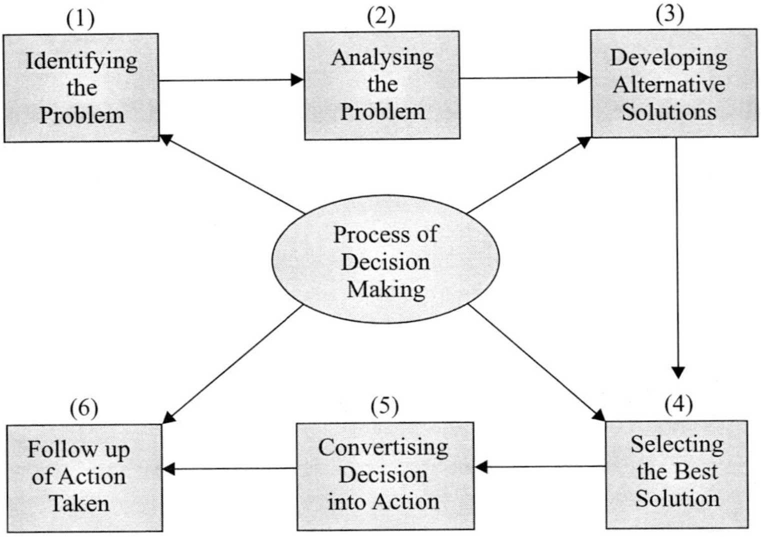

Decision-making Process

Some Factors Influencing Decisions

- ***Type of decision***
  - Routine, recurring, precedent related vs. unique, complex, strategic, nonrecurring
  - Well-defined vs. ill-defined
  - Single vs. multiple objectives
- ***Urgency, pressure, constraints***
- ***Attributes of environment and perceptions of these:***
  - Uncertainty, risk
  - Complexity, ambiguity
  - Conflict
- ***Decision-maker characteristics***
  - Physiological and psychological factors
  - Wants (partly physiological factors) vs "shoulds" (related to values)
  - Judgmental biases
    - Selective perception
    - Impression effects
    - Framing effects (including presentation and anchoring effects)
    - Escalation of commitment effects
    - Categorization effects
  - Creativity
  - Mental traps
- ***Who else is involved (as stakeholders and/or decision makers)***
- ***Ethics***

PHYSIOLOGICAL AND PSYCHOLOGICAL FACTORS

- We have emotional & physiological responses to conditions & events in environment
- Effects are continual & often unnoticed
- "Fight or flight" response is typical in uncomfortable or threatening situations
- Our senses can register much more than our minds can process or focus on
- Most of the time, quick common sense or intuition is adequate
- However, immediate emotional factors have a strong influence on thinking, often unrecognized
- Emotional distress blocks awareness and control, and promotes impulsive decisions
- Many decisions involve a desire to act emotionally and impulsively
- We may sacrifice long-term for short-term interests
- We need a balance of appropriate emotional response and rational thinking. Some things that can help:
  - ➢ Develop self-awareness and introspection
  - ➢ Anticipate emotional triggers and traps
  - ➢ Think through in advance of emotional triggers
  - ➢ Strive for awareness of feelings and emotional responses
  - ➢ Understand sources of feelings
  - ➢ Consider both short-term and long-term implications
  - ➢ Manage competition and tensions between what we "want" and what we "should do"
  - ➢ Identify and consider multiple options (we are more likely to address long-term implications when we believe that we have multiple, desirable options)
  - ➢ Use others to assist us.

JUDGMENTAL BIASES

Selective perception

- Perceiving what we expect, based on past experience and/or predictors
- Perceiving what we want something to be, usually grounded in strong attachment to a belief, position, or desired outcome. Strong emotional attachment to beliefs or positions creates filters that screen all statements and actions in terms of how they relate to these beliefs or positions.
- We tend to attribute good outcomes to internal causes (what we did) and bad outcomes to external interference (what others did or did not do)
- It helps to be aware of our expectations beforehand and to take multiple perspectives
- Using good communication skills can help, especially balancing advocacy with more inquiry, and making use of the "ladder of inference" model (we

will study and practice these skills later, including use of the web module *Communications, Improving*)

- Some examples of inquiry and introspection that can help reduce selective perception:
  - ➢ How would an objective third-party view the situation?
  - ➢ Am I seeking information to confirm my beliefs, rather than information that disconfirms what I expect or want?
  - ➢ Am I using only sources of information that are predisposed to my way of thinking?
  - ➢ Has my perception of the situation and/or problem altered as a result of new information?
  - ➢ Am I framing the decision in a way that would make it extremely difficult to convince me to change my mind while making it very easy to believe I was right all along?

Impression effects

- Primacy and recency effects
  - ➢ Primacy effect when initial events and impressions are believed to be more relevant and important than later ones
  - ➢ Recency effect when the most recent events and impressions are recalled clearly and believed to be more important than previous ones
- Either of these effects can cause bias, in contrast to considering events from first to last
- Is primacy or recency more powerful?
  - ➢ Moot if presentations of two sides and the decision are close together in time
  - ➢ Primacy effects are more powerful if the decision is to be at least a few days after both presentations (therefore, present first, if possible)
  - ➢ Recency effects are more powerful if there is to be a gap between presentations of the two sides and the decision follows shortly after the last presentation (therefore, present last, if possible)
- Halo effects occur when we extrapolate from one characteristic of a person to believe they have other characteristics (these can be desirable or undesirable characteristics)
- Halo effects tend to produce self-fulfilling prophecies. For example, if a supervisor believes an employee to be capable and an excellent performer, both the supervisor's perceptions of and the actual employee's performance tend to be positive (the Pygmalion effect). Conversely, if a supervisor believes an employee to be incompetent and a poor performer, both the supervisor's perceptions of and the actual employee's performance tend to be poor.
- It is important to be aware of these sources of potential bias and attempt to reduce impression effects in our decisions.

Framing effects (including presentation effects)

- Study web module on *Framing Decisions* plus exercises and cases we do in class
- Involves selecting & emphasizing certain aspects, while excluding or minimizing others
- Different parties will believe and/or advocate rather different frames for the same situation
- Frames can be ours or put forward by others (presentation effects)
- Can include anchoring (*e.g.,* the sales person who starts with, "this was originally selling for $500, but I'm able to offer it to a select group of individuals, including you, for only $300..."). Our tendency to anchor from an initial starting point can influence our decision.
- When we are trying to decide between the safety of a modest but sure gain against the possibility of a major loss, we tend to take the definite gain and avoid the possible disaster.
  - ➢ However, when faced with an unavoidable loss, we tend to grasp at opportunities that may salvage the situation.
- Another way of describing these tendencies: When faced with potential losses, we are more willing to take risks to avoid losses. However, when presented with possible gains, we are more willing to take the sure thing and less willing to take risks for higher gain.
- We tend to value what we have more than what we do not have. For example, when asked (*a*) how much we would pay someone to do a task for us and (*b*) how much we would have to be paid to do this same task for someone else, people demand 5 to 20 times as much for the latter choice (our own time) as we would pay for the former choice (someone else's time) (Bazerman, 1990)
- Some things that can help reduce bias from framing effects:
  - ➢ Frame problems in terms of objectives
  - ➢ Never automatically accept a frame as first given (by others or yourself)
  - ➢ Play devil's advocate to uncover reasons why you feel as you do
  - ➢ Consider events that occurred both before and after what comes to mind readily
  - ➢ Consider multiple possibilities to provide more anchor values
  - ➢ Reframe in multiple way

Categorization effects

- Are subtle yet pervasive and can be powerful
- Representativeness
  - ➢ Stereotyping
  - ➢ Perceive information as typical of the category
- Misperceive random event as a pattern (fail to realize that chance is not self-correcting). For example, the fact that an honest coin has turned up heads 10 times in a row does not change the probability of 0.5 that the next turn will be heads.

- Miss reality that a single event is always more probable than it plus another event. For example, the probability that a man who begs you for money on the street is homeless and an illegal immigrant is always less than the probability that he is homeless.
- Miss tendency of regression to the mean and allow an unusual event to bias predictions. For example, a pro basketball player who has been averaging 12 points per game made 32 points in the last game; do you predict that he will make considerably more than 12 points in the next game?
- Availability bias (*e.g.*, the probability of death from falling airplane pieces is 30 times the probability of death from a shark attack, yet we worry more about the latter because it is better publicized)
- Suggestion: distinguish data from inferences, assumptions, and conclusions (we will practice this more when we work on improving communication skills)
- General suggestions to reduce judgmental biases:
  - ➢ Consider different perspectives
  - ➢ Consider multiple alternatives
  - ➢ Strive for objectivity
  - ➢ Use inquiry
  - ➢ Use others to help
  - ➢ Separate facts from assumptions, inferences, and attributions
  - ➢ Identify and examine implicit assumptions

Effects of Risk Perceptions

- **Uncertainty**
  - ➢ Decisions become more risky as uncertainty increases
  - ➢ We can have uncertainty about potential outcomes, probabilities of their occurrence, and/or their controllability
  - ➢ We try to avoid, control, or ignore uncertainty

- **Potential gains & losses**
  - ➢ We prefer positive outcomes, so overestimate them
  - ➢ We tend to take more risk when facing sure losses, in an attempt to cut losses
  - ➢ We tend to avoid risk with a sure gain
- **Information framing**
  - ➢ We place more value on possibility of avoiding loss over accepting an assured loss
- **Personal involvement (relevance, responsibility, consequences)**
  - ➢ We tend to take greater risks with less personal responsibility and impact
- **Personal characteristics vary among decision makers**
  - ➢ Risk-taking propensity (some are comfortable with, even seek, risk)
  - ➢ Beliefs about control and competency

EXERCISES

Short Questions Answers

1. Describe Organization Behavior.
2. Elaborate the scope of Organizational Behavior.
3. What do you mean by Organization Development?
4. What do mean by Change.
5. Describe the process of Decision Making.
6. Define characteristics of Organizational Development.
7. Define Learning Organization
8. Define Action Learning.

Long Questions Answers

1. What are the contributing disciplines to the field of Organizational Behavior?
2. How Organization behavior helps in the development of organization?
3. Define various components of Organizational behavior.
4. Elaborate the theories of Organizational behavior.
5. Define Weber's ideal of Bureaucracy.
6. Define Modernization theory of Organizational behavior.
7. Define Scientific Management Theory.
8. What is the criticism of Weber's theory of Bureaucracy?
9. Elaborate the results of Hawthorne studies.
10. Define contingency theory of Organizational behavior.
11. What are the different models of Organization Development?
12. What do you mean by Learning Organization?
13. Elaborate the process of Change.
14. Elaborate the model of Organization Development.

EXAMINATION PAPER

B.Tech. (Sem. VI) Examination 2012
Principle of Management

Time : 2 Hours **Total Marks : 50**

1. Please read the following case let carefully and answers the questions given at the end.

The world's leading mobile phone company, Nokia Corporation, launched its 'take back' campaign in early 2009. The initiative, aimed at educating mobile phone users about the importance of recycling *e*-waste, had launched in a number of countries since 2005

Main issues behind these campaigns were:

(*a*) Sustainability

(*b*) Recycling

(*c*) Awareness campaigns

On January I, 2009, Nokia India, the Indian arm of world's leading mobile company, announced this 'take-back' campaign, on the occasion of the launch of the initiative, S. Shailendra, Managing Director,

Nokia India, said, "if every Nokia user recycled just one unused phone, the discarded phone will live on in something new, as the material is used to make item like park benches, utensils etc."

This campaign was first launched in 2005 in Finland before which it handled pilot project in the UK

And Sweden during the late IWs. It supported the concept of individual producer responsibility IPR)

And the take-back initiative was launched or an IPR initiative According to the company. Nokia was very environment conscious and followed the "Design for Environment" process where it took into account environmental aspects in every phase of product development right from the beginning of product life cycle.

Questions:

(*a*) Critically Analyze Nokia's take back initiative. (2)

(*b*) Do you think Nokia's Take-Hack initiative will receive good response in India? What should Nokia's Strategies to create awareness about recycling *e*-waste among Indian consumers?

2. Write short note on any four of the following: (4 × 4)

(*a*) Efficiency and effectiveness.

(*b*) Distinguish between contingency approach and system approach.

(*c*) Job enrichment.

(*d*) Explain all the steps involved in planning process?

(*e*) Define motivation and its process.

(*f*) What are the techniques of control?

3. Attempt any two of the following: (5 × 2)

(*a*) Fayol's general principles of management provide a basis for development of management theory, comment on the validity of statement by giving a brief account of Fayol's principles Which of these principles are more important and why?

(*b*) Explain nature of leadership? There is no one best style of leadership universally applicable to all situations? Comment.

(*c*) Coordination is the essence of managership. Explain how coordination is achieved?

4. Attempt any two of the following: (5 × 2)

(*a*) Explain Mc Gregor's theory X and theory Y, how theory Y an improvement over the traditional.

View that you can motivate employees by paying them more money.

(b) Explain the concept and mechanism of 360° appraisal?

(c) To manage is to forecast and plan to organize and to command to coordinate and to control. Discuss this statement

5. Attempt any two of the following: (5 × 2)

(*a*) Explain the types of management information system with their relevant importance for each Level of management.

(*b*) What is organizational structure? Explain different types of organizational planning structures?

(*c*) Compare and contrast various theories of management which approach would you suggest and why?

Index

■ ■ ■

Win Prizes

ATTENTION READERS

Win Prizes

We request for your frank assessment regarding various aspects of the book as given below :

Principles of Management

BY : Shikha Taneja/Renu Rana (UTU)

Please Fill up the Sheet in Neat and Clean Writing

(i) What topics of your syllabus, important for your examination point of view, are not covered in this book ?

....

....

....

....

(ii) What are the chapters wherein the treatment of the subject matter is not systematic or organised or updated ?

....

....

....

....

(iii) Have you come across misprints/mistakes/factual inaccuracies in the book ? Please specify the chapters and the page numbers.

....

....

....

....

(iv) What other books on the same subject have you found/heard better than the present book ? Please specify in terms of price and quality.

....

....

....

....

(v) Further suggestions and comments for the improvement of this book.

....

....

....

....

(PS : If need be please attach a separate sheet to write your views)

OTHER DETAILS

(i) Who recommended you this book ? Please tick (✓) in the box near the option relevant to you.

☐ Teacher ☐ Friends ☐ Bookseller

....

....

....

....

(ii) Name of the recommending teacher, his designation and address:

....

....

....

....

(iii) Name and address of the bookseller you purchased the book from :

....

....

....

....

(iv) Name and address of your institution (Please mention the University or Board as the case may be)

....

....

....

....

(v) Your name and complete address :

....

....

....

....

(vi) Course you are enrolled in :

....

....

....

....

The best assessment will be awarded each month : The award will be in the form of our publications as decided by the Editorial Board.

Please mail the filled up coupon at your earliest to :

S.K. KATARIA & SONS®
4885/109, Prakash Mahal, Dr. Subhash Bhargav Lane
Daryaganj, New Delhi-110002
☎ +91-11-43551243
Mobile : 91-9871775858
email : katariabook@gmail.com